Lecture Notes in Physics

Edited by J. Ehlers, München, K. Hepp, ...
R. Kippenhahn, München, H. A. Weidenmüller, Heidelberg
and J. Zittartz, Köln
Managing Editor: W. Beiglböck, Heidelberg

136

The Role of Coherent Structures in Modelling Turbulence and Mixing

Proceedings of the
International Conference
Madrid, Spain, June 25–27, 1980

Edited by J. Jimenez

Springer-Verlag
Berlin Heidelberg New York 1981

Editor

J. Jimenez
IBM Scientific Center
P. Castellana 4
Madrid (1). Spain

ISBN 3-540-10289-2 Springer-Verlag Berlin Heidelberg New York
ISBN 0-387-10289-2 Springer-Verlag New York Heidelberg Berlin

Printed in Germany

Printing and binding: Beltz Offsetdruck, Hemsbach/Bergstr.
2153/3140-543210

Sponsorship: UAM-IBM Scientific Center, Madrid
School of Aeronautics, U. Politecnica Madrid.

Scientific Committee

H.W. Liepmann (Chairman), California Institute of Technology, Pasadena, Ca., USA.
C. Dopazo, Universidad de Zaragoza, Spain.
H. Fiedler, Technische Universität, Berlin, Germany.
J. Jimenez, UAM-IBM Scientific Center, Madrid, Spain.
A. Linan, Universidad Politecnica, Madrid, Spain.
J. Mathieu, Ecole Central de Lyon, France.
P.G. Saffman, California Institute of Technology, Pasadena, Ca., USA.

Local Organization

C. Dopazo, E.T.S.I. Industriales, Zaragoza.
J. Jimenez, UAM-IBM Scientific Center, Madrid.
M. Kindelan, UAM-IBM Scientific Center, Madrid.
A. Linan, E.T.S.I. Aeronauticos, U. Politecnica, Madrid.
M. Rebollo, UAM-IBM Scientific Center, Madrid.

Financial Sponsors

IBM SAE, Madrid.
U.S. Army European Research Office, London.

PREFACE

In the past years there has been a growing recognition of the existence of large-scale structures in many turbulent flows. Qualitatively this has led to a new insight into the nature of turbulence and to a new way of doing experiments in the field. In the area of flow computation and modelling, however, the new ideas have not produced a corresponding number of important results, and the whole field has become clearly driven by experimental physicists and engineers.

This is probably inevitable after a major experimental breakthrough, and so it was time to bring together experimentalists and theoretists to discuss which had been the main achievements of the last years and to try to find out which will be the development in the future. This Conference was hosted by the IBM Scientific Center at the Universidad Autonoma of Madrid to provide a forum in which this discussion could take place. As such, the emphasis was on the interpretation of known results. Nevertheless, much new material is covered in these Proceedings.

One session concentrates on theoretical and computational problems. The next two are devoted to experimental evidence. One of them (Session II) tries to emphasize experimental methods, while the other one deals more with the presentation of results. The remaining session is devoted to the discussion of those applications, especially noise production in jets and turbulent combustion and mixing, in which turbulent flows play a dominant role as the underlying phenomenon and in which it is currently possible to see a connection with the concept of coherent structures. The session chairmen were asked to review the state of the art in the different fields. Their lectures are included at the begining of each session.

The IBM Scientific Center and the School of Aeronautics of the Universidad Politecnica of Madrid have been working for some years in the field of experimental turbulence and the application of computer image processing and pattern recognition methods to the processing of the data. It was therefore an honor for us to organize the Conference.

This was only possible through the generous support from IBM SAE and the European Research Office of the United States Army (USARSG). We wish to express our thanks to them.

Professor H.W. Liepmann from the California Institute of Technology acted as external advisor on the scientific programme. Without him this Conference could not have taken place.

I will also like to express my gratitude and appreciation to all the lecturers, chairmen and members of the Scientific Committee; they bore much of the burden of organizing and preparing the Conference. The help of everybody in the local Organizing Committee was invaluable. As in most cases, most of the remaining work was carried out by the secretarial staff of the Scientific Center. To them my sincere thanks.

Madrid, June 1980 Javier Jimenez

CONTENTS

SESSION I - THEORY

Chairman: P.G. Saffman, California Institute of Technology, Pasadena.

SESSION II - EXPERIMENTS

Chairman: J. Mathieu, Ecole Centrale de Lyon.

SESSION III - EXPERIMENTS

Chairman: H.E. Fiedler, Technische Universität, Berlin.

SESSION IV - APPLICATIONS

Chairman: A. Liñan, E.T.S.I. Aeronáuticos, Madrid.

LIST OF PARTICIPANTS

L.A. Aisa Miguel
Dpto. Mecanica de Fluidos G.XIX
E.T.S.I. Industriales B.
Diagonal 647, Barcelona 28 (Spain)

A. Barrero
E.T.S.I. Industriales
Sevilla, (Spain)

A.C.M. Beljaars
Koninklijk Netherlands Meteorologisch Instituut
Postbus 201
3730 Ae de Bilt
The Netherlands

H.C. Boisson
Institut de Mecanique des Fluides (IMFT)
2, Rue Charles Camichel
31071 Toulouse Cedex (France)

J.L. Bousgarbies
Laboratoire de Dynamique des Fluides
40 Ae. du Recteur Pineau
86022 Poitiers, (France)

L. Boyer
Lab. Thermophisique
Universite de Provence
Marseille

G.L. Brown
Graduate Aeronautical Lab.
California Institute of Technology
Pasadena, Ca. 91125
USA

A. Castellanos
Dept. Fisica Fluidos, Univ. Autonoma
Cantoblanco, Madrid - 34 (Spain)

D. Coles
Graduate Aeronautical Lab.
California Institute of Technology
Pasadena, Ca. 91125
USA

G.M. Corcos
Dept. of Mechanical Engineering
6101 Etcheverry Hall
University of California
Berkeley, Ca. 94720

A. Crespo
E.T.S.I. Industriales
Madrid (3)
Spain

D.G. Crighton
Dept. of Applied Mathematical Studies
University of Leeds
Leeds, LS2 9JT(England)

F. Cuesta Benito
Instituto Nal. del Frio (CSIC)
Ciudad Universitaria
Madrid - 3 (Spain)

B.A. Delcourt
c/o Von Karman Institute
72, Chee Waterloo
1640 Rhode-St-Genese, Belgium

J. Diaz-Roche
E.T.S.I. Navales
Madrid(3), Spain

C. Dopazo
E.T.S.I. Industriales Zaragoza.
Zaragoza, (Spain)

A. Dyment
Institut de Mecanique des Fluides de Lille
5 bd Paul Painleve
59000 Lille (France)

H. Eckelmann
Max-Planck Institut fur Stromungsforschung
Bottingerstr. 4/8
D34 Gottingen(F.R. Germany)

S. Einav
Faculty of Engineering
University of Tel-Aviv
Tel-Aviv (Israel)

H. Fiedler
Technische Universitat Berlin
Hermann-Fottinger Institut
Strasse des 17 Juni
D-1000 Berlin 12 (Germany)

K. Krisna Prasad
Dept. of Applied Physics
Eindhoven University of Tehcnology
Postbus 513, Eindhoven (Netherlands)

A. Gyr
Federal Institute of Technology
Institute of Hydromechanics and Water Resources Mangement
CH-8093 Zurich-Honggerberg (Switzerland)

M.A. Hernan
IBM Scientific Center
Paseo Castellana 4
Madrid (Spain)

F.S. Higuera
E.T.S.I.Aeronauticos
Ciudad Universitaria
Madrid (Spain)

A.K.M.F. Hussain
Director, Aerodynamics and Turbulence Laboratory
University of Houston
Houston, Texas 77004(USA)

J. Jimenez
IBM Scientific Center
P. Castellana 4.
Madrid(1), Spain.

M. Kindelan
IBM Scientific Center
P. Castellana 4.
Madrid(1), Spain.

D.D. Knight
Dept. of Mechanical, Industrial and Aerospace Engineering
Rutgers University, P.O. Box 909
Piscataway, New Jersey 08854(USA)

A. Leonard
NASA-Ames Research Center
Building 202-1
Moffet Field, Ca 94022 (USA)

O. Leuchter
Office National d'Etudes et de Recherches Aerospatiales (ONERA)
29 Ave. Division Leclerc
92.320 Chatillon (France)

H.W. Liepmann
Graduate Aeronautical Laboratories
California Institute of Technology
Pasadena, Ca 91125 (USA)

A. Linan
E.T.S.I. Aeronauticos
Ciudad Universitaria
Madrid (3), Spain

N. Lopez Rodriguez
Instituto Nal. del Frio, (CSIC)
Ciudad Universitaria
Madrid - 3 (Spain)

D.R. Manzano
E.T.S.I. Aeronauticos
Ciudad Universitaria
Madrid (3), Spain

R. Martinez-Val
E.T.S.I. Aeronauticos
Universidad Politecnica de Madrid
Madrid (Spain)

J. Mathieu
Lab. Machanique des Fluides
36, Route de Dardilly
69130 Ecully, (France).

F. Orti Daras
IBM Scientific Center
P. Castellana 4.
Madrid(1), Spain.

M.J. Ortiz
Dpto. de Fisica, Universidad de Alcala
Alcala de Henares, Madrid (Spain)

N. Peters
Institut fur Allgemeine Mechanich
RWTH, Aachen
West Germany

M. Rebollo
IBM Scientific Center
P. Castellana 4.
Madrid(1), Spain.

M. Rodriguez
E.T.S.I. Aeronauticos
Ciudad Universitaria
Madrid (3), Spain

A. Roshko
Graduate Aeronautical Lab.
California Institute of Technology
Pasadena, Ca. 91125 (USA)

A. Ruiz de Elvira
Dpto. de Fisica, Universidad de Alcala
Alcala de Henares, Madrid (Spain)

P.G. Saffman
Dept Applied Mathematics,101-50
California Institute of Technology
Pasadena, Ca 91125 (USA)

B. Steverding
USARGS (Europe)
223 Old Marylebone Rd.
London NW1, (England)

R. Takaki
Hermann-Fottinger-Institut, TU Berlin
Heerstrasse 131-135
1000 Berlin 19, (West Germany)

J.M. Vega de Prada
E.T.S.I. Aeronauticos
Ciudad Universitaria
Madrid (Spain)

M.G. Velarde
Dept. Fisica Fluidos, Univ. Autonoma
Cantoblanco, Madrid - 34 (Spain)

L. Virto Albert
Escuela T. S. de Ingenieros Industriales
Tarrasa (Spain)

J.M. Wallace
Dept. of Mechanical Engineering
University of Maryland, College Park
Md. 20742 (USA)

I.J. Wygnanski
Tel-Aviv University
School of Engineering
Ramat-Aviv, Tel-Aviv 69978 (Israel)

A.J. Yule
University of Sheffield
Dept. of Chemical Engineering and Fuel Technology
Mappin Street
Sheffield S1 3jd (England)

M. Yuste
Dept. Fisica Fluidos, Univ. Autonoma
Cantoblanco, Madrid - 34 (Spain)

Coherent Structures in Turbulent Flow

by P.G. Saffman
Applied Mathematics
California Institute of Technology

It is customary to say that the problem of turbulence remains the outstanding challenge of fluid dynamics, but in fact the concerted efforts of many scientists and engineers over the years have led to a considerable body of useful knowledge. The three processes of cataloguing, curve fitting and 'postdicting' are well under control. If there is a problem, it is partly because we now ask harder questions and are more concerned with prediction and control; or we hanker for the El Dorado of a general theory and the chimaera of simple understanding.

The purpose of this meeting is to consider if coherent structures are the road to the promised land, to decide if we are right to retreat from the statistical approaches of energy spectra, correlations, the Reynolds equation, etc., and to examine critically the prospects for a deterministic approach to turbulent flow. Put briefly, is the understanding and description of vortex dynamics the key to turbulence?

The concept of coherent structure or organized eddy is of course not exactly new. In a sense, the mixing length theories of Taylor and Prandtl, in which momentum or vorticity is convected by a blob of fluid, use a vague kind of physical entity as the basis for transport or mixing. More formally, Synge & Lin [1] proposed 37 years ago a model of isotropic turbulence as a random superposition of Hill Spherical Vortices (H.S.V's). The H.S.V. is an exact solution of the Euler equations (in fact it would be an exact solution of the Navier-Stokes equations except that the viscous stresses are not continuous across the boundary) in which the vorticity is confined to the interior of a sphere, the only non-zero component being the azimuthal with magnitude $(15U/2R^2)y$, where $U = |\underset{\sim}{U}|$ is the speed and R the radius of the sphere. Associated with the H.S.V. is a kinetic energy $(10/7)\rho\pi U^2 R^3$, and a hydrodynamic impulse $\underset{\sim}{I} = 2\rho\pi R^3 \underset{\sim}{U}$. The velocity field at a point r relative to the vortex center can be written in the form

$$\underset{\sim}{u}_{HSV}(\underset{\sim}{r},\underset{\sim}{U},R) = U\,A(r/R)\underset{\sim}{r} + B(r/R)\underset{\sim}{U}.$$

Then Synge & Lin expressed the velocity field of the fluid as

$$\underset{\sim}{u}(\underset{\sim}{x}) = \iiint W(R,\underset{\sim}{\xi},\underset{\sim}{U})\ \underset{\sim}{u}_{HSV}(\underset{\sim}{x}-\underset{\sim}{\xi},\underset{\sim}{U},R)\ dR\ d\underset{\sim}{\xi}\ d\underset{\sim}{U}$$

where W is a random function giving the probability density of finding at the point $\underset{\sim}{\xi}$ a H.S.V. of radius R moving with speed $\underset{\sim}{U}$. From this expression, can be synthesized all the statistical properties of $\underset{\sim}{u}$. This appears to be the first formal kinematic expression of a turbulent velocity field as the random superposition of laminar structures or coherent vortices.

It is ironic that the behavior $f(r) \sim 1/r^3$ for the asymptotic behavior of the longitudinal velocity correlation was predicted in this way in 1943 but ignored, while the statistical theory which assumed exponential decay considered the Loitsianskii invariant $\int_0^\infty r^4 f(r)dr$, which of course does not exist if $f(r) \sim r^{-3}$. Later work confirmed the correctness of the r^{-3} prediction and provided a belated recognition of the possible advantages of physical models over purely statistical considerations.

A later development was Townsend's [2] attempt in 1951 to construct a physical model for the dissipation eddies in isotropic turbulence as a random superposition of vortex sheets and tubes, stimulated in part by the success of Kolmogorov k,ε,ν scaling (k = wave number, not kinetic energy) for the velocity covariance and energy spectrum at Reynolds numbers too small for the assumptions of the theory to be valid, and in part by the strong implications from the Batchelor & Townsend [3] observations, confirmed later by numerous other measurements (see Van Atta & Antonia [4] for a recent survey) that flatness factors of high order are large and strongly Reynolds number dependent. This suggests that the energy dissipating regions are spatially intermittent, as predicted on theoretical grounds by Landau [5]. The basic structures employed by Townsend were the equilibrium viscous vortex sheet and tube (apparently due to Bergers and rediscovered several times) in an irrotational straining field

$$\omega_x = \omega_o\, e^{-\alpha y^2/2\nu},\ u = \alpha x,\ v = -\alpha y,\ w = -\int \omega_x\, dy$$

or

$$\omega_x = \omega_o\, e^{-\alpha r^2/2\nu},\ u = 2\alpha x,\ v_r = -\alpha r,\ v_\theta = \frac{\nu\omega_o}{\alpha}\,(1-e^{-\alpha r^2/2\nu})$$

where $r = (x^2+y^2)^{\frac{1}{2}}$. Random superposition of sheets and tubes (by a formula like that for H.S.V's) gives reasonable agreement with fine scale spectra observations. More sophisticated developments of Townsend's idea have been tried by several authors, e.g. Saffman [6] and Tennekes [7]. Saffman's appears to be the most detailed, and was an attempt to give physical significance to the Taylor micro-scale $\lambda = (5q^2)^{\frac{1}{2}}/\varepsilon$ and Kolmogorov length $\ell = (\nu^3/\varepsilon)^{\frac{1}{4}}$. The former was envisaged as an internal boundary layer between large eddies (e.g. H.S.V's) and the latter a scale produced by Taylor-Gortler instability of the internal boundary layers.

However, none of these physical or deterministic models can be regarded as being satisfactory or more than suggestive speculation. There is no real agreement on the kinematics of the structures, and they contain no dynamics. Also their predictions (e.g. for the dependence of flatness factors on Reynolds number) are inferior to those obtained by non-physical statistical dimensional analysis. Thus it cannot be said that the concept of coherent structures has been particularly useful or productive to date in the study of homogeneous turbulence. Its main value has been perhaps to suggest another point of view, in which the dynamics of vorticity is emphasized, spatial structure is important, and the possibility that Fourier components have strongly correlated phases is recognized. The concept found its real support in the studies of shear flow turbulence, most spectacularly in the turbulent mixing layer, and there is little doubt that the now classic work of Brown and Roshko was the impetus that changed a marginal activity into an active industry.

The remaining candidate to date for the role of building block or fundamental particle for a coherent structure is the vortex ring or more generally the vortex filament. The classical vortex ring is the thin torus filled with uniform vorticity, which propagates with speed

$$U_V = \frac{\Gamma}{4\pi R}\left(\log\frac{8R}{\delta} - \frac{1}{4} + O\left(\frac{a^2}{R^2}\right)\right),$$

where R is the radius of the ring, δ the radius of the core, and Γ the circulation about the axis. The vortex ring and H.S.V. are actually the opposite ends of a continuous family of finite cored vortex rings, which were computed by Norbury [8]. The vortex ring can also be regarded as the special case of a vortex filament, or thin tube of vorticity, in the shape of a circle. The velocity of a filament is given to leading order by the Biot-Savart law (or equivalent)

$$\underset{\sim}{u} = \frac{\Gamma}{4\pi}\oint \frac{d\underset{\sim}{s} \times \underset{\sim}{r}}{r^3}$$

where $\oint$ means that a cut-off, of length proportional to δ, has to be employed in the evaluation of the integral for a point on the filament. (A careful and thorough study of the equations of motion of a vortex filament, going to higher order than the Biot-Savart law and discussing in addition the equations for the internal structure, which can have an O(1) effect on the speed, was carried out by Moore & Saffman [9]).

The use of such vortex structures as models for the quasi-permanent coherent structures of a turbulent flow makes the study of their interactions and stability a matter of pressing interest. There has been some fascinating work on the interaction of vortex rings and filaments, showing remarkably strange behavior (e.g. Oshima & Asaka [10], Oshima [11]), which indicates that the dynamics of interacting vortex rings is a far from simple problem. In particular, the breaking and reforming of vortex lines (clearly noticeable for instance in condensation trails [12]) seems to play a vital role in the interactions of vortex rings, and we have no idea as yet of the basic processes involved. There has, however, been much work in the related problem of magnetic field line reconnection in magnetohydrodynamics which may be relevant, e.g. [13].

The three dimensional stability of pairs or arrays of line vortices needs further investigation in order to clarify the interaction properties of coherent structures. The long axial wavelength disturbances with $k\delta \ll 1$ (k = axial wavenumber, δ = core radius) can be studied using the Biot-Savart law (with a suitable cut-off for the self induction) to calculate the speeds of the deformed vortices. This was done by Schlayer [14] and Rosenhead [15] for the Karman vortex street, but these papers are obscure and it is not clear exactly what their results are. Simpler treatments of a pair of counter-rotating and co-rotating vortices were carried out by Crow [16] and Jimenez [17], respectively. These results are rather interesting. Crow found weak instability for sufficiently small k while Jimenez found that the configuration was always stable under the long wave approximation. These results are consistent with the apparent slow instability to three-dimensional disturbances of the Karman vortex street, which contains vortices of both signs, while the mixing layer, which contains coherent vorticity of just one sign, appears to be stable to long wave instabilities (other than the pairing instability) and is accordingly a more permanent structure. The observations of trailing vortices show that counter-rotating vortices break and reform into loops if the instability grows to finite amplitude, and a similar phenomenon can perhaps be expected in the Karman vortex street. The relevance or importance of the short wave parametric instability of vortex filaments with $k\delta = O(1)$, see Moore & Saffman [18], is not completely certain but may be related to the spanwise structures seen in the mixing layer.

Pierrehumbert [19] has recently calculated the two and three dimensional linear stability of the inviscid mixing layer described by the stream function

$$\psi = \frac{1}{2} \log(\cosh 2y - \rho \cos 2x) \quad .$$

As ρ goes from 0 to 1, this flow field changes from a hyperbolic tangent to an array of point vortices distance π apart. A good overall fit with experimental profiles is claimed for $\rho = 0.25$. This flowfield exhibits the subharmonic (wavelength 2π) pairing instability, with growth rate a slowly increasing function of ρ. Thus, it appears that finite core size may decrease slightly the pairing instability growth rate of a row of point vortices, but not by any significant amount. It was also found that the instability persists when the disturbance is three-dimensional, but is most unstable when two-dimensional and stabilizes when the spanwise wavelength is comparable to the vortex separation. The parametric instability of a straight filament corresponds to disturbance of wavelength π, in which all vortices are disturbed in the same way. The disturbance is neutrally stable when two-dimensional, and has a maximum growth rate when its spanwise wavelength is comparable to vortex separation. However, it is uncertain how typical these results are. Saffman & Szeto's [20] investigation of the two-dimensional stability of an array of uniform finite cored vortices showed stability to disturbances of wavelength π until the vortices reached a certain size when they became unstable, but their arguments were qualitative and they did not calculate actual eigenvalues.

The extent to which the different and various aspects of turbulent flow can be modelled or understood as a collection of H.S.V's, or vortex tubes and sheets, or vortex rings and filaments, or alternative large coherent flow structures, or large number of two-dimensional point vortices is one of the problems which this meeting will address. There is one significant point which should perhaps be made. Apart from the difficulty in dealing adequately with the dynamical interactions of the vortex structures, which raises some challenging mathematical problems, and incorporating viscosity, instabilities and 'fine scale turbulence', it may not be a too good approximation to assume that the vortex structures are surrounded by irrotational fluid. We perhaps need a theory for the structure of vortices in a background of weak vorticity. Some first steps have been taken by Kiya & Arie [21], [22].

An interesting comment has also been made by Taganov & Dudoladov [23] who have modelled the turbulent mixing layer by finite and point vortices. They suggest that two classes of vortex formation should be considered. The first class comprises structures which are essentially inviscid, and are relevant to jets, wakes, mixing layers, etc. While the second class contains those structures in which viscosity plays a large role in the momentum transfer and are modelled by low Reynolds number solutions. Those of the second class would satisfy no-slip boundary conditions at solid walls and be relevant to the viscous sublayer, etc.

The experimental study of the coherent structure is most firmly established in the mixing layer, boundary layer, jet and wake. Theoretical models are also developed most extensively for these flows, as we shall be hearing in the following talks, and considerable progress has been made in describing and understanding these flows in terms of vortex interactions. These are, however, free flows, apart from the boundary layer. There is also evidence of coherent or organized structure in bounded flows, which in fact predates the discovery of free flow phenomena, and it is worth suggesting that theoreticians should perhaps also direct their efforts to these. Coles [24] in 1961 described the intermittency and periodic turbulent flow structures in pipe flow and circular Couette flow. There has lately been renewed interest in the Taylor vortices at large Taylor number, e.g. Koschmieder [25]. To what extent the intermittent and periodic nature of these bounded flows can be explained in terms of coherent structures or vortex dynamics remains to be elucidated.

One aspect of vortex dynamics that will be mentioned explicitly here is the information that can be gathered from steady solutions of the Euler equations. The philosophy behind coherent structures is that at least the gross features of the flow are determined by the interaction of vorticity, and thus the possible equilibrium configurations are of interest. These have been calculated recently by Saffman & Szeto [20] and independently by Pierrehumbert & Widnall [26] for a linear array of uniform vortices which models the turbulent mixing layer. Equilibrium shapes were computed and it was found that solutions only exist if the vortices are not too large. For vortices with uniform constant vorticity, the area A has to be less than $.238L^2$, where L is the distance between centers. This provides support for a deterministic model of the evolution of the coherent structures in a shear layer, in which the vortices grow by the continuous ingestion of irrotational fluid until they reach such a size that they can no longer exist, at which stage tearing occurs and the array is reformed with a larger L and A but smaller A/L^2, and the process repeats. Of course, a lot of questions are left unanswered, the most important being the rate or frequency of the process. An alternative model is to recognise the instability of the array to the pairing instability in which neighboring vortices rotate around each other and coalesce. The objection here that there is no coalescence mechanism is answered by further work of Saffman & Szeto [27] on the steady states of two rotating vortices. They showed that for this flow there are no steady solutions when $L < 1.8A^{\frac{1}{2}}$, and that since the process of pairing will reduce the separation to less than this, it is expected that coalescence can take place. Both models make the same kind of quantitative prediction, giving values of a/b around 2 and δ_ω/L around 0.3. While there is clearly some uncertainty in deciding what is the basic physical process, it does appear that the processes of coalescence and mixing layer growth are consistent with the inviscid dynamics of vorticity.

The deterministic approach is certainly an exciting one, holding out the possibility of a fundamental advance in understanding. The search for quasi-permanent coherent structures and the study of their interactions promise to be challenging tasks. But there are clearly serious obstacles. It is not yet easy to make quantitative predictions. The most glaring omission is the inability to predict in any reasonable way the logarithmic law of the wall of the turbulent boundary layer. The steady Reynolds stress equation (and the associated 'invariant modelling') would certainly seem to be irrelevant, having as much physical content as the statement that the average vehicle has 3.4 wheels or the average family 2.7 children, but perhaps a conditionally sampled Reynolds equation or an integro-differential equation for the probability distribution function based on the pioneering work of Lundgren and Monin would be a useful subject for study. A recent step in this latter direction has been made by Onufriyev [28]. Such an approach would certainly not be inconsistent with intermittency if a bimodal probability distribution were employed.

One issue that needs to be carefully considered is the role of the computer in theoretical turbulence research. One day the computer may be able to answer all questions, or more precisely predict any number that is required (whether this constitutes 'understanding' is a subjective question), but this is still in the distant future. At the moment, the computer can churn out millions of random numbers, as in sub-grid modelling or following clouds of point vortices, or it can be used to solve specific equations for well posed models, such as those that give the structure of vortex rings or steady vortex arrays. It is hoped that this meeting will clarify its role in the coherent structure approach to turbulence.

The writing of this survey and my work on equilibrium structures was supported by the Department of Energy (Office of Basic Energy Sciences) and the Army Research Office (Durham).

References

[1] Synge, J.L. & Lin, C.C. On a statistical model of isotropic turbulence. Transactions Royal Society of Canada, Sec. III, p45-79. (1943)

[2] Townsend, A.A. On the fine-scale structure of turbulence. Proc. Roy. Soc. A208, p534-542. (1951)

[3] Batchelor, G.K. & Townsend, A.A. The nature of turbulent motion at large wave-numbers. Proc. Roy. Soc. A199, p238-255. (1949)

[4] Van Atta, C.W. & Antonia, R.A. Reynolds number dependence of skewness and flatness factors of turbulent velocity derivatives. Phys. Fluids 23, p252-257. (1980)

[5] Landau L.D. & Lifshitz, E.M. Fluid Mechanics, Pergamon. Footnote p126.

[6] Saffman, P.G. Lectures on Homogeneous Turbulence. Topics in Non-Linear Physics (Ed. N. Zabusky) Springer-Verlag p485-614. (1966)

[7] Tennekes, H. Simple model of the small-scale structure of turbulence. Physics of Fluids 11, p669-670. (1968)

[8] Norbury, J. A family of steady vortex rings. J. Fluid Mech. 57, p417-431. (1973)

[9] Moore, D.W. & Saffman, P.G. The motion of a vortex filament with axial flow. Phil. Trans. Roy. Soc. A272, p403-429. (1972)

[10] Oshima, Y. & Asaka, S. Interaction of two vortex rings moving side by side. Nat. Sci. Rep. Ochanomizu Univ. 26, p31-37. (1975)

[11] Oshima, Y. The game of passing through of a pair of vortex rings. J. Phys. Soc. Japan 45, p660-664. (1978)

[12] Scorer, R.S. & Davenport L.J. Contrails and aircraft downwash. J. Fluid Mech. 43, p451-464. (1970)

[13] Soward, A.M. & Priest, E.R. Fast magnetic field line reconnection. Phil. Trans. Roy. Soc. A284, p369-417 (1977)

[14] Schlayer, K. Über die Stabilität der Karmanschen Wirbelstrasse gegenüber beliebigen Störungen in drei Dimensionen. Z.A.M.M. 8, p352-372. (1928)

[15] Rosenhead, L. The spread of vorticity in the wake behind a cylinder. Proc. Roy. Soc. A127, p590-612. (1930)

[16] Crow, S.C. Stability theory for a pair of trailing vortices. A.I.A.A. J. 8, p2172-2179. (1970)

[17] Jimenez, J. Stability of a pair of co-rotating vortices. Phys. Fluids 18, p1580-1581. (1975)

[18] Moore, D.W. & Saffman, P.G. The instability of a straight vortex filament in a strain field. Proc. Roy. Soc. A346, p413-425. (1975)

[19] Pierrehumbert, R.T. The structure and stability of large vortices in an inviscid flow. M.I.T. Fluid Dynamics Lab. Rep. 80-1. (1980)

[20] Saffman, P.G. & Szeto, R. Structure of a linear array of uniform vortices. Stud. in App. Math. (to appear). (1980)

[21] Kiya, M. & Arie, M. Formation of vortex streets in shear flow. App. Sci. Res. 34, p313-339. (1978)

[22] Kiya, M. & Arie, M. Helmholtz instability of a vortex sheet in uniform shear flow. Phys. Fluids 22, p378-379. (1979)

[23] Taganov, G.I. & Dudoladov I.V. Application of results of analysis of transient vortex flows of ideal fluids to description of the turbulent mixing layer. Fluid Mech. Sov. Res. 7, p10-23. (1978)

[24] Coles, D. Interfaces and intermittency in turbulent shear flow. Mecanique de la Turbulence p229-250. (1961)

[25] Koschmieder, E.L. Turbulent Taylor vortex flow. J. Fluid Mech. 93, p515-527. (1979)

[26] Pierrehumbert, R.T. & Widnall, S.E. The structure of organized vortices in a shear layer. A.I.A.A. Paper 79-1560. (1979)

[27] Saffman, P.G. & Szeto, R. Equilibrium shapes of a pair of equal uniform vortices. (1980)

[28] Onufriyev, A.T. Model equation for the probability density in the semi-empirical theory of turbulent transport. Fluid Mechanics Sov. Res. 7, p143-152. (1978)

The Deterministic Description of the Coherent Structure of Free Shear Layers

G. M. Corcos
Department of Mechanical Engineering
University of California, Berkeley

ABSTRACT

The large coherent structures observed in turbulent mixing layers are tentatively identified with the unsteady flow which is generated in two dimensions by the same initial and boundary conditions. This flow is briefly described and means of calculating it are reviewed. Typical results of such calculations are illustrated. For large Reynolds numbers and the outer folds of the interfacial spirals, a boundary-layer-like treatment allows the calculation of the evolution of the diffusive attributes of the flow in terms of the rate of stretching of the material interface. This important quantity has been calculated as a Lagrangian function of time and initial location for a roll-up and pairing and for a moderate Reynolds number. It is suggested that interfacial stretching does not depend strongly on Reynolds number. The use of this convenient approach is illustrated in a simple example of a diffusion-limited reaction.

INTRODUCTION

The mixing layer between two uniform streams is a turbulent flow endowed with a well documented large scale coherent structure. What is most remarkable about this coherent motion is that it appears to be closely similar to the two-dimensional unsteady flow which arises out of the initial superposition of two uniform streams and a few weak two-dimensional perturbations.

We have learned a great deal more about the latter, purely two-dimensional flow that we have been able to measure in a real shear layer so that the evidence and the degree of similarity, while visually striking should perhaps still be viewed as incompletely established.

If, as we believe likely, the coherent structures of turbulent shear layers and the deterministically calculable unsteady flow, which arises in two-dimensions from the juxtaposition of two uniform streams are essentially one and the same, our methods of theoretical study of this turbulent flow should, in our opinion, be revised. In particular:

(a) The deviations from two-dimensionality of a real layer may be viewed as (possibly large) perturbations of the basic two-dimensional time-dependent flow in the sense that their interaction with the flow is evidently unable to destroy or even to control it.

(b) Therefore, such a two-dimensional flow should be used as the base flow for the turbulent mixing layer, no matter how the rest of the flow is calculated, i.e., whether one pursues the deterministic route further by deriving models of three-dimensional motions compatible with and grafted on the time-dependent base flow*, or if one prefers to approximate the three dimensional motion by using plausible functional or phenomenological relations between the known and the unknown parts of the flow. The outstanding simplification is that the base flow which is proposed as a substitute for the usual mean flow has an independent existence.

In this short survey, we will focus on the description of the two-dimensional unsteady flow, and in particular, we shall evaluate available methods of computing it, give examples of these configurations, suggest methods of generalizing them, and discuss a few applications. For a more comprhensive discussion these and other aspects of the problem, Patnaik, Sherman and Corcos (1976), Corcos and Sherman (1976) and Corcos (1979) may be consulted.

A Thumb-nail Sketch

As is now well known, a two-dimensional shear layer rolls up in a connected row of spirals whose spacing and diameter grow with streamwise distance from the origin. The first spirals are the result of the shear layer instability whose origin is described by linear theory. But, while this first instability is self-limiting,

*For a progress report on such an approach, see Corcos (1979). A substantial amount of additional work in this area is being carried out at Berkeley.

the configuration is unstable to subharmonic perturbations and the spirals pair or amalgamate and, as a result, both the circulation around a spiral and the spacing between two adjacent spirals increases.

As the original interface is distorted into a spiral, it is stretched on the average because its total length steadily increases. As a result, the region to which the original spanwise vorticity of the layer is confined thins out and the volume occupied by the vorticity at any time is a fraction of the volume of the fluid which appears to be sheared (the space within the nominal outline of the shear layer).

The focii of the spirals are points of accumulation of spanwise vorticity. Between them, there are stagnation points which are saddles, the neighborhood of which is depleted of spanwise vorticity. They are the site of intense deformation, consistently positive in the direction of the interface.

As we shall see, around the focii of the spirals, the successive coils become not only dense, but also wide, because they suffer compression, and they form a solid core in which vorticity gradients and gradients of diffusive attributes of the layer are in general far more gradual.

We have mentioned that coils of vorticity coalesce. According to both computations and to observations, the preferred mode involves two coils at a time. When two coils pair, the saddle which was in between (an irrotational stagnation point) is destroyed. But it survives long enough to contribute, during the pairing and for a certain time afterwards, an intense strain in between the two vortex centers. This strain in turn causes energetic interdiffusion of vorticity between the two previously separated vortex centers. Nevertheless the pairing vortices while pressed against one another retain a separate identity for a substantial time after pairing so that a stationary probe, past which paired vortices are convected would be likely to detect a large oscillation in velocity near the apparent center of the new vortex.

Computation of the Flow

By now, a number of independent numerical simulations of the two-dimensional shear layer have been carried out, by Ashurst (1977), Patnaik, Sherman and Corcos (1976), Peltier, Halle and Clark (1978), Delcourt and Brown (1979) and others. They include simulation of the vorticity distribution by a large number of point vortices and simulation of the Navier-Stokes equation by finite differences.

While all these simulations are very suggestive of the coherent structures observed in the laboratory, the finite differences solutions of the diffusive equations for the temporally developing instability and for moderate Reynolds numbers has been carried out in greatest detail. Its main drawback is that practical flows develop streamwise rather than temporally, and that a mapping only gives an approximation to the former in terms of the latter.

I shall only discuss temporal simulation by finite differences, with which our group has had substantial experience.

In these computations, the flow develops in time on a grid in which the streamwise boundary conditions are periodic. The initial flow is the superposition of a simple shear layer and one or several eigenfunctions of the linear stability problem. One of the eigenfunctions corresponds in general to the wave number of the fastest growing wave according to linear theory, since that wave is known to dominate the early instability in a natural flow. The other eigenfunctions correspond to subharmonics of that wave.

While the cyclical boundary conditions limit the number of subharmonics which can be studied simultaneously with the fundamental with good resolution, the solution for the simultaneous growth of a fundamental and two subharmonics (wave length ratios 1, 2, 4) is currently carried out at reasonable cost.

Resolution limits the maximum value of the Reynolds number. We are routinely using an initial Reynolds number $U\delta/\nu = 100$, where

$$\delta = \frac{\Delta U}{2(\partial U/\partial z)_{max.}} \qquad \text{and} \qquad U = \Delta U/2$$

Our computations provide as printed outputs many details of the flow such as the two components of velocity, the vorticity and the pressure at all grid points, the integral of the Reynolds stress, the rate of dissipation, the momentum thickness, the maximum elevation of the limiting closed streamline (cat's eye), the area enclosed by that streamline, the rate of increase of the fluctuation kinetic energy. Plots are provided at regular intervals of streamlines, and constant vorticity contours. Other quantities of dynamic importance are also calculated as a by-product of the computation. Some of these will be discussed later.

As an illustration, figures (1 abcd) reproduce plots of vorticity contours and of selected streamlines during the later stages of a first pairing between fundamental and its first subharmonic and figures (2 abcd) show similar plots for an initial condition which includes fundamental, first and second harmonic perturbations. For the latter case, the times chosen on the figure corresponds to the second pairing.

It is, of course, also possible to compute the evolution of a passive scalar which diffuses from one steam to the other (e.g., temperature, if one stream is slightly warmer than the other). This is illustrated in Figure (3 ab).

Development in time vs. streamwise development

Can our time-developing layers be used to model streamwise development?

There is no reason of course why a mapping of one onto the other should satisfy the governing equations precisely. It is known for instance that the initial infinitesimal growing waves are non-dispersive for the instability in time of x-wise periodic waves, while they are weakly dispersive for the streamwise instability of time-wise periodic waves.

Nevertheless, such a mapping may be useful for many approximate purposes. For instance, one may postulate that if x, z, t are coordinates and time in the temporal problem x', z', t', the corresponding coordinates and time in the streamwise case, $\overline{U}$, the average velocity of two streams, and ω is the spanwise vorticity,

$$\omega(x, z, t) = \omega'(x', z', t')$$

where

$$z' = z \qquad z = z'$$

$$x' = \overline{U}t \qquad x = x' - \overline{U}t'$$

$$t' = t - {}^{x}/\overline{U} \qquad t = x'/\overline{U}$$

The vorticity equation expressed in the coordinates of the streamwise problem is

$$(2 - {}^{u'}/\overline{U}) \frac{\partial\omega'}{\partial t'} + \overline{U} \frac{\partial\omega'}{\partial x'} + w' \frac{\partial\omega'}{\partial z'} = \nu \left\{ \frac{1}{\overline{U}^2} \frac{\partial^2\omega'}{\partial t'^2} + \frac{\partial^2\omega}{\partial z'^2} \right\}$$

This equation is clearly wrong, but it indicates that the accuracy of the mapping increases as $\overline{U}/\Delta U$ increases. A measure of the error of the mapping is the difference at any point between the velocity V associated with ω in the temporal calculation and the velocity V' induced by the distribution $\omega' = \omega$ at any corresponding point x'*. We are presently making such a comparison in connection with the question: What role do the paired downstream vortices have in exciting subharmonic perturbations upstream, i.e. is there a strong feedback?

For other purposes, a grid computation with boundary conditions relevent to a streamwise-growing-layer is indicated. The main difficulty with this version of the problem is that while it is elliptic, the downstream boundary condition is unknown. However several variations of a scheme which neglects streamwise diffusion over the last streamwise interval can be derived and the distribution of vorticity which would be found downstream of the grid can be idealized with varying degrees of realism, e.g. by use of the transformation suggested above.

Such a computation would be particularly valuable for the study of the early nonlinear period of development. For instance, it might well shed welcome light on the nature of the competitive growth of two periodic perturbations of the layer slightly different (incommensurate) in frequency around the frequency for maximum initial growth rate.

Extension to large Reynolds numbers

It is clear that where the coils are thin and distinct, the dominant gradients of vorticity, concentration, etc. are normal to the coil and further, that the shear across a coil has only a minor effect on diffusion across it. Thus, for computation

---

*This suggests that the mapping is the first term of an expansion of the streamwise-developing vorticity in powers of $\varepsilon = \Delta U/U$.

of the flow quantities in the neighborhood of a coil, an adequate approximation to the advective operator is

$$\frac{\partial}{\partial t} + \vec{u} \cdot \nabla \cong \left(\frac{\partial}{\partial t}\right)_{\vec{s}o} - \gamma\,(\vec{s}o,t)\,\eta\,\frac{\partial}{\partial \eta} \tag{1}$$

where the time derivative on the right is taken as Lagrangian (s_o is the location of a material point at the original interface), V_s is the velocity along the material interface whose local direction is $\vec{s}$, s is the local coordinate along this interface and η a coordinate normal to s. This Lagrangian operator casts advective problems as local problems solved along the characteristic $s = so + \int V_s\, dt$. The approximations leading to eqn (1) require not only a high Reynolds number, but also a sufficiently positive value of the Lagrangian integral $\int_o^t \gamma(t',so)dt'$, in order that diffusive layers be narrow and centered on the material interface. Its use requires the knowledge at all times both of the location of identified points of the material interface and of the Lagrangian function $\gamma(s_o, \tau)$, the interfacial strain rate which is also the logarithmic extension rate of an interfacial segment.

From numerical calculations for several Reynolds numbers and somewhat weaker (i.e., less direct) experimental evidence, we suggest that the material interface location and the function γ are both only weakly affected by the Reynolds number* and we propose that their values computed at a Reynolds number = 100 as part of our finite difference solutions can reasonably be used for all Reynolds numbers.

The Interfacial Evolution

Figures 4 give an idea of the computed evolution of interfacial markers for a simple roll-up and figures 5 for a roll-up and a first pairing. Figures 6 and 7 show the corresponding Lagrangian functions $\gamma(so, \tau)$ for the two cases.

The results indicate that while on the average, the length of the material interface increases substantially (by a factor close to 4 for a roll-up from $\tau = \frac{tU}{\lambda o} = 0$ to $\tau = 2.0$; by a factor of about 8 for a roll-up and pairing from $\tau = 0$ to $\tau = 2.50$), the Lagrangian strain function γ is anything but monotonic with either s_o or time. The total strain of individual elements of the interface vary with initial location by a factor of about 130 in the first case and by more than 500 in the second. Some sections of the initial interface are more consistently compressed than stretched and therefore any diffusive thickness associated with such segments would increase at any Reynolds number rather than decrease. In general, a sample solution of an interfacial diffusion problem of the type

$$\frac{\partial c}{\partial t} - \gamma \frac{\partial c}{\partial \eta}\,\eta = D\,\frac{\partial^2 c}{\partial \eta^2}$$

This is less true for the initial roll-up than for subsequent pairings.

yields a diffusive layer thickness, $\delta(s_o, t)$ which, for a given value of t, increases almost monotonically from the saddle (the middle of the braid) to the focus in a roll-up; a substantial part of the region surrounding the focus in the core would be occupied by overlapping rather than distinct diffusive layers at any Reynolds number. During pairing, the behavior of γ is still more complex. In particular, the region along the interface midway between two merging vortex cores is subject to an especially large positive strain which enhances the inter-diffusion of the two vortex centers. We have not yet analyzed these recent results carefully. But it already seems clear that the boundary layer-type approximations which we have mentioned for high Reynolds numbers are only applicable to the braids and to the external folds of the spiral. On the other hand, for many applications in which total rate of mixing or rate of diffusion-limited reaction is desired, these are the regions which make by far the greatest contribution to these rates.

An application: Diffusion-limited reaction

The knowledge of the interfacial properties of the shear layer can readily be used in problems involving diffusion and chemical reactions between the two streams. We use a simple example as an illustration. It is an extension of that offered by Marble and Broadwell (1977).

The two streams have different chemical compositions and can react with each other. The binary reaction is fast enough to be diffusion-limited. Let K_1, K_2 be the mass fraction of the two reacting species, K_3 that of the product. All three satisfy the equation

$$\frac{\partial K_i}{\partial t} - \gamma z \frac{\partial K_i}{\partial z} = D_i \frac{\partial^2 K_i}{\partial z^2} \tag{2}$$

under the conditions discussed earlier. Assume $D_i = D$ for all three species. Neglect the effect which the reaction may have on the motion. The diffusion layers are assumed distinct from one coil to the other so that K_1 and K_2 reach asymptotic values away from the reaction surface. Let z_o (T, s_o) be the normal distance of the reaction surface from the material interface. If f is the stoichiometric ratio, Marble and Broadwell (1977) show that

$$\frac{\partial K_1}{\partial z}(zo, t) \Big/ \frac{\partial K_2}{\partial z}(zo, t) = -f \tag{3}$$

K_1 is found for $z > z_o$, K_2 for $z < z_o$ and, at $z = z_o$, $K_1 = K_2 = 0$. K_3 is produced at the reaction surface and is found on either side of it. Mass conservation gives at the surface the equation

$$\left(\frac{\partial K_1}{\partial z} + \frac{\partial K_3}{\partial z}\right)_{z = z_o^+} - \left(\frac{\partial K_2}{\partial z} + \frac{\partial K_3}{\partial z}\right)_{z = zo^-} = 0 \tag{4}$$

at $z = z_o$, the fluid normal velocityis due to the net mass flow across the surface, i.e.:

$$D\left(\frac{\partial K_1}{\partial z} + \frac{\partial K_3}{\partial z}\right)_{z = zo^+} = -\rho w\,(z_o)$$

Now, take the reaction surface as the origin, y = 0 and assume the similarity form

$$K_i = f_i\left({}^{y}/\delta_D\,(s_o,t)\right) = f_i\,(\eta)$$

with boundary conditions

$$\begin{pmatrix} K_1 \\ K_2 \end{pmatrix} = 0 \quad \text{for } y = 0; \qquad \begin{array}{ll} K_1 \to K_{1\infty}, & \eta \to \infty \\ K_2 \to K_{2\infty}, & \eta \to -\infty \end{array}$$

then one finds that

$$K_1 = \frac{K_{1\infty}}{1-\alpha}\left[\text{erf}\left(\frac{\pi^{1/2}}{2}\,\eta + \beta/\pi^{1/2}\right) - \alpha\right] \tag{5a}$$

$$K_2 = \frac{K_{2\infty}}{1+\alpha}\left[\text{erf} -\left(\frac{\pi^{1/2}}{2}\,\eta + \beta/\pi^{1/2}\right) + \alpha\right] \tag{5b}$$

Where $\alpha = \text{erf}\ \beta/\pi^{1/2}$; equation (3) leads to

$$\alpha = \frac{1-\phi}{1+\phi}$$

where ϕ is the equivalence ratio,

$$\phi \equiv \frac{K_{1\infty}}{K_{2\infty} f} \tag{6}$$

The local volume consumption rates are

$$V_1 = D\,\frac{\partial K_1}{\partial y}\bigg|_{y = o} = \frac{K_{1\infty} D}{\delta_D}\;\frac{\phi+1}{2\phi}\;e^{-\beta^2/\pi} \tag{7a}$$

$$V_2 = -D\,\frac{\partial K_2}{\partial y}\bigg|_{y = o} = \frac{K_{2\infty} D}{\delta_D}\;\frac{\phi+1}{2}\;e^{-\beta^2/\pi} \tag{7b}$$

where:

$$\delta_D^2 = \delta_D^2\,(t = o,\, s_o)\,\exp.\left[-\int_o^t 2\gamma dt'\right] + \pi D \int_o^t \exp.\left(-\int_{t'}^t 2\gamma dt''\right) dt' \qquad (8)$$

The total reactant consumption up to time t is found by integrating first in time for an identified segment of the reaction interface and then over the whole initial reaction surface. For reactant 1, this is for instance,

$$C_1 = D\, K_{1\infty} \frac{\phi + 1}{2\phi}\, e^{-\beta^2/\pi} \int_o^{\lambda o} \left[\int_o^t \left(\frac{1}{\delta_D(so,t')} \int_o^{t'} \exp \gamma\, dt''\right) dt'\right] ds_o \qquad (9)$$

where λ_o is the initial length of the reaction surface considered. δ_o is given by equation (7) and γ by the computation discussed earlier.

The rate of generation of product (3) is the sum $V_1 + V_2$ and one finds easily that

$$K_3 = K_{30}\left(1 - \frac{K_1}{K_{1\infty}}\right); \quad y > 0 \qquad (10a)$$

$$K_3 = K_{30}\left(1 - \frac{K_2}{K_{1\infty}}\right); \quad y < 0 \qquad (10b)$$

where $K_{30} = K_3(y = 0) = K_{2\infty}\,(f + 1)\,\frac{1 + \alpha}{2}$.

We note, from the form of eqs. (8) and (9) that production takes place almost exclusively where $\int_o^\tau \gamma\, d\tau > 0$, so that the solid cores where the treatment above is inappropriate because $\int_o^\tau \gamma d\tau < 0$ there in general, makes only minimal contributions to the production; on the other hand the products should occupy the core almost exclusively. For a real turbulent shear layer, it would seem realistic to expect the stochastic nature of the large scale (i.e. two-dimensional) structures to have little effect on the reaction rate. On the other hand the small, three-dimensional scales which are observed to grow rapidly and to create a mixing transition (Konrad, 1977, Breidenthal, 1978), need to be taken into account. One should replace the molecular diffusion coeeficient D, by a local eddy-diffusivity related to γ and to its Lagrangian history. At this time we have little experimental data and insufficient theoretical models to define this relation.

A speculation about the origin of the randomness of the coherent structures in the shear layer

As was demonstrated by Brown and Roshko (1974), the coherent structures which occur in an experimental shear layer evolve with a substantial measure of randomness. Thus at a fixed point, the frequency of passage of vortex cores is broadly distributed around a mean, the location of the nth pairing fluctuates etc. Yet the model which we have offered for these structures is fully deterministic. The difference can conceivably come from two distinct sources. The first one is that in the calculation we control the phase, wave length and amplitude of all initial disturbances (fundamental and subharmonics) while the background fluctuations upon which the sequential instabilities feed in a natural flow are evidently random. The second possible cause, which is more or less axiomatic in usual discussions of turbulence is that the flow is effectively not unique in the sense that an infinitesimal perturbation in, say, the boundary conditions leads rapidly to a finite variation in the flow field. The difference is subtle but real, since in the first case a proper, say, statistical, definition of the initial conditons may lead via stable solutions, to an accurate statistical description of the solution, while in the second, there is no obvious way to construct the statistics. We suggest that a flow in which the only sequential instabilities are subharmonic is unlikely to be sensitive to small perturbations during its finite development and is in fact neutrally stable to such influences. In principle, then, we suggest that the random properties of the coherent structures may be calculable from the statistics of the initial perturbations. We have as yet no evidence to support this suggestion except the insensitivity of our numerical calculations to small round-off errors.

F.S. Sherman and S.Y. Lin have both considerately contributed to the calcualtions discussed above.

References

Ashurst, W.T. (1977) Numerical Simulation of Turbulent Mixing Layers via Vortex Dynamics. SANDIA Laboratories Report SAND 77-8613.

Breindenthal, R.E. (1978) A Chemically Reacting Shear Layer. Thesis, Cal. Inst. Tech.

Brown, G.L. and Roshko, A. (1974) On Density Effects and Large Structure in Turbulent Mixing Layers. J. Fluid Mech. 64, 775.

Corcos, G.M. (1979) The Mixing Layer: Deterministic Models of a Turbulent Flow. University of California, Berkeley, College of Eng. Report No. F. M-79-2.

Corcos, G.M. and Sherman, F.S. (1976) Vorticity Concentration and the Dynamics of Unstable Free Shear Layers. J. Fluid Mech. 73, 241.

Delacourt, B.A. and Brown, G.L. (1979) The Evolution and Emerging Structure of a Vortex Sheet in an Inviscid and Viscous Fluid Modelled by a Point Vortex Method. Second Symposium on Turbulent Shear Flows, July 2-4, 1979, Imperial College, London, England.

Konrad, J.H. (1977) An experimental Investigation in Two-Dimensional Turbulent Shears Flows with Applications to Diffusion-Limited Chemical Reactions. Thesis, Calif. Inst. Tech.

Marble, F.E. and Broadwell, J.E. (1977) The Coherent Flame Model for Turbulent Chemical Reactions. Project SQUID Report, Contract No. 4165-52 and 8960-18.

Patnaik, P.C., Sherman F.S. and Corcos, G.M. (1976) A Numerical Simulation of Kelvin-Helmholtz Waves of Finite Amplitude. J. Fluid Mech. 73, 215.

Peltier, W.R., Halle, J. and Clark, T.L. (1978) The Evolution of Finite Amplitude Kelvin-Helmholtz Bellows. Geophys. and Astrophys. Fluid Dyn. 10, 53.

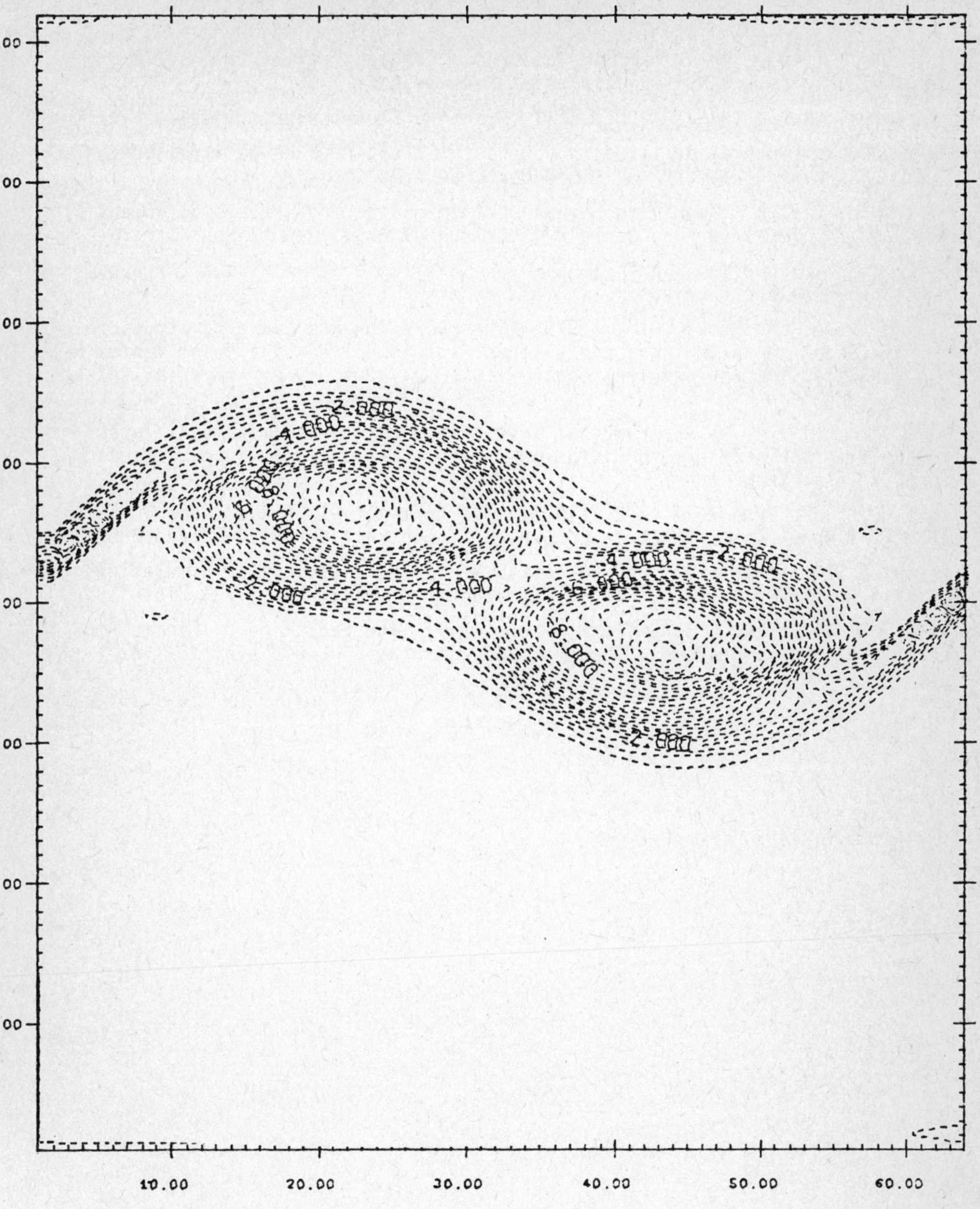

Fig. 1a Vorticity contours. Roll-up and pairing. $\alpha_1 = 0.43$; $\alpha_2 = 0.215$; $R_e = 100$.

$\tau = \frac{tU\alpha}{2\pi\delta} = 2.$ Ω max ≈ 13.5

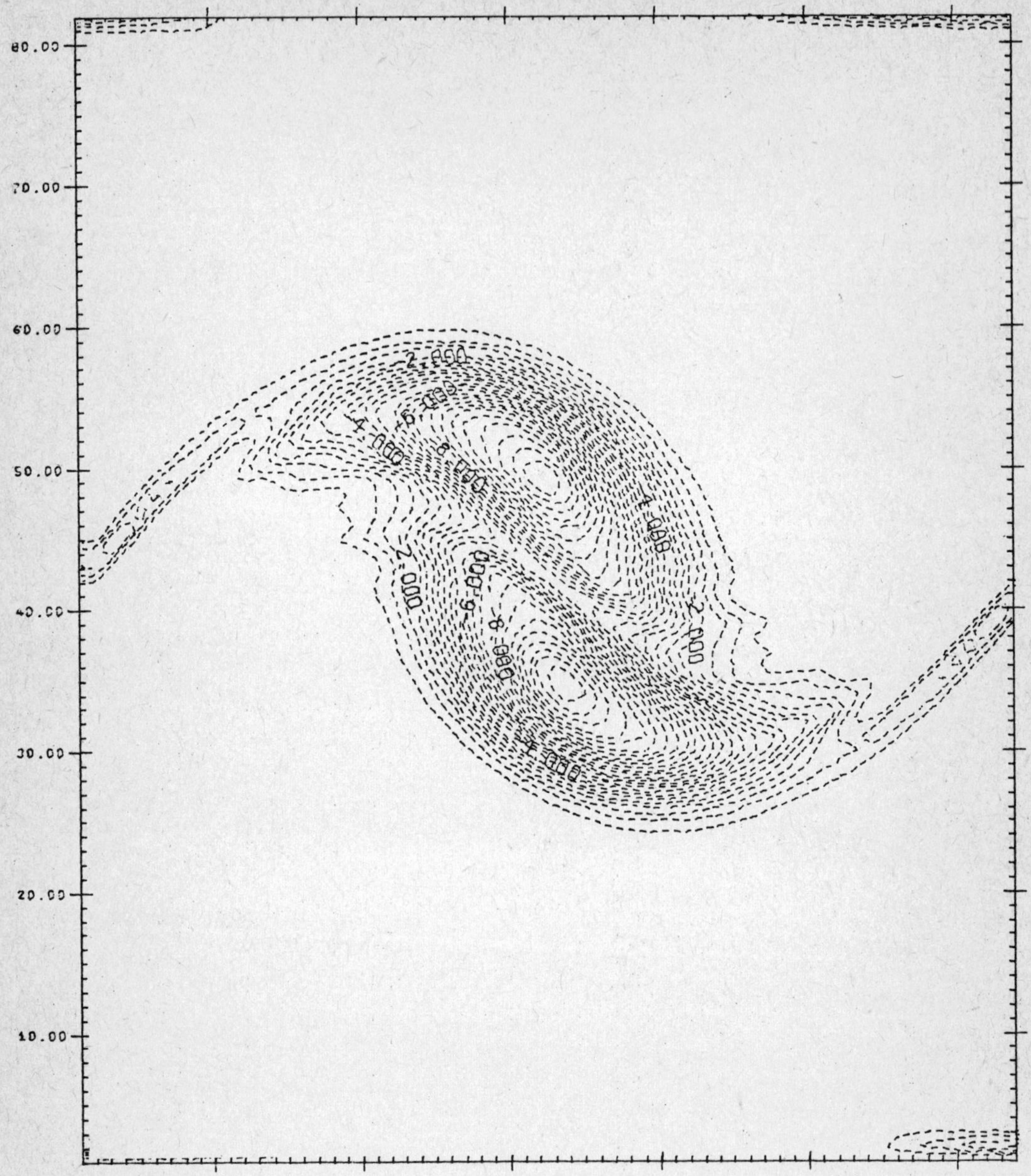

Fig. 1b Vorticity contours. Same case as 1a. $\tau = 2.5$

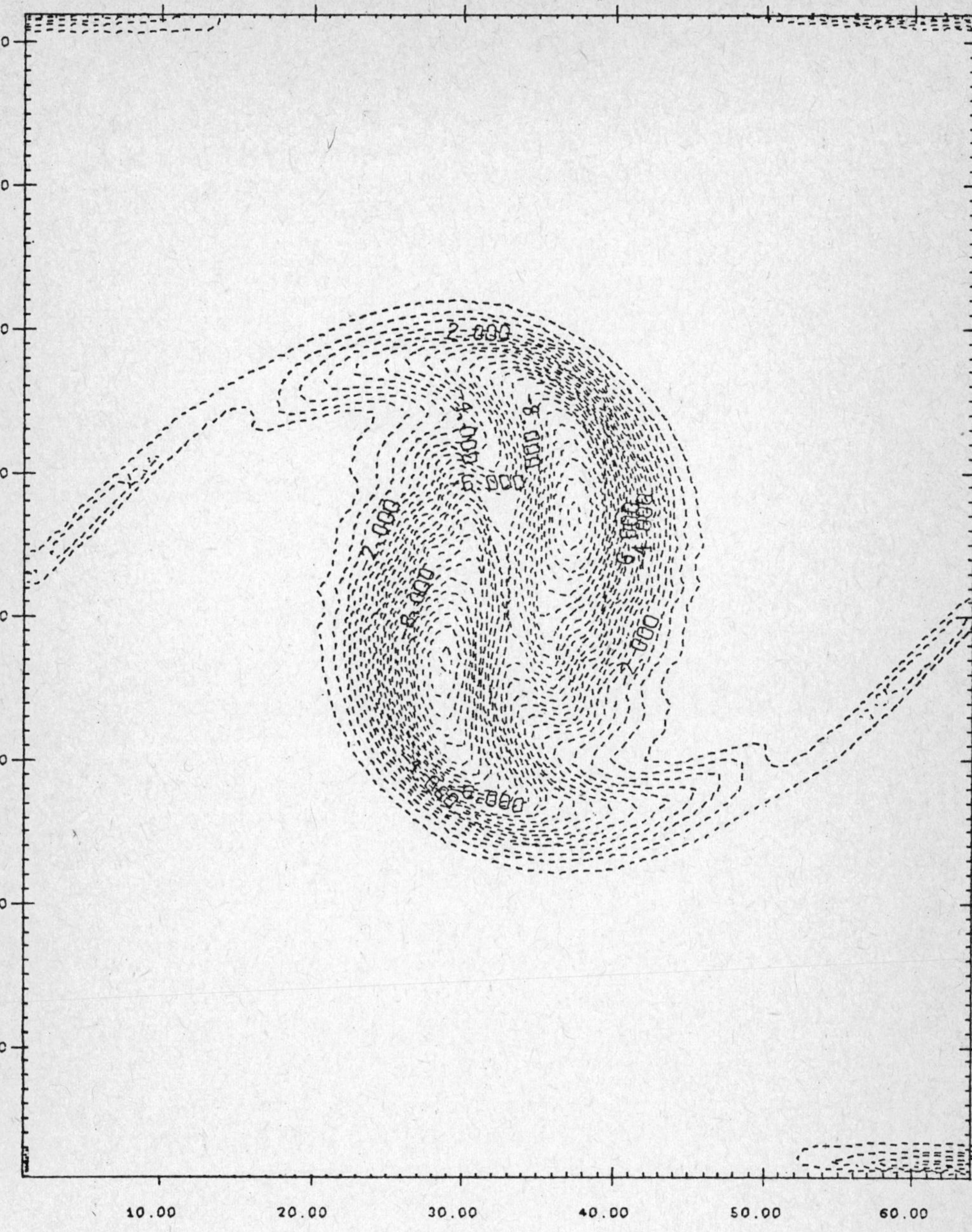

Fig. 1c Vorticity contours. Same case as 1a. $\tau = 2.75$

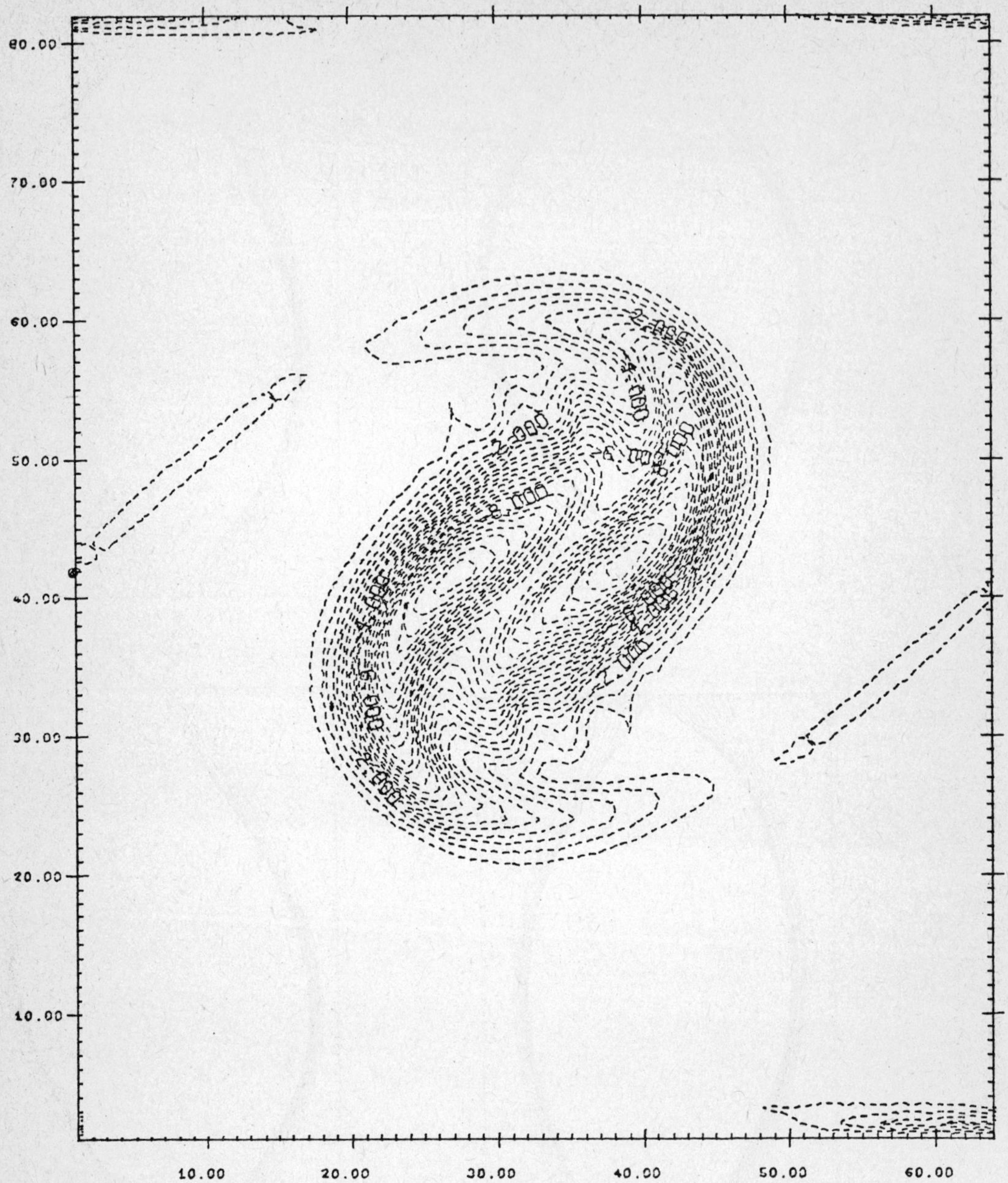

Fig. 1d Vorticity contours. Same case as 1a. $\tau = 3.0$

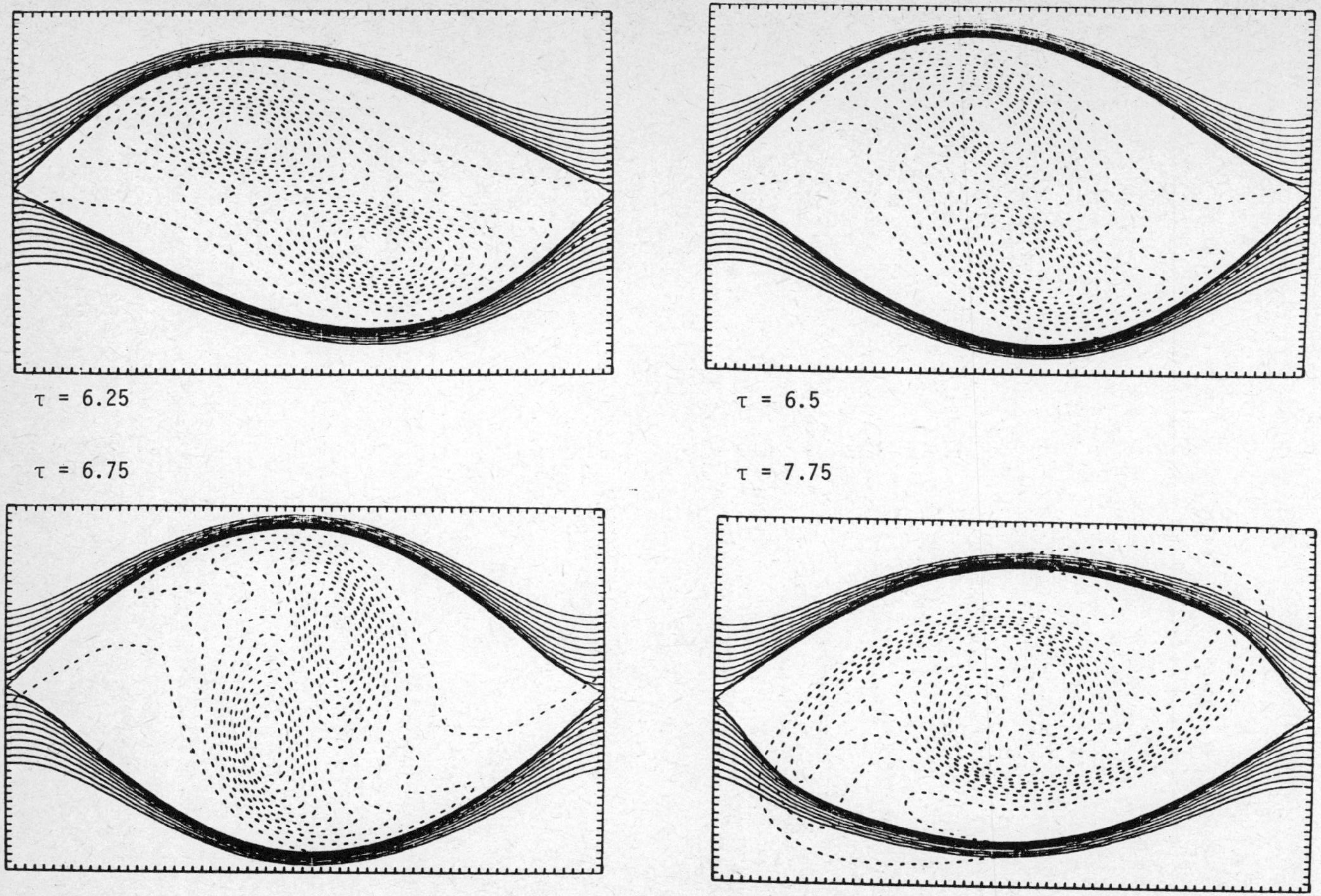

Fig. 2 Vorticity Contours (---) and streamlines (—). Roll-up and two successive pairings. $\alpha_1 = 0.43$. $\alpha_2 = 0.215$; $\alpha_3 = 0.1075$; $R_e = 50$. The first pairing has left no trace of the shorter wave.

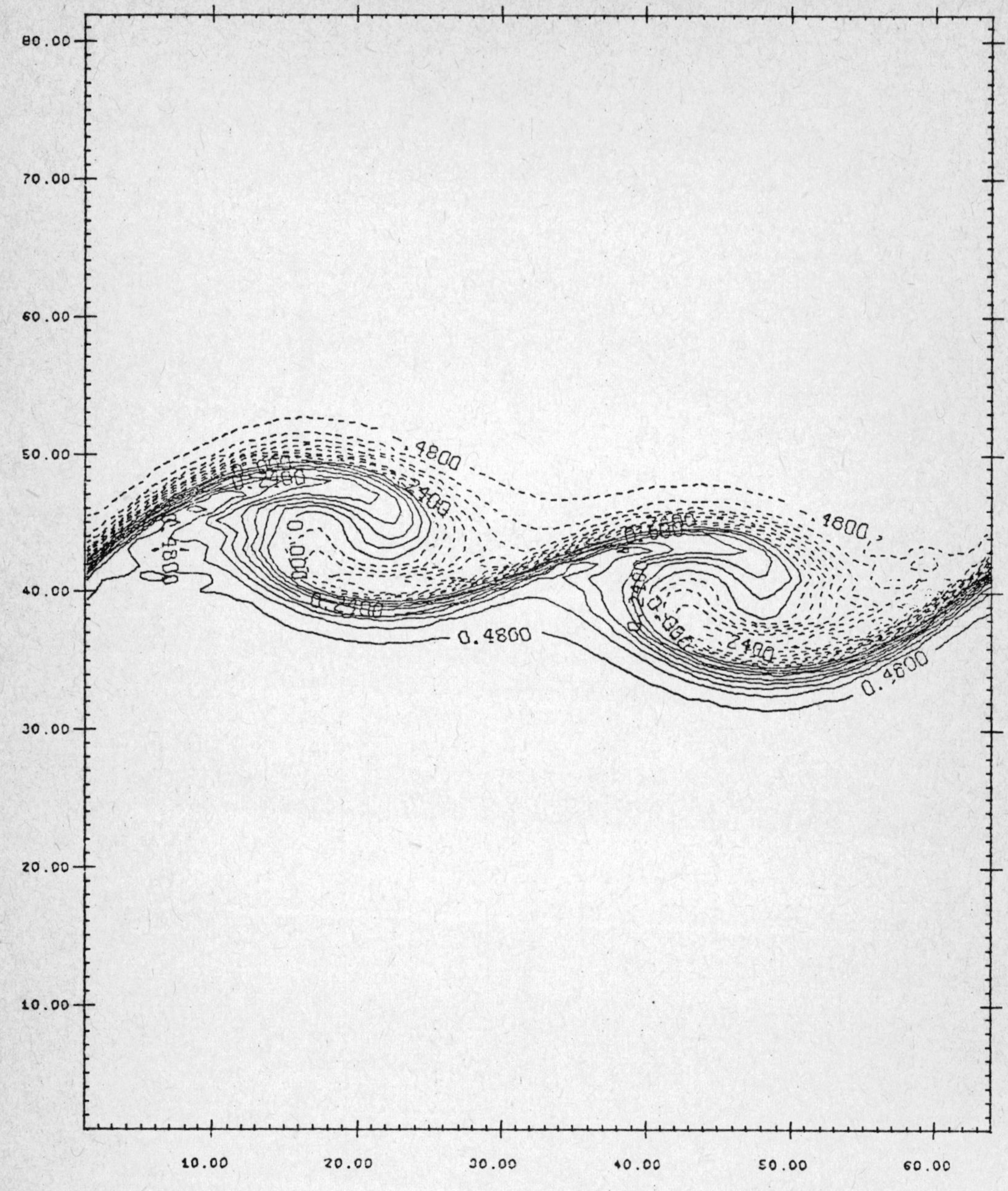

Fig. 3a Diffusion of a passive scalar. Pr = 0.72. Lines of constant concentration. Same case as 1abcd. Free stream values of scalar {± 0.500. τ = 1.50

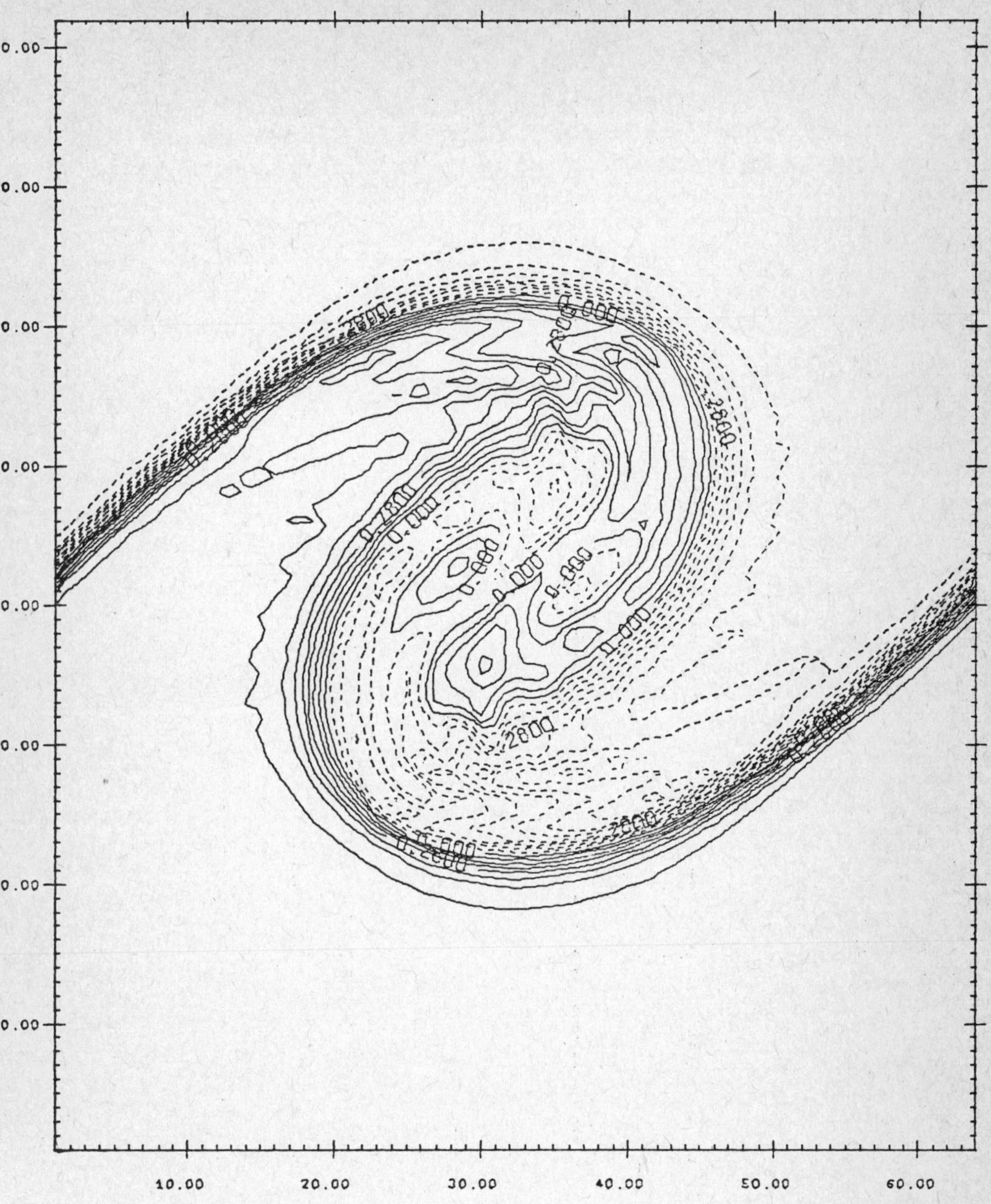

Fig. 3b Same case as 3a. τ = 3.00

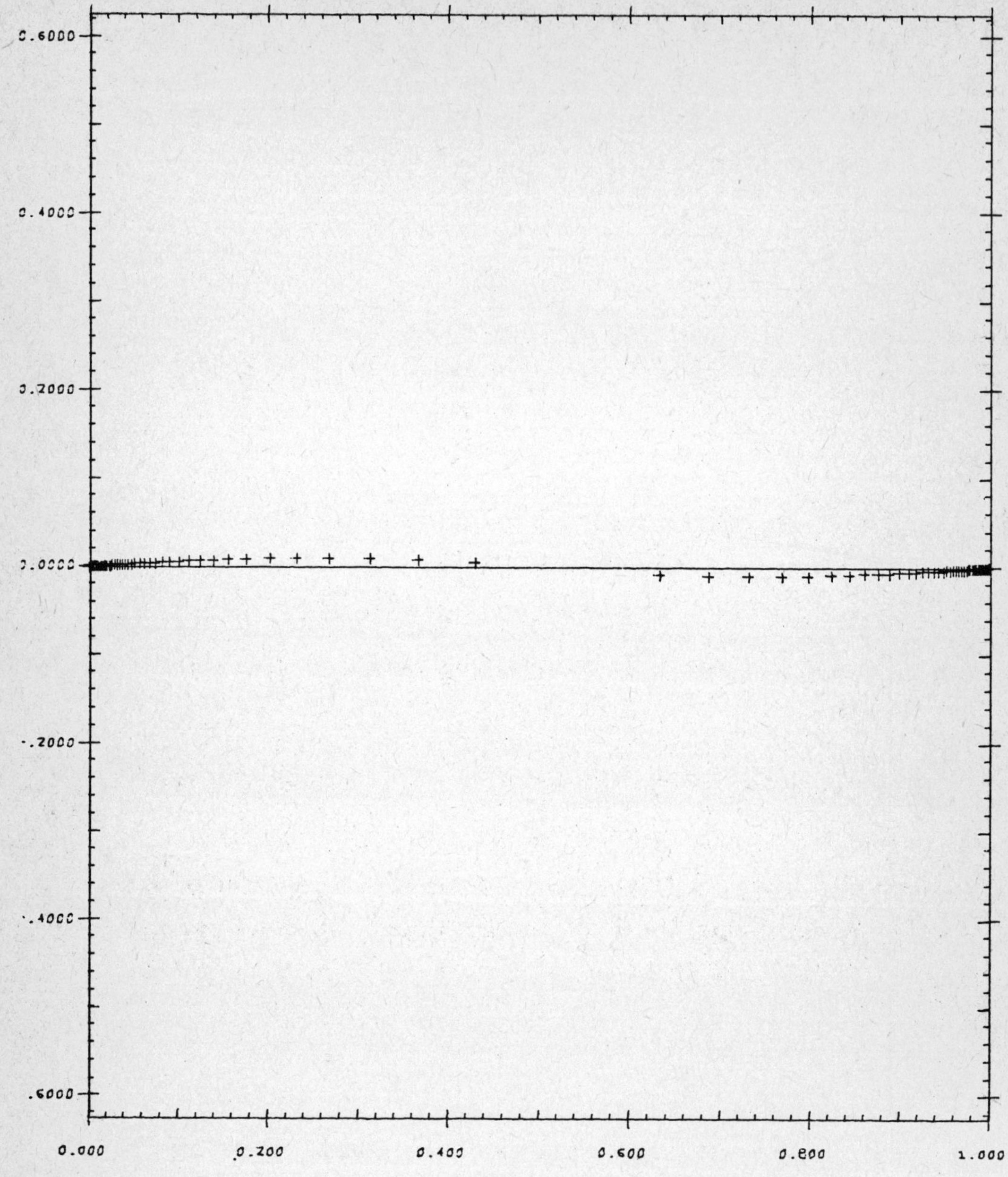

Fig. 4a The evolution of the material interface. Crosses are non-diffusive markers. Roll-up; α = 0.43. R_e = 100 τ = 0

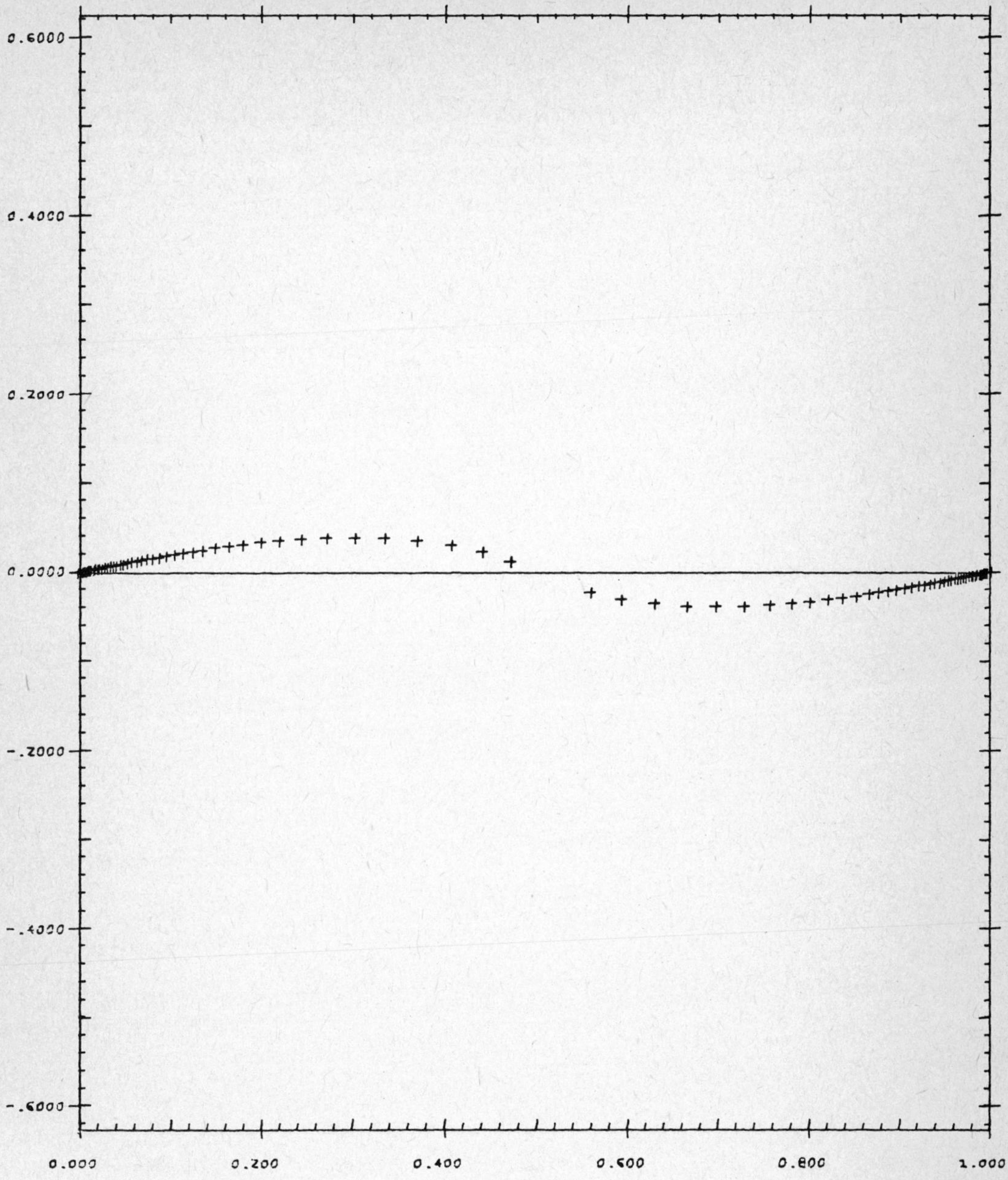

Fig. 4b Same as 4a; $\tau = 0.50$

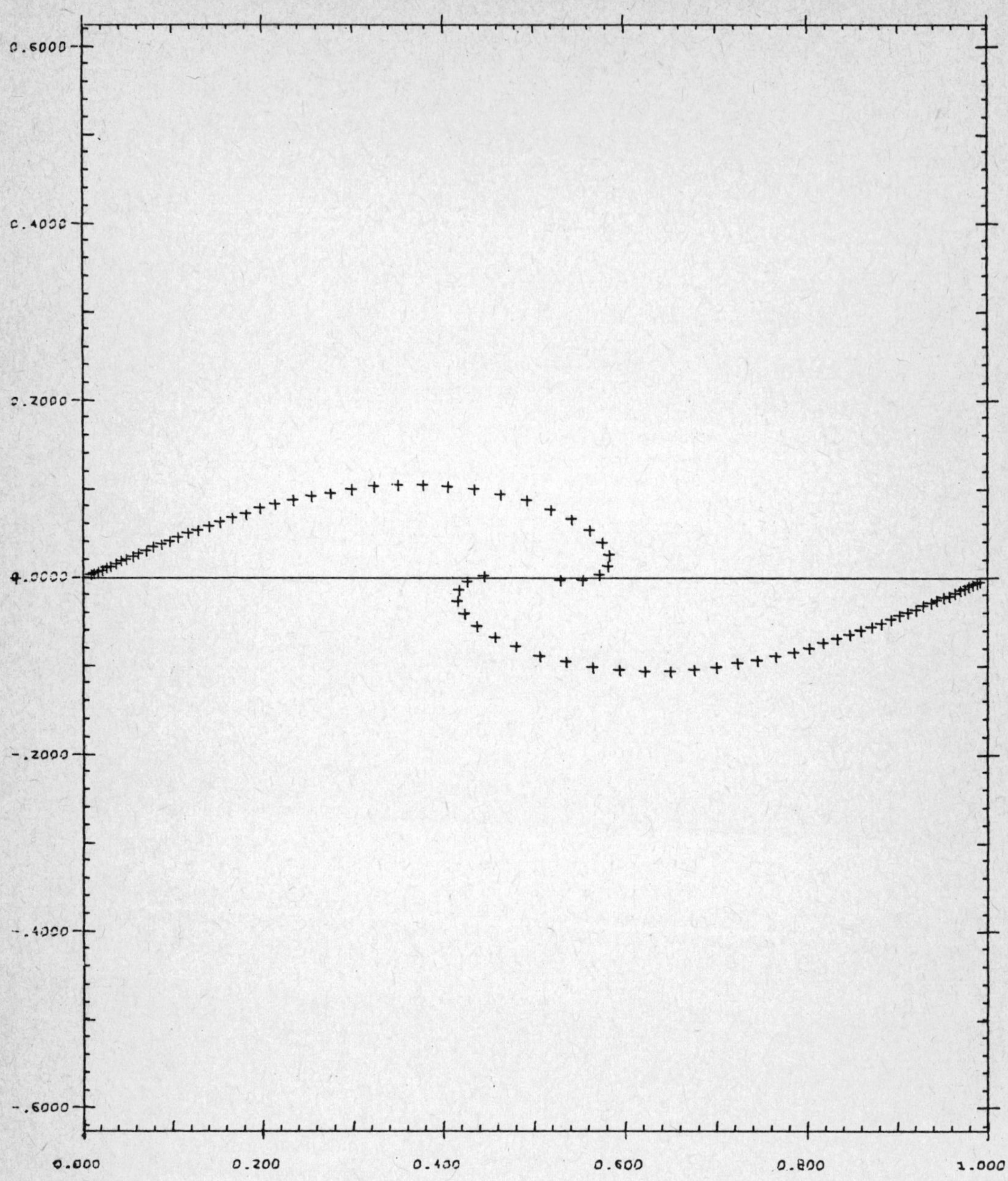

Fig. 4c Same as 4a; $\tau = 1.0$

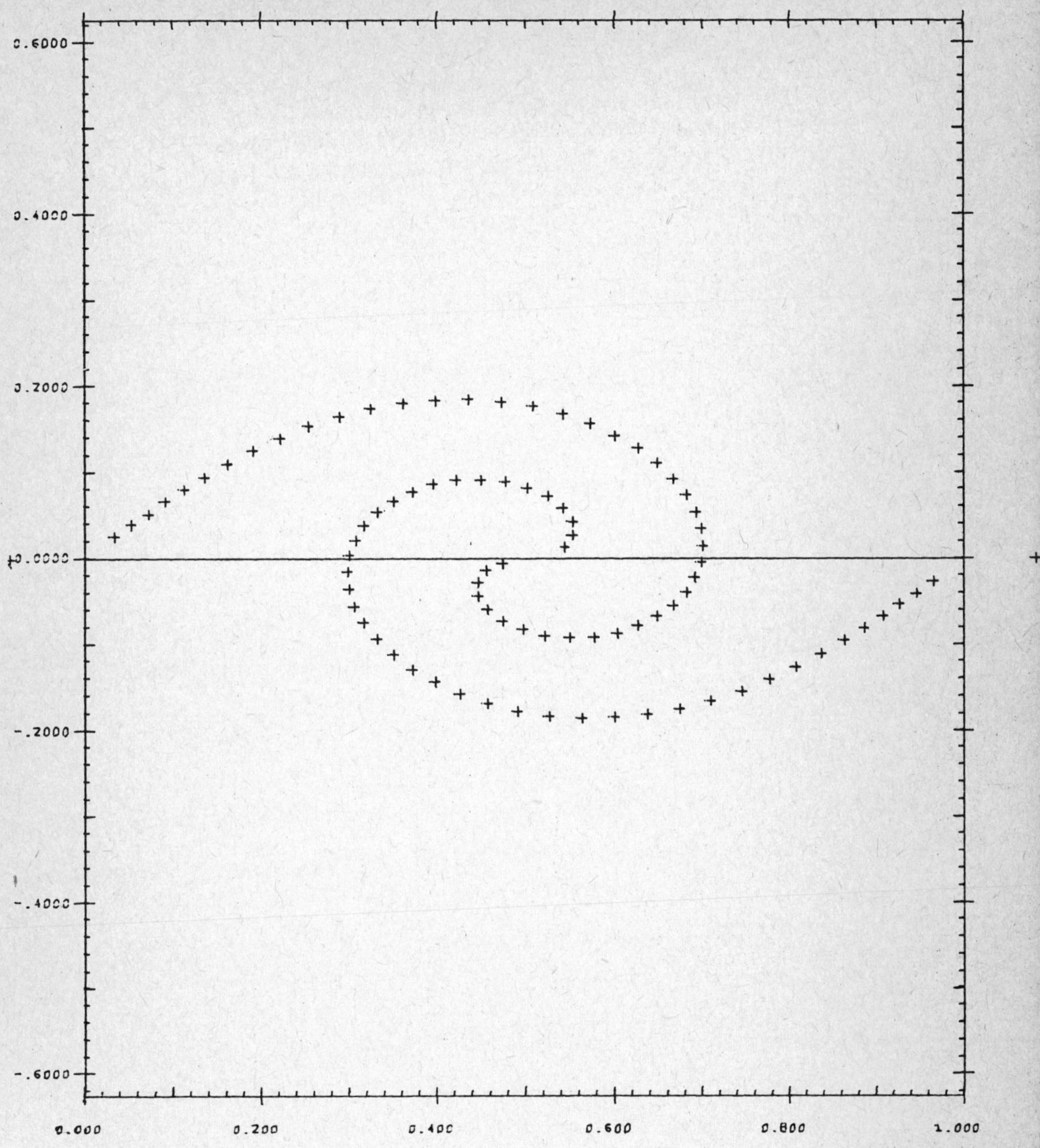

Fig. 4d Same as 4a; $\tau = 1.50$

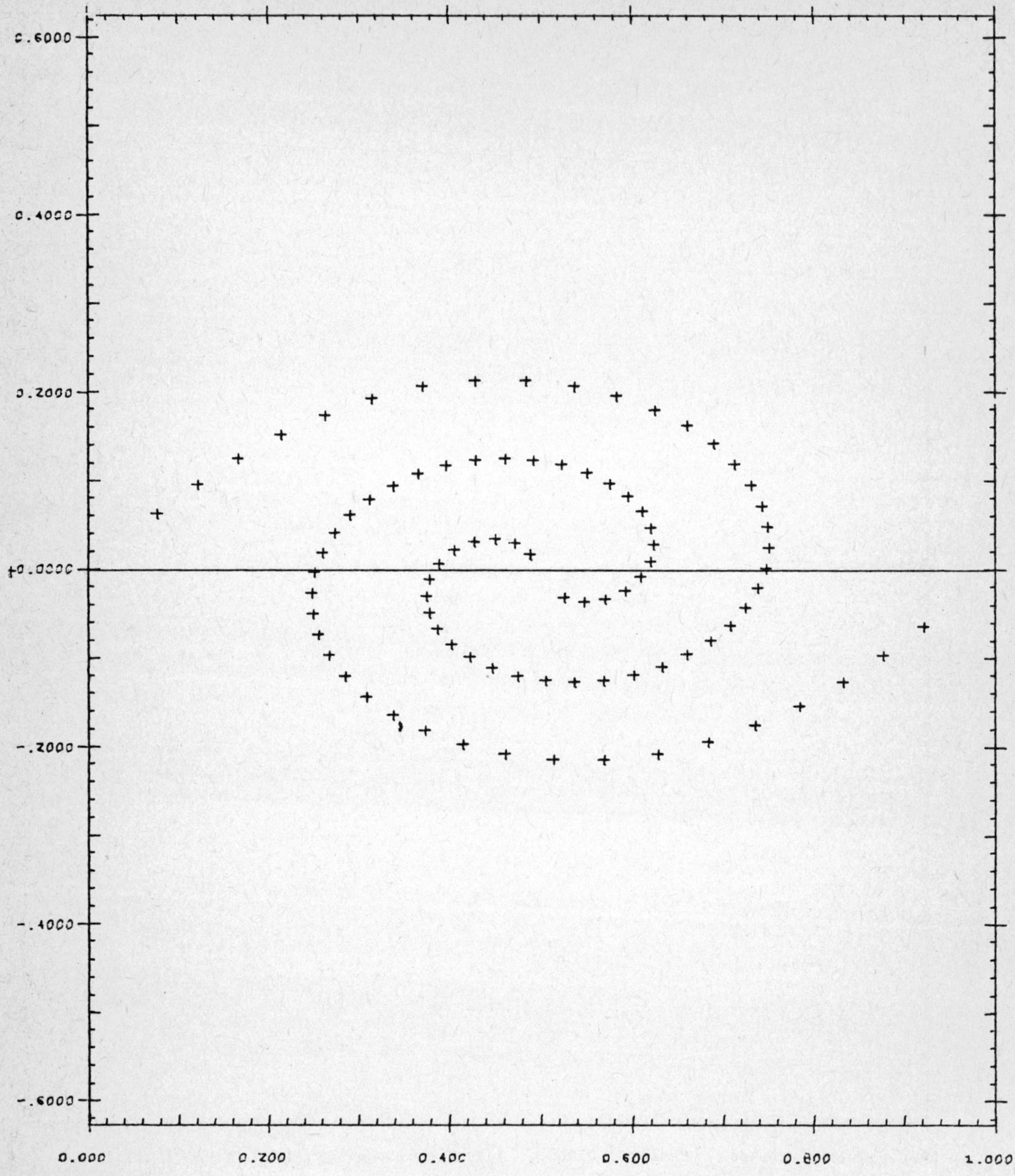

Fig. 4e Same as 4a; $\tau = 1.75$

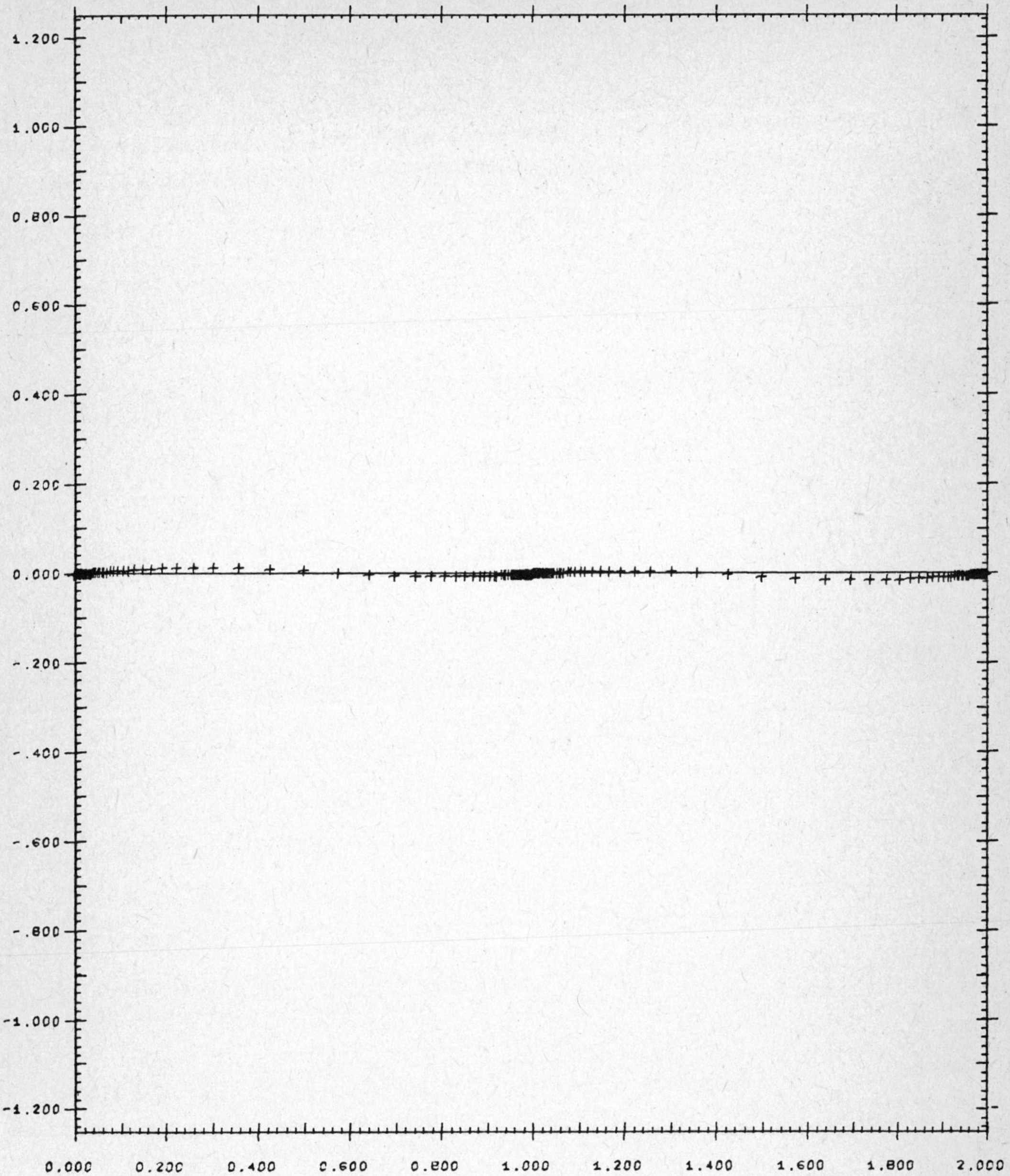

Fig. 5a The evolution of the material interface. Crosses are non-diffusive markers. Roll up and pairing. $\alpha_1 = 0.43$; $\alpha_2 = 0.215$. $R_e = 100$; $\tau = 0$

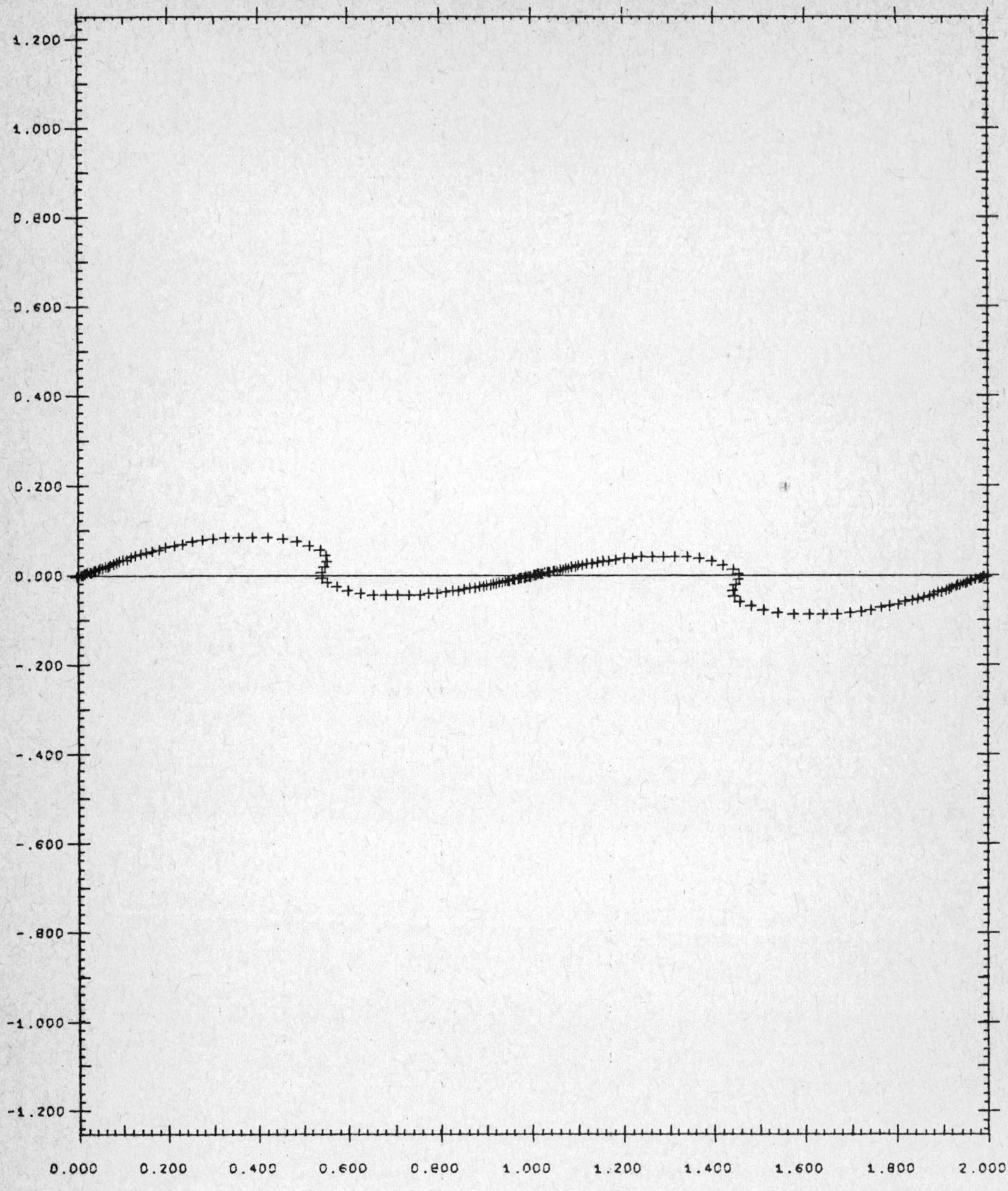

Fig. 5b Same as 5a; $\tau = 0.75$

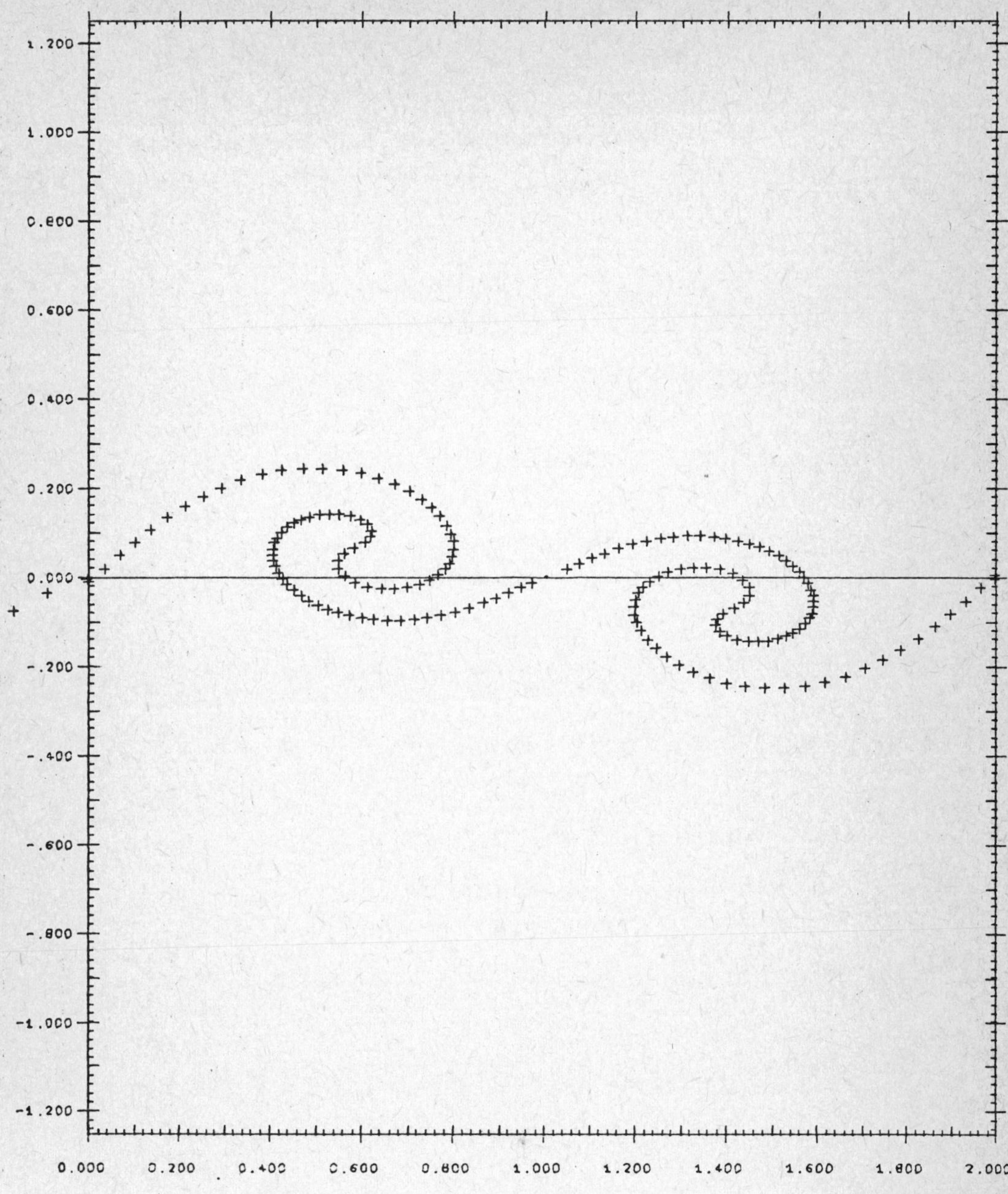

Fig. 5c Same as 5a; $\tau = 1.50$

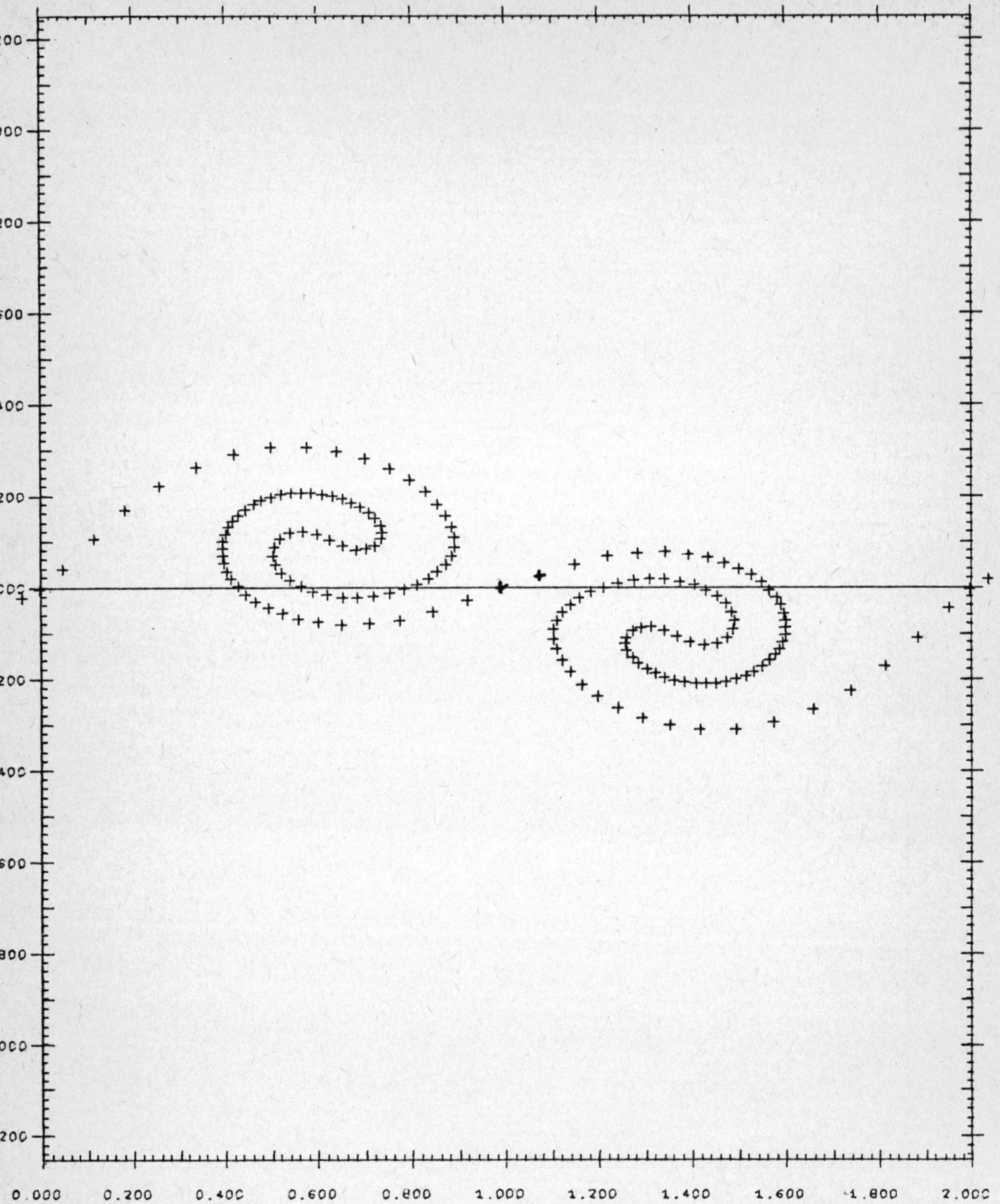

Fig. 5d Same as 5a; $\tau = 1.75$

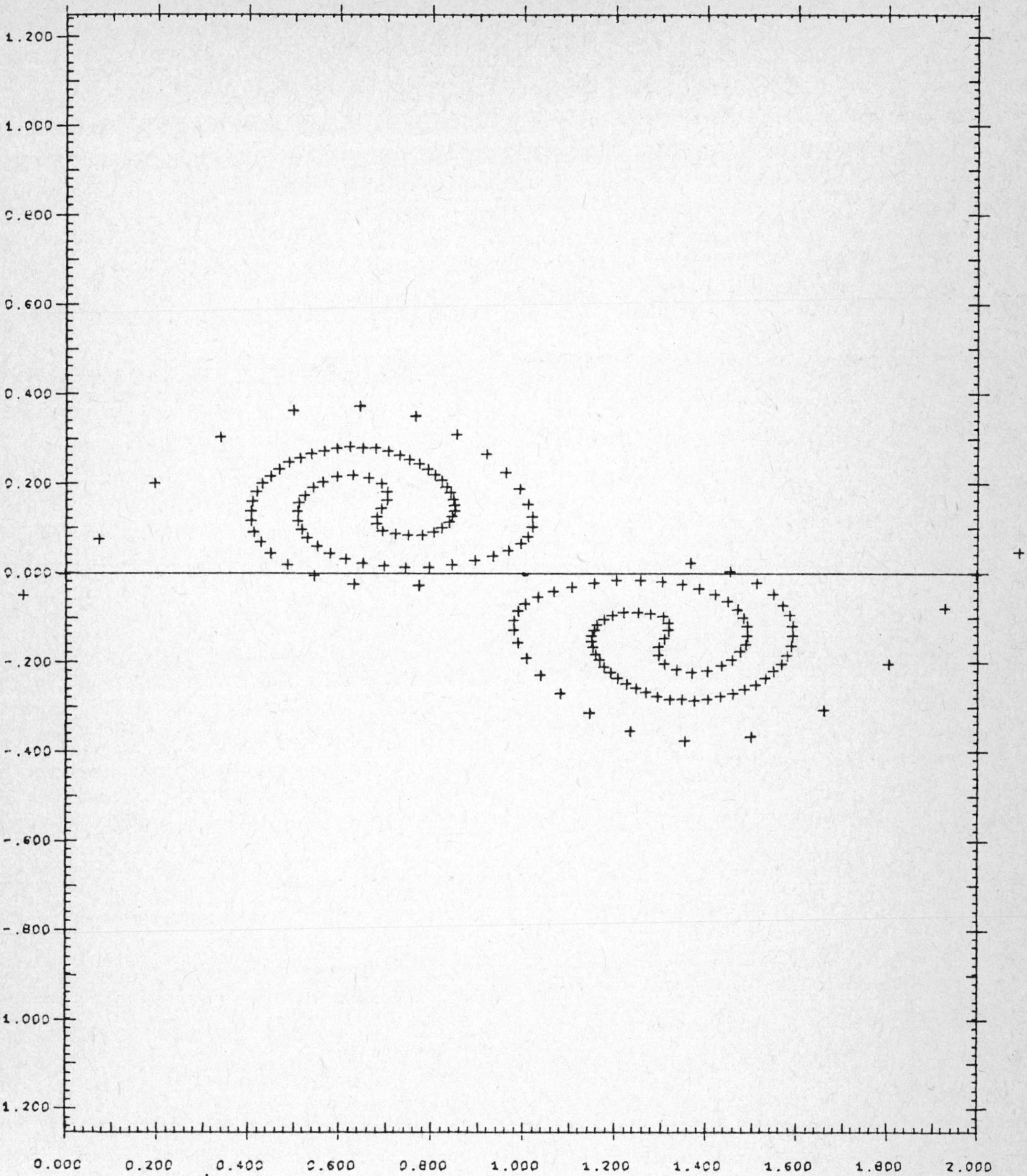

Fig. 5e Same as 5a; $\tau = 2.00$

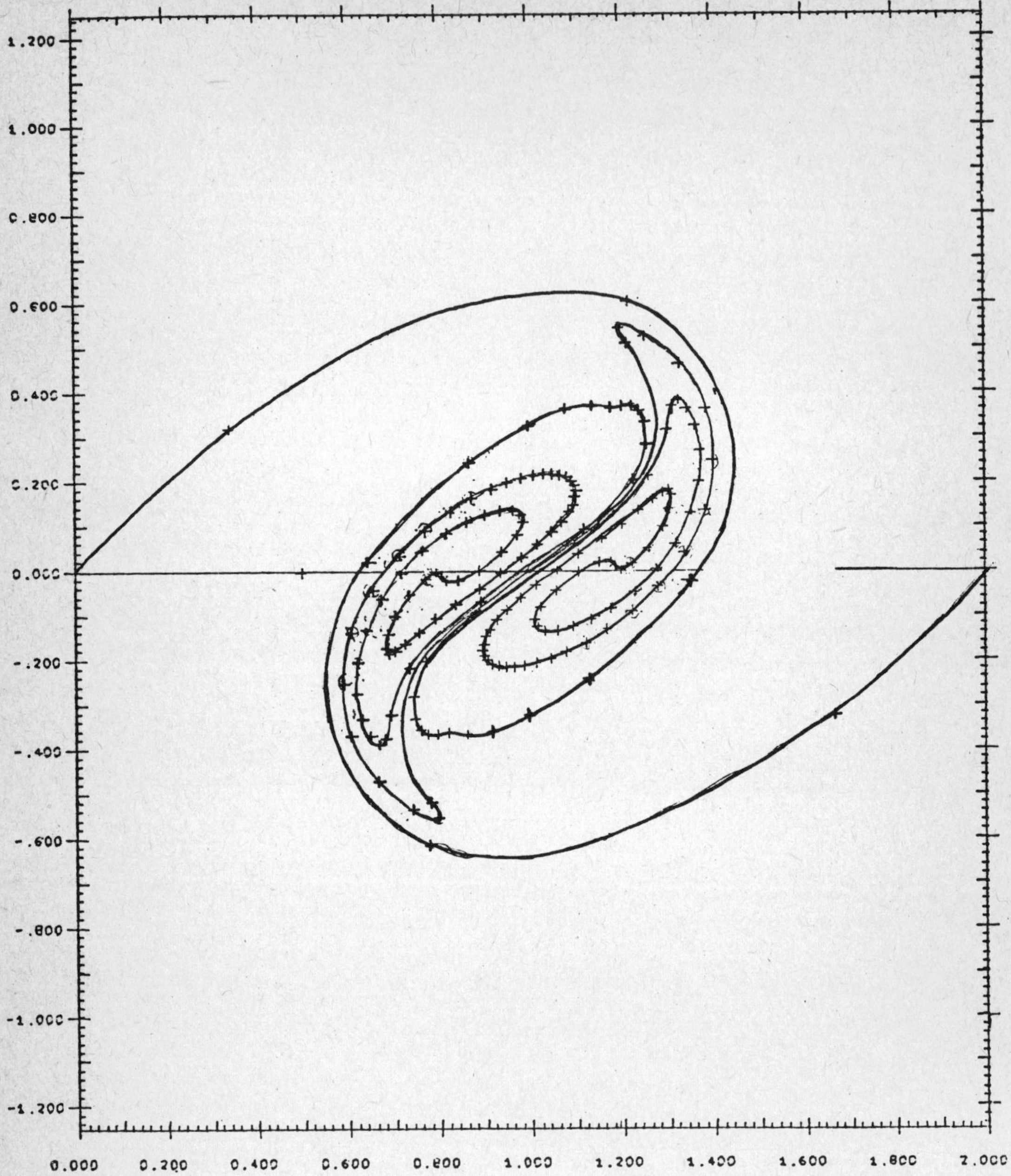

Fig. 5f Same as 5a; $\tau = 3.0$

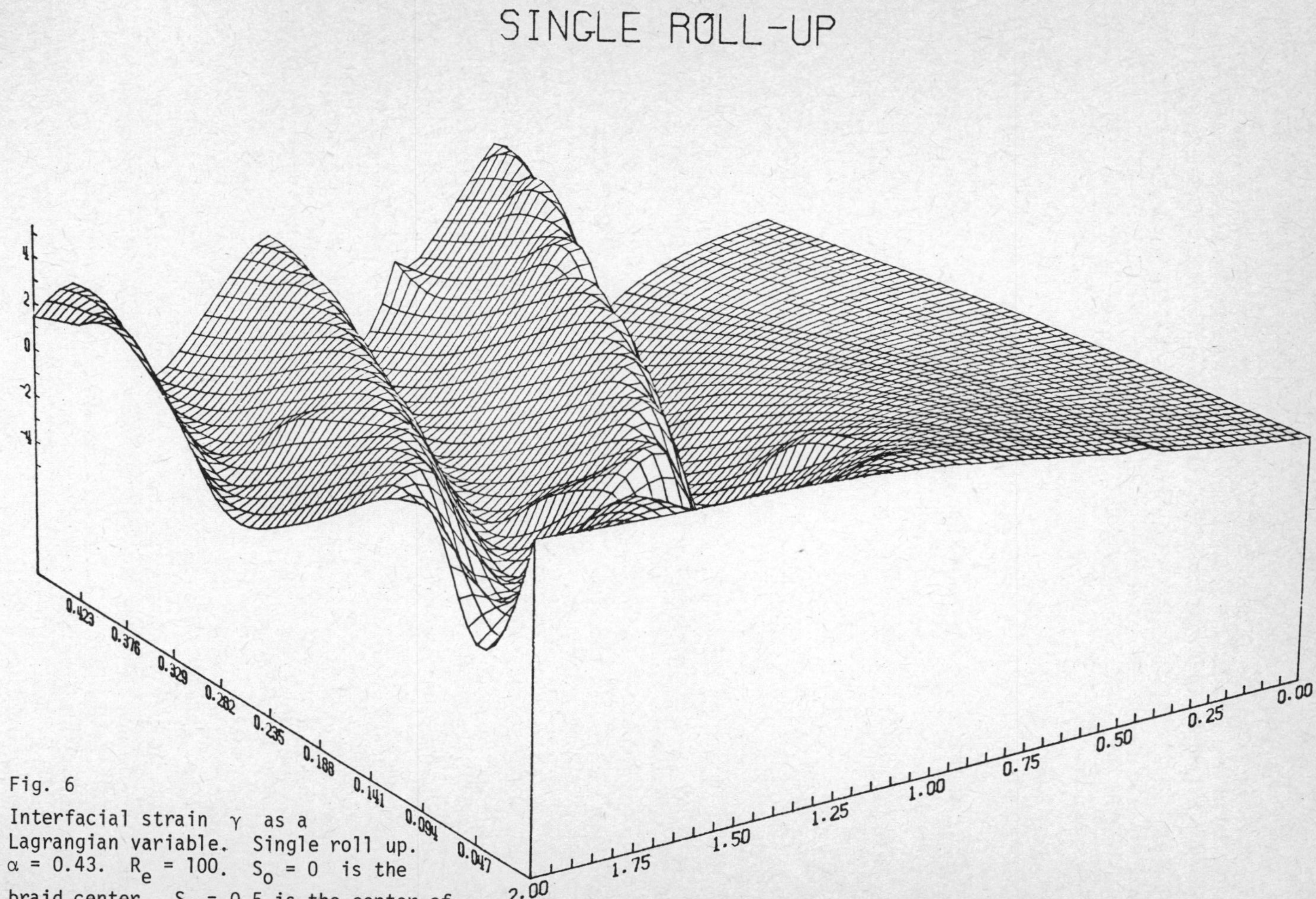

Fig. 6

Interfacial strain γ as a Lagrangian variable. Single roll up. $\alpha = 0.43$. $R_e = 100$. $S_o = 0$ is the braid center. $S_o = 0.5$ is the center of the Vortex core.

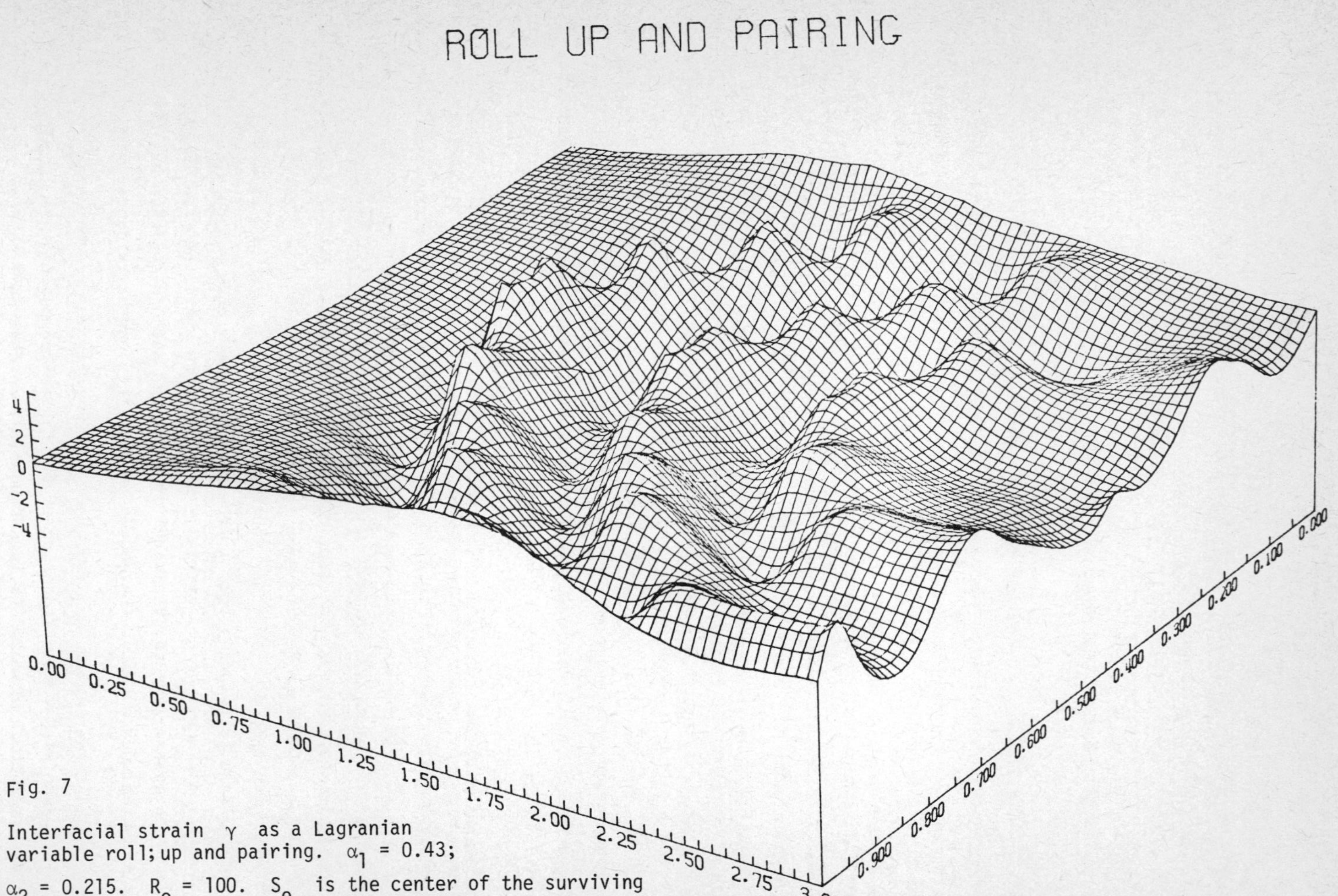

Fig. 7

Interfacial strain γ as a Lagranian variable roll; up and pairing. $\alpha_1 = 0.43$; $\alpha_2 = 0.215$. $R_e = 100$. S_o is the center of the surviving braid. S_1 is the center for the pairing vortices.

SHEAR LAYER MODELS AND COMPUTER ANALYSIS OF DATA

Javier JIMENEZ
UAM-IBM Scientific Center
Castellana, 4, Madrid-1, Spain.

R. MARTINEZ-VAL and M.A.HERNAN
Universidad Politécnica, Madrid.

Introduction

Modelling a turbulent flow involves more than predicting its technologically interesting properties. While it is probably true that a better model will predict better the separation of a boundary layer, the converse is not necessarily true; quite as important is the amount of insight gained. This is difficult to define. Appart from a certain amount of esthetic appeal that makes you feel that you have "understood" a certain situation in a certain case, insight has something to do with minimum complexity and with the facility of including new features starting from a given model.

Thus, once we are able to use Prandtl's lifting-line theory to compute the behaviour of a slender wing, we can hope to understand which are the effects of body-wing interaction, even if for it we need to put the problem in a computer. However if we use a large computer simulation just to study the behaviour of the wing profile, even if the results are probably much better than those of the simple theory, chances are that the complete problem will soon be outside the bounds of available computer power or programming complexity. The lifting-line theory is a less accurate predictor but gives more insight.

A different question is how to construct good models. An useful approach is to break the big problem into smaller modules and to try to understand them one by one. The problem then is to validate the pieces. If one wants to keep the submodels understandable and simple it is often necessary to purposely neglect some effects, even if they are known to be important for the final predictions. Therefore these predictions cannot be used to check whether the model is good, and we need experimental results relating only to the partial aspect under study. This has been traditionally difficult to do, specially in turbulence where the experimental signals are complex. Instead, the usual approach has been to see what quantities can be measured and to tailor the models so that they can be validated with those quantities.

This has changed with the introduction of the digital computer as an integral part of the data reduction process. It is now possible to analyze a complex signal and isolate from it, using pattern recognition techniques, specific features relating to a particular aspect of the flow and to use these features to validate submodels.

Two examples are given in this paper, both relating to the plane mixing layer. This flow is a good test case for modelling and experiments. Its two-dimensionality, at least in the initial region, makes it easier to understand than almost any other turbulent flow and, in addition, a large amount of experimental and theoretical work has been done on it in the last few years.

There are two different aspects of the mixing layer. First, the flow field. At the exit from the splitter plate we have the continuous creation of a thick vorticity sheet that immediately develops a wavelike instability resulting in the formation of vorticity "knots" at regular intervals. This linear phase is well understood and the theoretical predictions agree well with experiments [1, 2]. There is less agreement in the subsequent nonlinear evolution. It seems clear that the vorticity concentrations keep growing and form the seeds of the large coherent structures that are observed later. Since total vorticity is essentially conserved at large Reynolds numbers the growth of the knots, or "cores", can only happen by the amalgamation of several of them into bigger structures. Two mechanisms have been proposed for this amalgamation; historically the first observations were described as "pairings" [3], in which two vortices rotate around each other and become somehow entangled to form a single entity. If we momentarily think of the cores as point vortices this process is easy to understand. A row of vortices is unstable and the dominant mode of instability results precisely in the common rotation of pairs of contiguous vortices. The subsequent entanglement does not result from this model but, in some sense, a pair of rotating cores can be considered a bound state with twice as much vorticity.

A different process, "tearing", was proposed in [4]. Here the finite size of the cores is essential; under some circumstances the cores themselves desintegrate, their vorticity is set free and eventually the neighbouring cores absorb it and grow.

The dynamics of vorticity evolution is a complex nonlinear phenomenon which is not easy to study without the use of simplified models. We address this matter in the second example in this paper in which a particular aspect, the mechanism of amalgamation, is modelled; in fact it is neglected. The predictions, or rather the parameters to be used in the model are then studied by a computer analysis of anemometry measurement in a mixing layer.

This evolution has a parallel in the behaviour of the concentration field. This is a much simpler problem; once the velocity field has been described (or assumed) the transport of a passive scalar property is a linear process which can be computed easily. Also the experimental evidence is more complete; the beautiful films of the mixing regions, specially those by the GALCIT group, contain enough information to completely define the two-dimensional evolution of the concentration field. In the next section we try to check some basic ideas; by just assuming that each concentration eddy contains a region of active vorticity and that these regions evolve more or less independently of one another, and therefore obey similarity rules depending only on their own size, we deduce a series of statistical parameters on the lifetime of each eddy. The different consequences of pairing and tearing are also discussed.

The point of view in this model is essentially Lagrangian, as opposed to Eulerian, in that the evolution of each eddy is followed individually. As a consequence the predictions are also Lagrangian and have to be extracted with care from the visualization films; this we do using digital image processing. The computer is trained to recognize the eddies, draw an x-t diagram and measure such quantities as eddy areas, positions and lifespans. The results are compared with the theoretical model.

The concentration field

In modelling the homogeneous mixing layer we will always make the assumption of temporal rather than spatial growth. The velocities of the top and bottom streams will be taken to be $\pm 1/2$, so that the vorticity in the layer is stationary in the average. Since the spreading angle of physical layers is small and since we will be neglecting in general the interaction of an eddy with all but its most immediate neighbours, this approximation is not critical. The transformation to the spatial case, when needed, will be done by the obvious Galilean transformation with a convection velocity U_c.

After the initial instability phase we assume that the vorticity is concentrated in a row of relatively compact cores. These cores act at large distances like point vortices and tend to roll one of the streams into the other, producing a spiral structure of the concentration field that grows with time [5]. This process continues as long as neighbouring structures do not interfere with each other, in effect exhausting the supply of available unmixed fluid; at this moment the structures ("eddies") will stop growing unless an amalgamation rejuvenates the situation by providing more growing space for each eddy.

For an incompresible fluid the amount of material entrained in an eddy is proportional to the eddy cross section, which can easily be measured on the visualization films. What happens to this area in an amalgamation depends on the details of the process;in a pairing the area after amalgamation will be the sum of the areas before, plus possibly a fraction entrained during the process. In a tearing an eddy will just disappear and the neighbouring ones will absorb its fluid, increasing its area proportionately.

Consider now the characteristic lengths in the flow. In the absence of viscosity and of long range interactions (splitter plate, etc.) the only length available to an eddy is its wavelength, the distance to its neighbours, and this should be equal

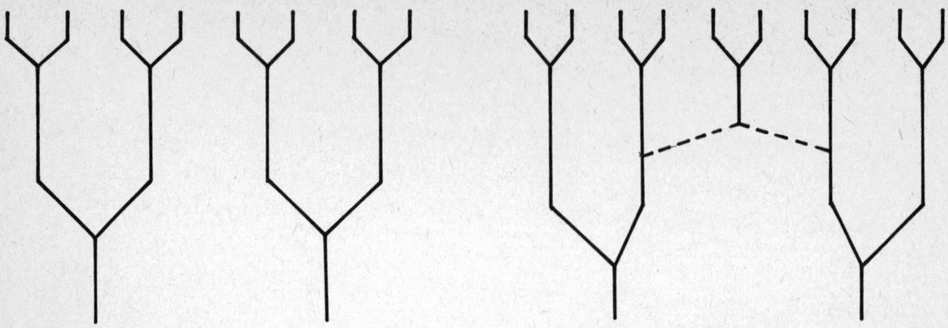

Fig. 1.- Sketch of x-t diagram for eddy motion. a: Only pairing; b: Pairing and tearing present.

in the average to the vorticity contained in the core, normalized with the velocity difference. Consider a simple case in which the only interaction is by pairing (fig. 1.a). After each pairing the vorticity in the average eddy doubles, and so do all scales; therefore the area should increase by a factor of four

$$S_{0(n+1)}/S_{0n} = 4 \qquad (1)$$

Consider a more complicated situation in which there is a certain amount of tearing (fig. 1.b), so that at the end of its lifetime each vortex has increased it vorticity by a fraction τ, received from the teared nearby eddies. By the same argument as before the area increase in one generation will be

$$S_{0(n+1)}/S_{0n} = 4(1+\tau)^2, \qquad (2)$$

and by measuring this ratio we can get an estimate for τ.

This area increase can be divided in two parts, the increase during the roll-up and that during the pairing. This separation can not be deduced from the similarity arguments proposed here but is interesting since the two opposing views have been forwarded: that growth is due to pairing [3] and to roll-up [5]. We will come back to this problem when analyzing the experiments.

There is one further prediction that does not depend on the details of the growth mechanism. After one generation the eddy scales increase by a factor of $2(1+\tau)$; the lifetime of each generation will increase accordingly and for an eddy born after the n-th pairing the lifetime will be

$$T_n = 2^n(1+\tau)^n\, T_0.$$

Going back to the spatial case, during this time the eddy has moved a distance U_cT_n,

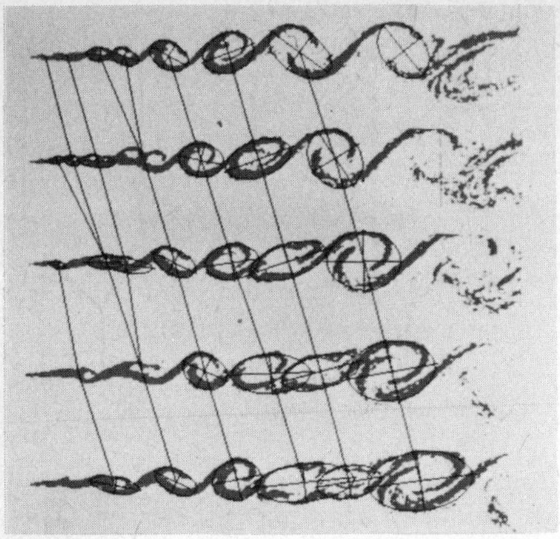

Fig. 2.- The x-t diagram for five frames as interpreted by the computer analysis program.

so that the distance from the n-th pairing to the splitter plate is

$$x_{0n}=U_c \sum_{k=1}^{n-1} T_K=U_c T_0(2^n(1+\tau)^n-1)/(1+2\tau)$$

and the normalized lifetime after a few amalgamations is

$$\frac{U_c T_n}{x_{0n}} = 1+2\tau, \qquad (3)$$

which can also be used to estimate τ.

Both the lifetime and the evolution of eddy cross section can be measured directly from the visualization films. Doing this by hand, however, presents several problems, arising basically from the large number of frames to be processed; there is the shear amount of repetitive work involved in measuring several thousand objects in a consistent manner, and the need to keep a unifom criterium all this time in such subjective matters as the exact boundaries of an eddie or the moment of amalgamation. This is why we decided to let a computer do the measurement. The actual processing is described elsewhere [6,7] but the relevant results and the outline of the method are given here.

A continuous sequence of frames (~400) from a film of a homogeneous mixing layer, kindly provided to us by Prof. Roshko, is digitized in a scanner reducing them to numerical matrices in which each element is proportional to the optical density of the film in the neighbourhood of a given point. This two-dimensional digital signal is then enhanced, filtered and simplified until each frame is classified into zones of mixed and unmixed fluids. The outline of these zones is extracted and the computer is given a series of rules by which it can segment the outlines into individual eddies, to each of which it fits an ellipse which is taken to represent the eddy shape and position. It is now an easy task to follow the motion of these ellipses accross different frames and to organize them into coherent histories.

Each history begins either with the appearance of a new eddy that can not be assigned an ancestor in a preceding frame, or as the merging of two (or several) previous histories. Similarly, histories end either when there is no reasonable continuation in a subsequent frame, or when they merge with a neighbour in an amalgamation (see fig. 2). The process by which a history suddely disappears to merge with appreciable probability with two different neighbours was not forbidden in principle, but was found only seldom (20 times is nearly 700 histories). When

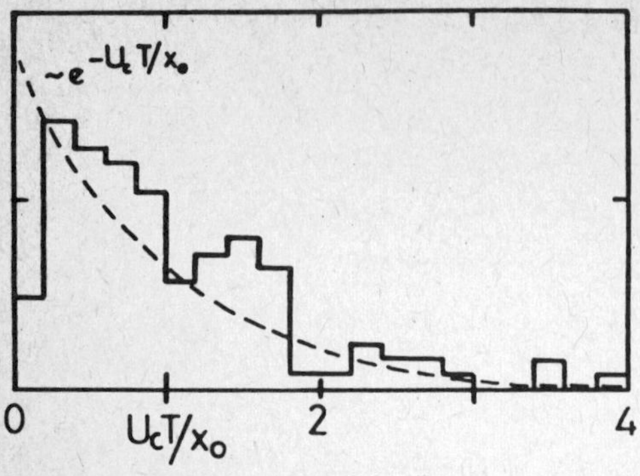

Fig. 3.- Histogram for nondimensional lifetime of eddies. Aver: 1.05 $\pm$ .77, Sample: 114 histories.

those cases where inspected visually, they were seen to correspond in all cases, but one, to small errors in the recognition procedure and to have in any case a clearer interpretation as pairings to one of the neighbours. The remaining case is our only documented candidate for a fast tearing.

The lifespan statistics are given in figure 3, in which only those histories starting and ending with amalgamations have been considered, and which should therefore correspond to equation (3). The average normalized lifetime is 1.05, implying a tearing frequency very close to zero; also the histogram is approximated reasonably well by an exponential distribution with parameter equal to one, confirming the value of the average. A similar histogram was presented in [8] but, even if the distribution was also exponential, the average was lower (.45) and therefore inconsistent with equation (3).

Figures 4(a-b) give histograms of area increase during roll-up and pairing. Again only histories beginning and ending with pairings have been considered, and equation (2) should hold. A slight correction might be needed since some of the interactions involve more than two eddies, but the effect is small and is not taken into account.

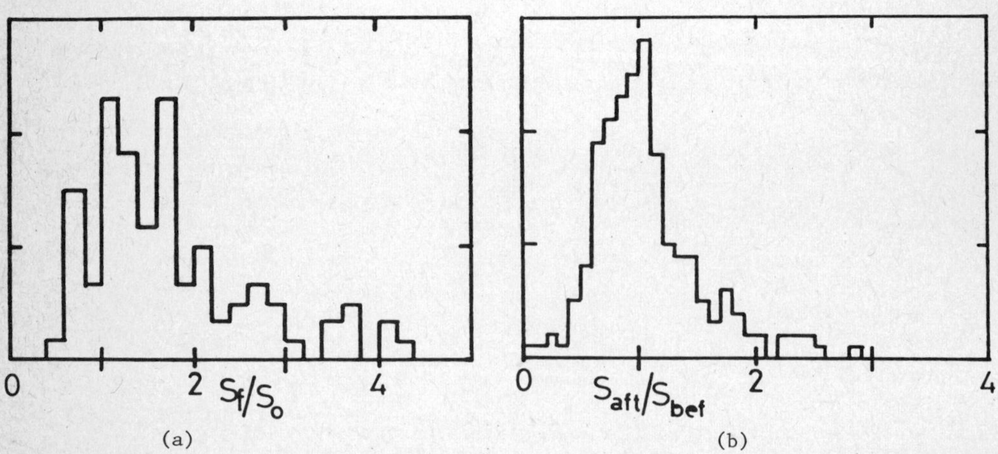

Fig. 4.- Histograms for eddy area growth. a: Area ratio from beginning to end of lifespan; Aver.: 1.75 $\pm$ .87, Sample: 91 histories. b: Ratio of total area before and after amalgamation; Aver.: 1.17 $\pm$ .45, Sample: 210 amalg.

Actually, out of 210 amalgamations studied, 196 are pairings, 14 triplings, and none is a quadrupling. It is interesting that in the average most of the area increase happens during the roll-up (1.75) and relatively little fluid is entrained in the pairing (1.17), in accordance with the idea that the entraining is due mainly to the active vorticity in the eddy cores. The average eddy increase in a generation is obtained by multiplying the average increase ratios in these two processes, times the number of vortices involved in a pairing, so that

$$S_{0(n+1)}/S_{0n} = 2 \times 1.75 \times 1.17 = 4.1,$$

which, compared with equation (2) implies again that essentially no tearing is present.

A more direct counting of pairing versus tearing is possible. Choosing the center part of the layer to avoid complications due to end effects, we counted directly the number of histories beginning and ending in different ways. In this region 96 histories began as a consequence of pairing while 20 started out of nowhere. These last ones were considered recognition errors, an interpretation confirmed by visual inspection. On the 145 histories ending in a pairing, a similar error rate would give 31 mistakes, while the actual number of histories ending without descendency was 36. The excess of 5 histories (out of 181) is a new estimate of the frequency of tearing. Actually, after displaying several cases in which histories ended without pairing, we found that, appart from obvious mistakes, there is a process accounting for these eddy disappearances.

In it, an eddy, usually small with respect to its neighbours, stops growing and is slowly stretched, or bleeded, by them. The process usually takes several frames and is different from the more or less rapid desintegration of an eddy of which we mentioned a single example above. Its frequency is consistent with the estimate of a few percent that is implied by the measurements of area growth and lifespan.

Thus, in all, the results of the computer analysis of the visualization films confirm the idea that eddies grow mainly by roll-up, and almalgamate by pairing. Tearing is observed, but it appears to be a relatively rare event of little importance in the global dynamics of the layer.

Modelling the vorticity field

The results presented above show that from the point of view of the concentration interface the plane mixing layer can be modelled by a more or less one-dimensional row of vorticity blobs, one within each of the concentration eddies.

Unfortunately this does not tell us much about the internal structure of the blobs. In fact, any concentration of vorticity whose effective diameter is less than the size of the concentration eddies will generate around itself a closed streamline and behave from the outside much like a point vortex. Besides, pairing is a major difficulty; in the above discussion it was assumed that when instability of the row brings two cores near one another they somehow merge into a single one.

Two isolated point vortices will not merge but move around each other in a permanent orbit. Tearing, on the other hand, is explained because cores of finite size, when subject to too much strain, become unstable and desintegrate. As cores move downstream in the layer, they grow in size until the strain produced by the collective effect of the vortex row is enough to destroy them [4] . However, the instability mechanisms involved in pairing and tearing are basically the same and the time scales involved are at most similar, while for compact cores pairing is always more unstable. Therefore we would expect to find in a mixing layer mostly pairing interactions with an occasional tearing, as is indeed the case in the experiment reported in the last section.

This still does not explain the merging of the cores involved in a pairing. One possibility, is that the strain induced by the close approach is again able to destroy the cores, which then reform together. However it can be shown [5] that the maximum strain in a vortex row occurs when the cores are at their undisturbed position and that the strain actually decreases when they come together in a pairing; therefore if the cores do not desintegrate before they begin to pair they will not do so afterwards.

Another possibility is viscosity, but the effect will be dependent on Reynolds number and will imply that for low viscosity the cores may orbit for long times without merging, even after the concentration eddies have coallesced completely. Numerical simulations of the mixing layer usually include some viscosity, although seldom in a well controlled way. In some cases [9,10] merging of the vorticity structures is very apparent in the results, while in others [11,12] vortex cores are seen rotating around each other for a long time without merging. Because of the different integration schemes used it is difficult to say whether the effect is real or a computational artifact.

A curious fact is that the easiest and probably one of the most succesfull numerical modellings of the mixing layer actually involves no pairing at all. A large number of point vortices, with some kind of short-range cutoff,are used to simulate an initial vortex sheet which is perturbed slightly, and the motion of the points are used to follow the subsequent behaviour of the layer [10,13,14]. The cutoff seems to be important to avoid numerical difficulties, but the details of how it is done do not seem to affect strongly the outcome.

The results of one such simulation are shown in figures 5 and 6. A hundred point vortices are distributed uniformly along a segment on the x axis and their initial positions are perturbed slightly before starting the integration. The whole

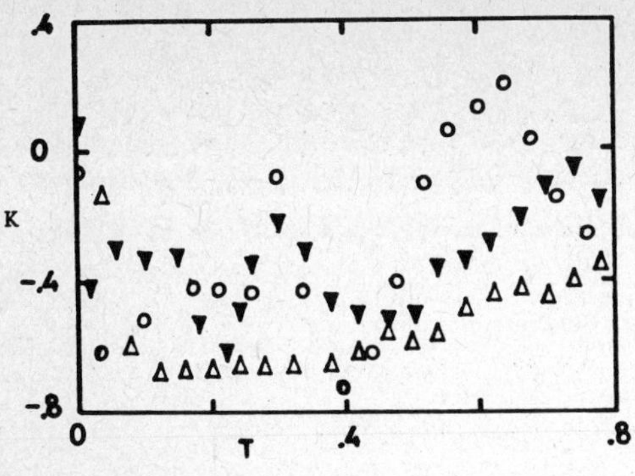

Fig. 5.- Evolution of kurtosis of lateral distribution of vortices in simulation. $K=\mu_4/\sigma^4-3$.

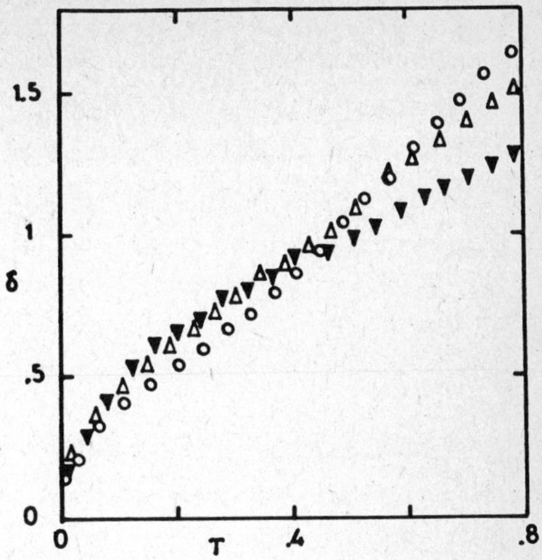

Fig. 6.- Vorticity thickness growth in point vortex simulation.Different symbols are different simulation parameters.

arrangement is assumed to be periodic in x and the short range cutoff is provided by using a look-up table to compute the induced velocities (for a similar method see [15]). Each of the curves in the figures is an average over five independent random initial conditions.

As soon as the integration begins, the vortices start to scatter laterally. Since all vortices have the same strength the probability density of this distribution is proportional to the average vorticity profile accross the layer and the average velocity can be obtained by simple integration.

Because of the symmetry of the equations of motions, if the odd central moments of this distribution are initially zero, they remain so forever and the distribution is determined by its even moments. The evolution of kurtosis is shown in figure 5; even if it shows wide oscillations it seems to drop very fast to about -.5 and to stay around that value. That implies that the vorticity profile is almost gaussian, but slightly flatter, in agreement with experimental measurements. Disregarding the deviation from gaussianity the velocity profile will be an error function and the vorticity thickness $(\partial u/\partial y)^{-1}$ will be proportional to the standard deviation

$$\delta = \sqrt{2\pi}\ \sigma \ . \tag{4}$$

This is represented in figure 6 for three different combinations of numerical parameters and it is seen to grow linearly in all cases, with growth rates (.11-.18) that again agree well with experimental evidence.

Since at least the mean velocity profile of the mixing layer is described so well by

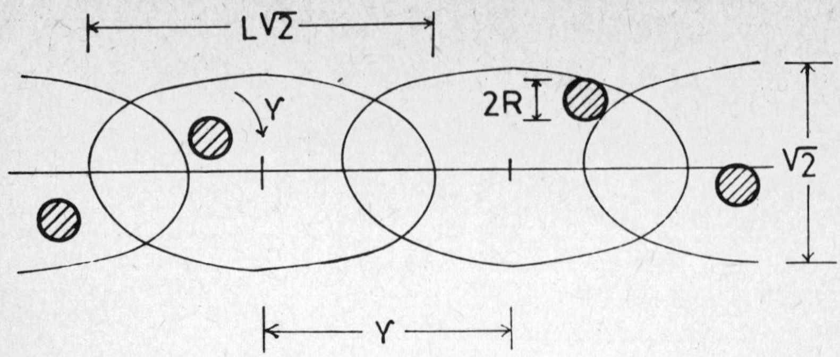

Fig. 7.- Geometry and scatter ellipses for the vortex cloud model.

a model that neglects such pressumably basic features as vortex core sizes, amalgamations and large scale structure, it is interesting to explore how far can other average properties be computed from a simple description of the layer as a cloud of vortices with the correct lateral scatter.

Consider an arrangement consisting of a row of solid vortex cores with strenght γ and radius R distributed more or less uniformly along the X axis. If the units are taken so that the velocity jump is one, the average longitudinal distance among vortices is also γ. The position of the n-th core is a random variable distributed normally around the center point $(n\gamma,0)$ with variances which are 1/2 in the transverse (y) direction and $L^2/2$ in the longitudinal one, so that the density function is

$$P_n(x,y) = p(x-n\gamma,y) = \exp\,(-y^2-(x-n\gamma)^2/L^2).$$

To simplify calculations the positions of all vortices are assumed to be mutually independent, and the spirit of the model will be to think of R and γ as small numbers (see fig. 7). Somewhat surprinsingly the resulting velocity distributions are found to depend not only on R and γ but on the longitudinal scatter L, and, while vortex size and strength are measurable physical properties, the scatter is a parameter of the model that has to be computed theoretically.

Happily the total energy associated with the vortex distribution is also a function of L and can be used to decide which value to use. For point vortices the excess energy (per vortex) with respect to the uniform row is proportional to [16]

$$E = -\iiiint_{-\infty}^{\infty} \sum_{n\neq 0} \log \frac{(x-x')^2+(y-y')^2}{n^2\gamma^2}\, P_0(x,y)P_n(x'y')dx\, dy\, dx'\, dy', \qquad (5)$$

which can be manipulated into the form

$$E = -2\pi^{-3}\gamma^2 L \iint_0^{\infty} \log \frac{\sin^2\xi+\mathrm{sh}^2\eta}{\xi^2+\eta^2} \exp \frac{-\gamma^2(\xi^2/L^2+\eta^2)}{2\pi^2}\, d\xi\, d\eta, \qquad (6)$$

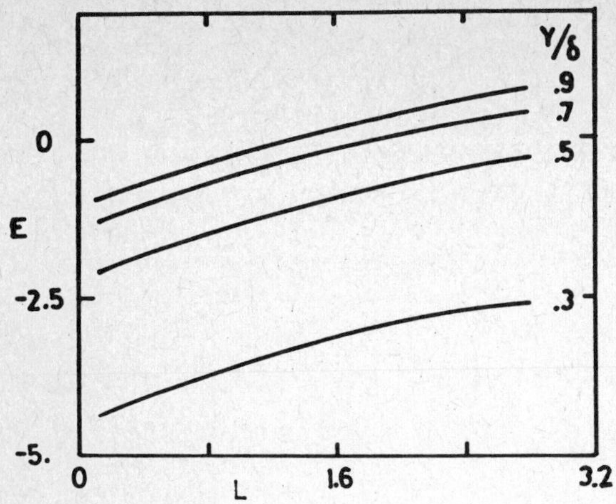

Fig. 8.- Excess cloud energy due to vortex scatter.

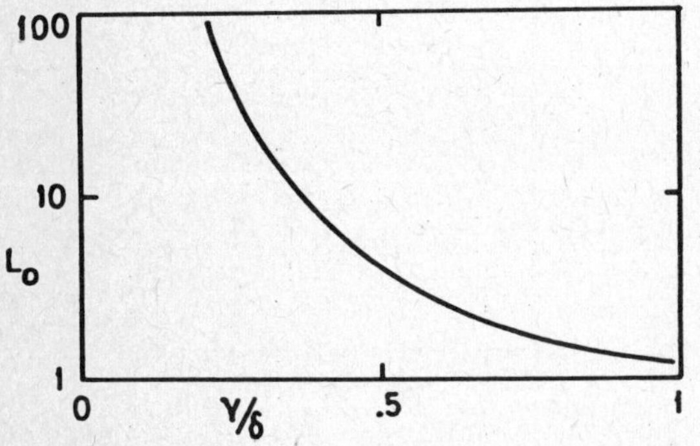

Fig. 9.- Longitudinal vortex scatter for zero energy, as a function vortex strength.

and then evaluated numerically. The effect of R on this energy is slight and has not been taken into account. The results are given in figure 8; the energy is a monotonically increasing function of L which can be shown to increase without limit for $L \to \infty$, and the minimum energy occurs for L=0. Intuitively, as the scatter of the cores gets bigger, the probability of two vortices getting very close to one another gets higher and this makes the average energy larger. We can then expect that the most probable configuration is that with the minimum energy and that we should use L=0 in our model.

Another possibility is to use the value of L that makes the excess energy equal to zero, as in a purely inviscid calculation in which the initial condition is very close to an uniform row, we expect to end up in such a state. The values of L satisfying that condition are given in figure 9 and all calculations in this section are presented for the two cases: zero and minimum energy. However it should be remembered that the natural evolution of a vortex row will not produce in general uncorrelated vortex positions and that it may not be consistent to force energy conservation at the same time as neglecting the correlations.

The average velocity profile is independent of L and γ and to the lowest order in R is an error function. The finite size of the cores has the effect of thickening slightly the layer so that the vorticity thickness is given by

$$\delta \simeq \pi^{1/2} \ (1+R^2/4) + O(R^4). \qquad (7)$$

The turbulent intensities depend critically on the core radius. For point vortices the integral of the square of the velocity diverges when the integration path

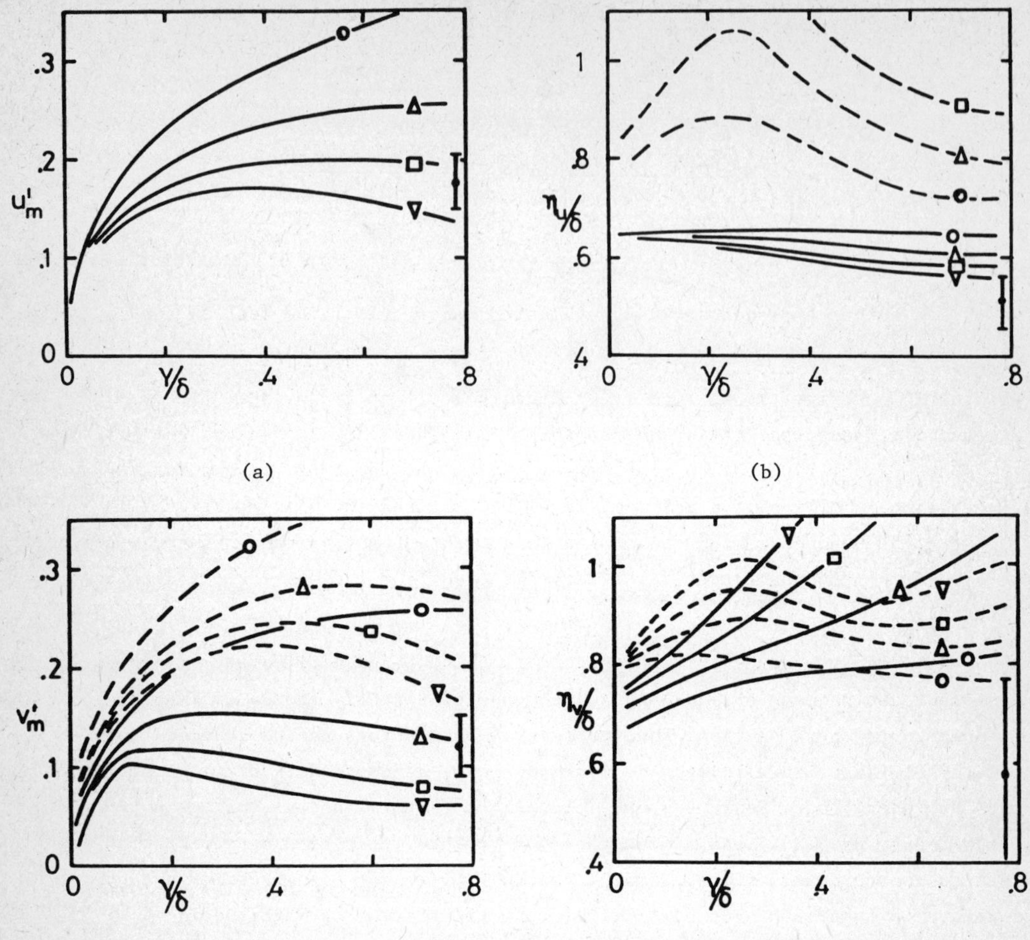

Fig. 10.- Results of vortex cloud model. O: R/γ=.1, Δ:.3, B:.5, ∇:.7; Continuous line is for minimum energy (L=0), dashed is for zero energy. a.- longitudinal turbulence level; b.- u' profile half-width; c.- transversal turbulence level; d.- v' profile half-width.

crosses a core, and all intensities are infinite. Consider the longitudinal fluctuations for finite cores. The velocity induced by a vortex is

$$u(x,y)=\begin{cases}\dfrac{\gamma}{2\pi}\,\dfrac{y}{x^2+y^2}\,; & x^2+y^2 \geq R^2\\[2ex] \dfrac{\gamma}{2\pi}\,\dfrac{y}{R^2}\,; & x^2+y^2 \leq R^2,\end{cases}$$

and the variance produced on (x,y) by a single vortex is

$$\sigma^2(x-n\gamma,y) = \iint_{-\infty}^{\infty} u^2(x-\xi,y-\eta)p(\xi-n\gamma,\eta)d\xi\, d\eta - [\iint_{-\infty}^{\infty} u(x-\xi,y-\eta)p(\xi-n\gamma,\eta)d\xi\, d\eta]^2 \quad (8)$$

Since the positions of the vortices are uncorrelated, the variances are additive and the fluctuation level is found by adding the individual variances and averaging the result over x,

$$u'^2(y) = \frac{1}{\gamma}\int_0^{\gamma} \sum_n \sigma^2(x-n\gamma,y)dx = \frac{1}{\gamma}\sum_n \int_{-n\gamma}^{(-n+1)\gamma} \sigma^2(x,y)dy = \frac{1}{\gamma}\int_{-\infty}^{\infty} \sigma^2(x,y)dy, \quad (9)$$

which can now be expanded in R and evaluated numerically. The results are given in figure 10.

Figures 10 a,c give the maximum (y=0) longitudinal and transversal fluctuation intensities (note that u' is independent of L). The other two graphs characterize the shape of the fluctuation profiles by giving the value of y for which the fluctuation is half of the maximum. In all four cases there is a vertical error bar in the right hand side of the plot showing a representative range of the experimental values for the quantities involved, collected from the available literature [17-21].

Any model with two adjustable parameters is bound to predict something and no judgement can be made until R and γ are measured directly from a mixing layer; this is done in the next section. However several conclusions can be deduced from figure 10. First, in all cases de minimum energy assumption (L=0) works better than the zero energy one. Next the peak intensities, both u' and v', are reasonably well predicted by $R/\gamma\sim.3-.5$ and a wide range of γ; this is in good agreement with the results in the next section. Finally the half width for both profiles are overestimated by about 30% but still within the right order of magnitude.

An important failure of the model is in the prediction of the Reynolds stresses. The prediction is zero, and this is a fundamental consequence of the parity of the velocity components of a vortex and of the hypothesis of zero correlation. No stresses obviously mean no layer growth and no momentum diffusion. In the numerical experiments, in which the rate of growth is accurately represented, there is a large scale clumping that introduces a correlation of vortex positions and looks strikingly like the coherent structures in real layer. In this sense it seems that the large structure is responsible for the dynamics of the mixing layer even if most of the commonly measured quantities u, u' and v' are predicted well by a purely kinematic description in terms of small vortices randomly distributed.

Measuring the core parameters

In the last section we have found another case in which, to check a given model, we need some elaborate processing of the experimental data to extract the required information. The properties to be measured are now the radius and circulation associated with the vorticity cores in the mixing layer and the appropriate signal is a hot wire velocity reading.

Some similar studies have been reported before. Browand and Weidman [22] and several other authors have mapped the vorticity distribution in the large structures by phase averaging and find single peaks in the center of the structure. However, averaging is intended precisely to smooth the differences in the internal arrangement of the different eddies and will give at most the probability distribution of the position of the small vortices, if they exist. Another attempt was done in [23] where individual eddies are studied, with no averaging, but the probes are positioned far outside the mixing layer and the effect of the small vortices is integrated by the distance.

To be able to see vortices which are small with respect to the large structure we need to observe them from distances which are smaller than the vorticity thickness. Think of a vortex passing so near our probe that locally its influence dominates the flow, but still far enough that we stay outside the rotational core; we will see something similar to the velocity induced by a point vortex and, from this velocity, we should be able to deduce the properties, circulation and distance, of the passing core.

Unfortunately, in most cases the situation is more complicated; either there are several cores at roughly the same distance from our probe, so that our signal is complex and difficult to analyze (figure 11-a), or else the path of the cores intersect the probe and the signal is strong but does not follow the behaviour of a simple point vortex (fig. 11-c). Occasionally however a core will just skirt the probe at just the right distance, and produce a clean peak that can be detected and analyzed (fig. 11-b).

We have used these peaks to study the vortex cores in the shear layer. We used a plane continuous air half jet discharging in a large room. The stream velocity was 50 m/s with a turbulence level of .3%, and the nozzle section was 130x70 mm. The boundary layer at the splitter plate was laminar, although close to transition, and its peak turbulence level reached 1-2%. The apparatus is described in detail in [24]

A hot wire anemometer was used to take longitudinal velocity and turbulence profiles at several sections and, in addition, long records (20 s) were obtained at several point for use in the peak recognition studies. Two examples of the signal are given in figure 12. The first record corresponds to the region of linear instability in which compact cores have pressumably not yet formed. The second is further downstream, after one or two large scale amalgamations, and is more complex. Large peaks appear in it that seem to have roughly the same scale as the ones in the linear record, but they happen less often and, in fact, it looks as if every second peak is missing, suggesting that the effect of the amalgamation has been not to produce bigger cores but to displace half of them out of sight from our probe.

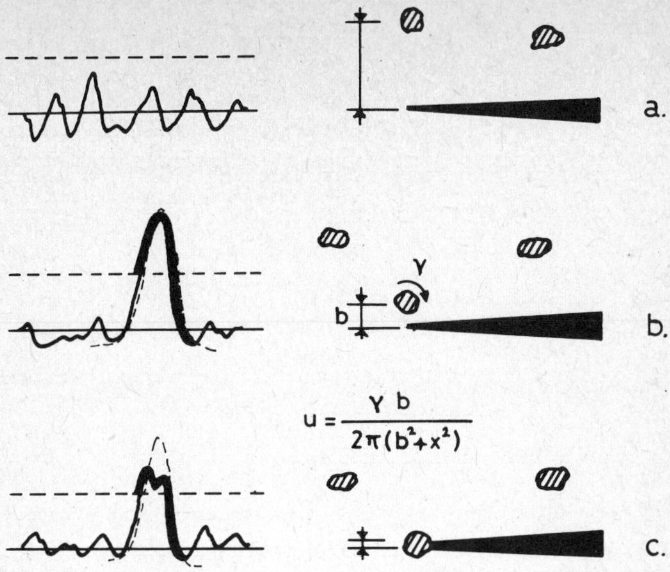

Fig. 11.- Relation between different relative positions of passing cores and signals detected by probe.

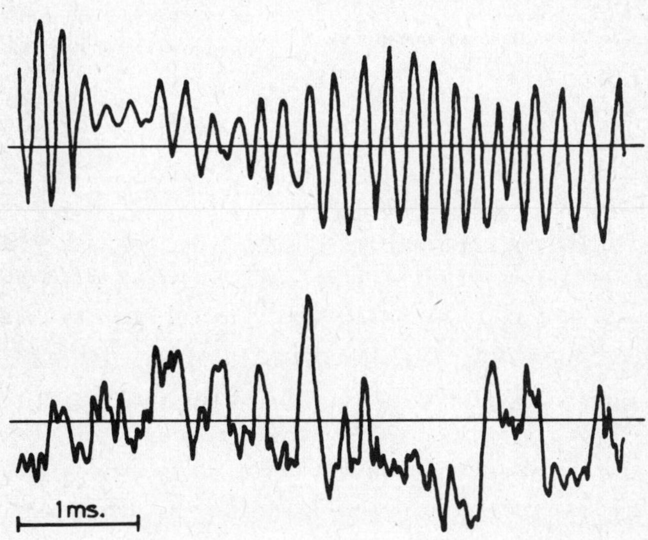

Fig. 12.- Examples of signals used in the peak recognition procedure. Top record is in the linear transition region (x=10 mm), bottom record after one amalgamation step (x=30).

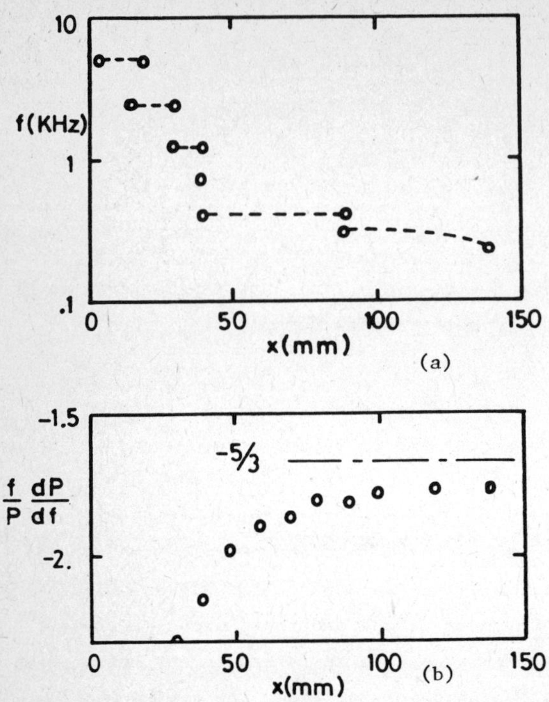

Fig. 13.- Spectral evolution of mixing layer vs downstream coordinate. a: Frequency of dominant spectral peak; b: Slope of inertial power spectrum range.

The development of the layer, as seen from th power spectrum of the velocity fluctuations, is documented in figure 13. In 13-a the frequencies of the dominant spectral peaks are given as function of downstream coordinate. They show periodic halvings that should correspond to the average position of the large scale amalgamation events. Figure 13-b shows the evolution of the slope of the inertial range in the (log-log) spectrum. Pressumably the moment where this slope attains the Kolmogorov value of -5/3 marks the beginning of full flow three-dimensionality and well developed turbulence. The records on figure 12,as well as all the results to be presented below are taken in the region in which the three-dimensional regime has not yet been entered.

The details of the peak recognition process are given elsewhere [25]. Essentially a long record is obtained, stored in digital tape, transformed to velocity and thresholded to identify the very large peaks. The thresholder detects every instance in which the velocity exceeds a certain value and then falls again below a different, lower, one.

Once a peak is found it is isolated from the record; its end point is the first minimum after crossing the downgoing threshold, and is taken to define the level of the "base" velocity, i.e. the velocity which would be observed if the core under consideration were removed from the layer. After substracting this level, the peak should be approximated by the velocity induced by a point vortex (figure 11-b).

$$u = \frac{\gamma/2\pi b}{1+(U_c t/b)^2}, \qquad (10)$$

where U_c is the unknown convection velocity of the vortex core. This equation is fitted to the shape of the peak and only those instances that show a reasonably good fit are accepted as real cores, thus rejecting all but the clear-cut cases. The acceptable error and the different thresholds are adjusted empirically so as to make the final results as independent of them as possible.

By this procedure we can get values of b/γ and U_c/b but, to get absolute circulations, we need to make hypothesis on U_c. We have used the assumption that the cores move with the average flow velocity at their transversal co-ordinate which, linearizing around the probe position, is

$$U_c = U_p + (\partial U_p/\partial y)b, \tag{11}$$

where U_p is the average velocity measured by the probe. This hypothesis permits us to compute values of γ and b for individual cores and to compile statistics for them.

Its accuracy is only approximate; inspection of the velocity records obtained at different transverse probe positions shows that the peaks are indeed wider at the low speed side of the layer, indicating that they are convected slower, and the ratio of the widths corresponds roughly to the ratio of the average velocities. A more quantitative estimate of the influence of the hypothesis in the accuracy of the final results can be had from figure 14. There we can see the values of the average vortex circulation obtained at three different sections, with the probe at different transverse positions. Since there is no "a priori" reason why cores should look different when seen from different points of view, the drift apparent in the figure is probably a measure of the inadecuacy of assumption (11) and a measure of the final errors to be expected.

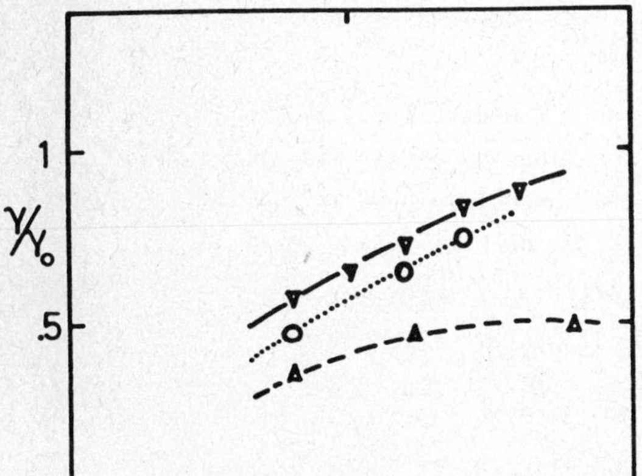

Fig. 14.-.Cross-stream variation of measured average circulation per core vs. velocity at probe position. Δ: x=20 mm, O: 40 mm, ∇:60 mm.

These errors have another consequence: equation (11) has a different effect on the accuracy of γ depending on whether the core is positive ($b<0$) or negative. Analysis shows that the resulting errors tend to cancel for positive peaks, so that the results are quite accurate in this case, while for negative peaks they reinforce each other, giving unacceptable results. Only positive peaks will be used below.

In summary, by observing the strong peaks in the velocity records, and subject to assumption (11), we are able to measure the strength and distance of the passing cores. The parameter b/γ is independent of (11)

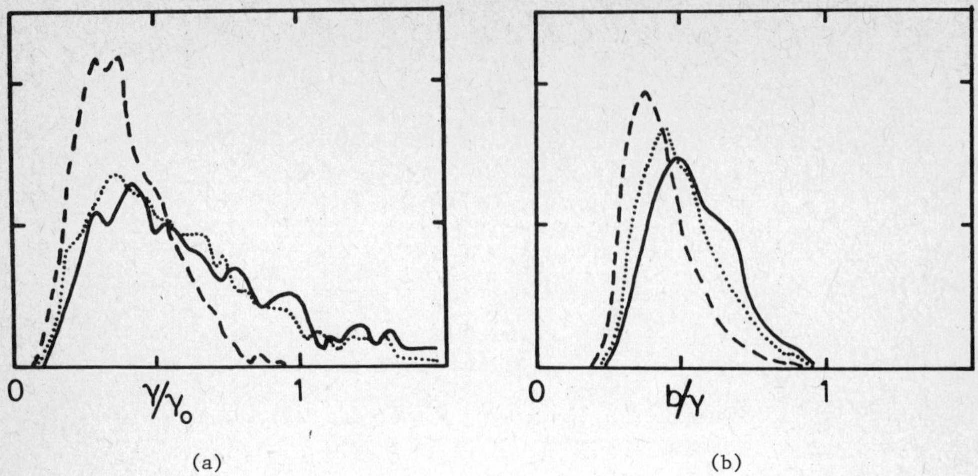

Fig. 15.- Histograms for core strength (a) and normalized distance to probe (b).
---:x=20 mm, ...:40 mm, -: 60 mm.

and the measurements should be accurate, while the absolute circulation and distance show large errors and should be taken only as indicative values.

The final histograms at three different downstream stations are given in figure 15. The vortex strength is shown normalized with the circulation γ_0 corresponding to the wavelength of the initial instability of the shear layer. It is rather surprising that in all cases the most probable value is smaller than one, and is actually close to .5. It is noteworthy in this respect that the structure of the most unstable mode predicted by the linear stability theory consists precisely of two vorticity concentrations, one "on top" and the other on the "bottom" of the initial perturbation, each containing half the vorticity of one wavelength, and rotating around each other in a pairing motion (fig. 10 in [2]). Even if it is not clear whether this solution continues to be valid after the initial moments, it is tempting to speculate that the cores observed in fig. 15 are those initial vorticity concentrations, many of which have not disappear by merging even at the most downstream station. Actually the circulation distributions at the second (x=40) and third (x=60) stations are almost identical, even if by then the large scale structures have suffered at least two and four amalgamations steps.

The distance data is easier to understand. At each station the histogram shows that no cores are detected either very near or very far from the probe. This was expected since vortices passing too far are simply lost in the general background while those passing too near intersect the probe and are rejected by the fitting algorithm. In fact the lower edge of the histograms can be used to estimate the radius of the cores. From figure 15 this radius increases slightly downstream but is always in the range $R/\gamma \sim .3.-4$. This corresponds nicely to some simple theoretical modelling; if we assume that the initial vortex sheet roll-up happens without strong

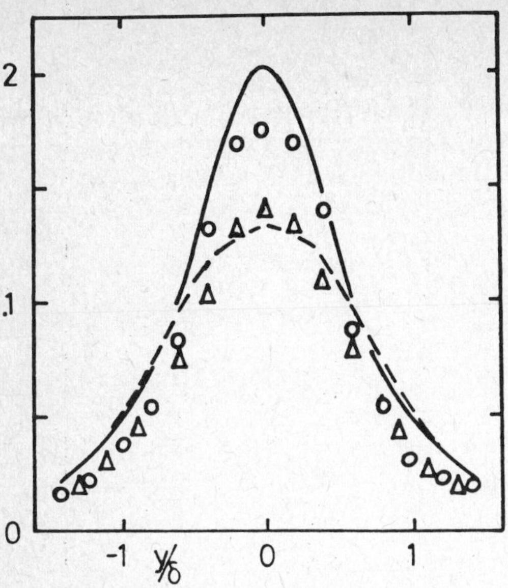

Fig. 16.- Comparison between uncorrelated vortex model (lines) and experimental data (symbols). In model, $R/\gamma=.4$, $\gamma/\delta=.3$, L=0. Experiments are from [20] .— and O: u'; --- and Δ: v'.

interaction among different wavelengths the moment of inertia of the vorticity distribution is conserved and the final radius must satisfy the constrain

$$48\,\pi\int_0^R \hat{\omega}(r)\; r^3\; dr = \gamma^3 .$$

The value of R implied by this equation depends on the particular distribution $\omega(r)$ chosen for the vorticity in the core, but in most cases falls in the range $R/\gamma\sim.3\div.5$, which agrees well with the experimental data.

Finally it is interesting to know which fraction of the passing vorticity is detected by the algorithm. This is easy to find since the total vorticity is known from the velocity profile and can be compared with that contained in the detected cores. The result varies at different stations but is usually in the range of 5-11%, which is actually high considering the complexity of the signals treated.

The purpose of the experiment was to try to validate the model according to which the mixing layer can be represented by a cloud of small vortex cores with a high lateral scatter. We are now in the position to introduce the values obtained here in the model developed in the last section, and check the prediction.

The average core strength can be found from figure 15 and increases only slowly downstream. Therefore the parameter γ/δ decreases, beginning at .7 for x=20 and varying between .4 and .3 in the downstream stations. It pressumably keeps decreasing slowly, although eventually the three-dimensionality might alter the picture completely.

At the last station (x=60) we get $\lambda/\delta=.3$ and $R/\gamma=.4$. Figure 16 shows the tubulence profiles both as predicted from the model using these parameters and taken from the experimental data in a different mixing layer [20]; even if part of the agreement is probably accidental, considering all the approximations involved, the general fit is surprisingly good.

Conclusions

We have shown first of all that pattern recognition methods, applied to classical signals from turbulent flows, can give information on aspects well beyond the standard statistical properties. Moreover the kind of results obtained can be changed by adjusting the processing techniques instead of the experimental situations. In this respect it is important to realize that the film used in the first part of this paper was not taken with the idea of digital image processing in mind, although it was obviously well fitted for that application, and that the anemometry results were obtained in a standard apparatus in a standard way and, in fact, were the very same records previously taken to study the spectral evolution of the mixing layer.

This intensive analysis of the experimental signals is a new application of the computer to experimental science that goes beyond the introduction of digital instrumentation and control that has been such an important feature on research laboratories in recent years. We believe this sort of analysis will become more widespread and important in the future.

Secondly, the results regarding the structure of the mixing layer are also interesting. The driving force behind the evolution of the concentration field can be described well by a series of vorticity blobs, convected with the average velocity and amalgamating periodically. The amalgamation mechanism seems to be the pairing motion of two eddies, after which the two concentration features merge into a single one; there is little fluid entrainment in this phase.

It also appears that the structure of these vorticity blobs is not simple and that they contain several individual vortex cores moving on bound orbits inside a common "bag". If this is true the merging of two blobs is relatively easy to understand since, instead of the desintegration and merging of two cores, all that we have to do is to tangle together two sets of already loose orbits.

The experiments in which this "vortex bag" idea is based were done in the region in which the mixing layer is still quite two-dimensional. Three-dimensionality can be expected to change the picture, since it is difficult to think of two highly kinked vortex lines coming together without wrapping around each other and essentially merging.

The size of the vortex cores used in this model and detected in the experiments correspond to what is usually identified in the power spectrum as the inertial subrange. In some sense, therefore, the random motion of these small cores is what is usually identified as turbulence and it should be interesting to see how far the study of turbulence can be taken in terms of this idea. We have shown that some properties, notably the turbulent intensities, can be modelled correctly disregarding completely the large scale order ordering of the cores, while others, such as the Reynolds stresses, need the explicit introduction of correlations. We suggest that this type of analysis might be useful in isolating which processes are important for different aspects of the flow.

The research of one of the authors (Martinez-Val) was supported in part by the Spanish Comisión Asesora de Investigación

REFERENCES

1. Freymuth, P., 1966, J.Fluid Mech., 25, p. 683-704.
2. Michalke, A., 1964, J.Fluid Mech., 19, p. 543-556.
3. Winant, C.D. & Browand, F.K., 1974, J.Fluid Mech., 63, p. 237-255.
4. Moore, D.W. & Saffman, P.G., 1975, J.Fluid Mech., 69, p. 465-473.
5. Jimenez, J., 1980, J.Fluid Mech., 96, p. 447-460.
6. Hernan, M.A. & Jimenez, J., 1979, Turb. Shear Flows, London, p. 7.7.-7.13.
7. Hernan, M.A. & Jimenez, J., 1980, 8th World Comp. Congress, IFIP, Melbourne.
8. Roshko, A., 1976, AIAA Paper 76-78.
9. Corcos, G.M., 1979, University of California, Berkeley, Report FM-79-2.
10. Acton, E., 1976, J.Fluid Mech., 76, p.561-592.
11. Riley, J.J. & Metcalfe, R.W., 1980 AIAA Paper 80-0274.
12. Aref, H. & Siggia, E.D., 1980, J.Fluid Mech.(to be published).
13. Ashurt, W.T., 1977, SANDIA Lab Report SAND77-8613.
14. Jimenez, J., 1977, Lecture Notes in Physics, 75, p. 147-161.
15. Delcourt, B. & Brown, G.L., 1979, Turbulent Shear Flows, London, p. 14.35-14.40
16. Batchelor, G.K., 1967, An Introduction to Fluid Mechanics, Cambridge, p. 530.
17. Liepmann, H.W. & Laufer, J., 1947, NACA TN-1257.
18. Spencer, B.W. & Jones, B.G., 1971, AIAA Paper 71-613.
19. Wygnansky, I. & Fiedler, H., 1970, J.Fluid Mech., 41, p. 327-362.
20. Pui, N.K. & Gartshore, I.S., 1979, J.Fluid Mech., 91, p. 111-130.
21. Hussain, A.K.M.F. & Zedan, M.F., 1978, Phys. Fluids, 21, p. 1100-1112
22. Browand, F.K. & Weidman, P.D., 1976, J.Fluid Mech., 76, p. 127-144.
23. Koochesfahani, M.M., Catherasoo, C.J., Dimotakis, P.E., Gharib, M. & Land, D.B., 1979, AIAA J., 17, p. 1347-1351.
24. Jimenez, J., Martinez-Val, R. & Rebollo, M., 1979, Final Report USAERO 79-G-079
25. Martinez-Val, R., 1980, Ph.D. Thesis, Madrid (in Spanish).

THEORETICAL INVESTIGATION OF INTERACTION AND COALESCENCE OF LARGE SCALE STRUCTURES IN THE TURBULENT MIXING LAYER

D.D. Knight and B.T. Murray

Department of Mechanical, Industrial and Aerospace Engineering
Rutgers University
New Brunswick, New Jersey 08903 USA

ABSTRACT

The dynamics of the interaction of the large scale eddies in the temporally-developing homogeneous turbulent mixing layer are investigated using two different models for the fine scale Reynolds stresses. The results indicate the growth, interaction and coalescence of an eddy pair. Detailed quantitative comparison with experimental data is provided at a particular stage in the eddy pairing.

I. INTRODUCTION

The recent observations of the remarkable coherence of the large scale structures ("eddies") in turbulent shear flows have stimulated the development of novel and exciting avenues of turbulence research. In a wide variety of flows, the large scale eddies have been found to exhibit a reasonably well-defined physical structure characteristic of the particular shear flow (Laufer 1975, Roshko 1976). A brief sampling of flows studied includes the turbulent boundary layer (Kline, et al. 1967), mixing layer (Brown and Roshko 1974, Winant and Browand 1974), two-dimensional turbulent wake (Papailou and Lykoudis 1974, Cantwell 1976), and axi-symmetric jet (Higuchi 1976).

From a conceptual standpoint, these observations suggest that the turbulent motion is comprised of "quasi-deterministic" unsteady large scale eddy motions and fine grained random fluctuations. This idea has, of course, been suggested on numerous occasions (Dryden 1948, Liepmann 1962, Roshko 1976). Recent experimental observations, however, have emphasized this concept, and new theories have been proposed regarding the importance of the large scale eddies in the phenomena of entrainment (Winant and Browand 1974, Dimotakis and Brown 1976, Hussain and Zaman 1979), mixing (Konrad 1976, Breidenthal 1978), combustion and pollutant formation (Yule, et al. 1977).

Theoretical turbulence research, however, has been slow to adopt this view of the structure of turbulent shear flows. As indicated in a recent review (Reynolds 1976), the overwhelming majority of phenomenological turbulence models do not explicitly incorporate the concept of coherent large scale motion. Wihin the past five years, however, substantial interest has developed in theoretical treatment of coherent structures. The turbulent mixing layer has been the focus of most of the research, due primarily to its favorable geometric features and the relative simplicity of the observed large scale structures. Two primary approaches have been employed, namely numerical calculations using collections of point vortices and finite-difference computations. In the former category, the results of Ashurst (1977), Couët et al. (1979) and Delcourt and Brown (1979) have displayed the interaction and amalgamation of large scale eddies in qualitative agreement with experimental flow visualization results. In the work of Delcourt and Brown (1979), the computed growth rate for one class of initial conditions indicated close agreement with experimental data.

In the category of finite-difference calculations, Mansour et al. (1978) considered the temporally-developing homogeneous mixing layer using a subgrid scale turbulence model. Computations were performed for both the two-dimensional mixing layer (for which the scales of the flow resolved by the calculation are assumed to be independent of spanwise position), and the fully three-dimensional mixing layer. In the two-dimensional case, the initial condition consisted of a mean velocity given by the conventional error function profile plus a perturbation associated with two line vortices. Large scale eddy interaction and pairing was observed. Throughout the interaction, the mean velocity and turbulence intensity profiles were found to remain self-similar. The growth rate was observed to be linear in agreement with experiment, although the magnitude was dependent on the choice of the initial conditions. In the three-dimensional case, two different initial conditions were employed for the velocity field, namely 1) the error function profile and a random three-dimensional disturbance, and 2) the error function profile, two line vortices and a random three-dimensional disturbance. Self-similarity was observed for the mean velocity profile, although the turbulence intensity never achieved self-similarity. A linear growth rate was found, whose magnitude was dependent on the initial conditions.

The dynamics of growth and decay of a single eddy in the temporally-developing mixing layer was considered by Gatski and Liu (1980) using the turbulence model of Hanjalic and Launder (1972) for the fine scale turbulent motion. The initial condition consisted of the mean velocity profile plus a small perturbation which represented the most unstable disturbance based on an inviscid stability analysis of the mean velocity profile. A large eddy structure was observed to grow, extracting energy from the mean flow while simultaneously losing energy at a smaller rate to the fine grained turbulence. After a finite time, the growth of the eddy ceased and energy was both transferred back to the mean motion (a phenomenon which may be interpreted as "negative" turbulence energy production) and lost to the fine scale turbulence.

The interaction of large scale coherent structures in the temporally-developing two-dimensional mixing layer was considered by Knight (1979) using the turbulence model of Saffman (1977) for the fine scale turbulent motion. The initial condition consisted of the mean velocity profile and a small perturbation comprising the most unstable disturbance and its first subharmonic obtained from an inviscid stability analysis of the mean velocity profile. Eddy interaction and pairing was observed to occur, with a strong energy transfer from the mean flow to the coherent motion during eddy coalescence and a reverse energy transfer from coherent scales to the mean flow ("negative" turbulence production) following the amalgamation. The correlation coefficient of the coherent motion displayed near unitary values during an intermediate stage in the eddy pairing, in quantitative agreement with the measurements of Browand and Weidman (1976).

The temporally-developing two- and three-dimensional turbulent mixing layer was also investigated by Riley and Metcalf (1980) using a direct pseudo-spectral numerical method at an initial Reynolds number $R_\lambda = u'\lambda/\nu \approx 50$ where u' is the maximum value of root-mean-square fluctuating longitudinal velocity and λ is the Taylor microscale. In the three-dimensional calculations, four different initial conditions were considered for which the mean velocity profile was perturbed by random three-dimensional fluctuations and/or a coherent two-dimensional perturbation consisting of the most unstable disturbance based on an inviscid stability analysis of the mean velocity. In the absence of a coherent perturbation, the initially random vorticity field developed a large coherent structure which displayed spanwise non-uniformity. The predicted growth rate was within 10% of the correlation of Brown and Roshko (1974), and most of the mean statistical quantities (e.g., mean velocity profile and integrated turbulence kinetic energy) developed in a self-similar manner. The inclusion of both the random fluctuations and coherent perturbation in the initial condition resulted in an enhanced growth rate during the early stages of the eddy development, followed by a substantial reduction in growth rate. In addition, the maximum values of the mean longitudinal turbulence intensity and Reynolds stress were reduced by inclusion of the coherent perturbation, in qualitative agreement with the experimental observation of Oster (1980) and Wygnanski et al. (1979) in a forced turbulent mixing layer. In the two-dimensional calculations, four different initial conditions were employed in which the mean velocity profile was perturbed by coherent perturbations consisting of the most unstable disturbance and/or its first subharmonic. When both harmonic and subharmonic disturbances were present, it was observed that the growth rate and phenomenon of amalgamation depended upon the initial relative phase of the harmonic and subharmonic perturbations, in qualitative agreement with the earlier calculations of finite-amplitude Kelvin-Helmholtz instability in a stratified fluid by Patnaik et al. (1976).

In this study, attention is focused on the interaction of the large scale eddies in the two-dimensional turbulent mixing layer. A particular point of interest is the modeling of the fine scale turbulence, which has been treated by a variety of approaches in previous investigations including omission of any explicit fine scale model in the relatively low Reynolds number calculations of Riley and Metcalf (1980), subgrid scale modeling in the large eddy simulations of Mansour et al. (1978), and full Reynolds stress equation models in the work of Gatski and Liu (1980) and Knight (1979). Conditional phase-averaged measurements at high Reynolds number in a variety of turbulent flows (Cantwell 1976, Kovasznay 1977, Latigo 1979) indicate that roughly half of the total Reynolds shear stress is attributable to the "incoherent" or phase-random turbulent fluctuations, suggesting therefore that proper treatment of the fine scale turbulence is important. In this paper, results are presented for the turbulent mixing layer for two different fine scale turbulence models, namely a full

Reynolds stress equation model and a simple eddy viscosity model. The results using the latter approach are also compared with experiment.

II. FORMULATION OF PROBLEM

A. Temporally vs. Spatially Developing Mixing Layer

For purposes of computational efficiency, the temporally-developing homogeneous turbulent mixing layer is considered. As indicated in Fig. 1a, a shear layer is formed by two opposing infinite parallel streams of equal speed U_∞. In this configuration, the mean convective velocity of the large eddies is approximately zero. Consequently a single eddy pairing can be considered using a computational domain approximately twice the eddy size. The common experimental configuration (Fig. 1b) is the spatially-developing turbulent mixing layer, formed by the merging of two parallel streams of velocity U_1 and U_2 downstream of a splitter plate. Since the mean convective velocity of the large eddies is approximately $(U_1 + U_2)/2$, numerical simulation of an eddy pairing in the spatially-developing configuration requires a computational domain whose streamwise extent is several times the typical eddy size.

There is no precise mathematical transformation between the temporally- and spatially-developing cases. Qualitatively, however, the configurations are similar when related by the Galilean transformation

$$x' = x + (U_1 + U_2)t/2$$

This interpretation has proven useful to the understanding of various phenomena in the spatially-developing mixing layer (e.g., the dependence of spreading rate (Brown and Roshko 1974) on the ratio U_2/U_1). The freestream velocity in the temporally-developing case is thus defined by $U_\infty = (U_1 - U_2)/2$.

B. Definition of Averages

Based on the aforementioned experimental observations, the instantaneous flowfield is viewed as a combination of large scale ("phase coherent") motion and fine-grained ("phase incoherent") random fluctuations. A flow variable $\mathcal{F}(x,y,z,t)$ is therefore decomposed according to (Hussain and Reynolds 1970, Hussian 1977, Hussain 1979).

$$\mathcal{F}(x,y,z,t) = f(x,y,t) + f''(x,y,z,t) \tag{1}$$

where f(x,y,t) is the conditional or phase-averaged value of $\mathcal{F}$ (denoted also by $\langle \mathcal{F} \rangle$), and f" is the random contribution. The phase-coherent motion is assumed to be unsteady and spatially two-dimensional based on flow visualization results (Brown and Roshko 1974), Winant and Browand 1974, Dimotakis and Brown 1976, Konrad 1976). Recent observations (Konrad 1976, Breidenthal 1978), however, suggest that a secondary spanwise flow may have an important additional effect on the mixing characteristics of the flow. The random contribution f"(x,y,z,t) is unsteady and spatially three-dimensional.

The concept of conditional or phase-averaged measurements has been successfully realized in a wide variety of "natural" (e.g., Browand and Weidman 1976, Cantwell 1976) and "perturbed" turbulent shear flows (e.g., Hussain and Reynolds 1970, Kendall 1970, Zaman and Hussain 1979, Wygnanski _et al._ 1979). The instantaneous components of the velocity $(\mathcal{U},\mathcal{V},\mathcal{W})$ in the x, y and z directions, respectively, are thus

$$\begin{aligned} \mathcal{U} &= u(x,y,t) + u''(x,y,z,t) \\ \mathcal{V} &= v(x,y,t) + v''(x,y,z,t) \\ \mathcal{W} &= w(x,y,t) + w''(x,y,z,t) \end{aligned} \tag{2}$$

Because of the assumption of two-dimensionality of the coherent motion, $w = 0$. The conditionally-averaged velocity components (u,v) may be further decomposed into a mean and coherent fluctuation according to

$$\begin{aligned} u(x,y,t) &= U(y,t) + u'(x,y,t) \\ v(x,y,t) &= V(y,t) + v'(x,y,t) \end{aligned} \tag{3}$$

where

$$\begin{aligned} U &= \bar{u} = \frac{1}{L}\int_0^L u(x,y,t)\,dx \\ V &= \bar{v} = \frac{1}{L}\int_0^L v(x,y,t)\,dx \end{aligned} \tag{4}$$

and L is the streamwise extent of the computational domain. From conservation of mass, $V = 0$. The instantaneous velocity field thus has the familiar representation (Hussain and Reynolds 1970)

$$\begin{aligned} \mathcal{U} &= \underbrace{\underset{\text{MEAN}}{U(y,t)} + \underset{\text{COHERENT FLUCTUATION}}{u'(x,y,t)}}_{\text{TOTAL COHERENT MOTION}} + \underset{\text{RANDOM FLUCTUATION}}{u''(x,y,z,t)} \\ \mathcal{V} &= v'(x,y,t) + v''(x,y,z,t) \\ \mathcal{W} &= w''(x,y,z,t) \end{aligned} \tag{5}$$

C. <u>Equations of Motion and Turbulence Models</u>

For purposes of simplicity, cartesian tensor notation is employed with $(x_1,x_2,x_3) = (x,y,z)$, $(u_1,u_2) = (u,v)$ and $(u_1'',u_2'') = (u'',v'')$. Application of the conditional average to the conservation equations yields (Hussain 1977)

$$\frac{\partial u_i}{\partial x_i} = 0 \tag{6}$$

$$\frac{Du_i}{Dt} = - \frac{\partial p}{\partial x_i} + Re^{-1} \nabla^2 u_i - \frac{\partial}{\partial x_j} <u''_i u''_j> \tag{7}$$

where

$$\frac{D}{Dt} = \frac{\partial}{\partial t} + u_j \frac{\partial}{\partial x_j}$$

In the above, ∇^2 is the Laplacian operator, and the Einstein summation convention is employed. The equations have been non-dimensionalized using U_∞ and a length scale L_0 (taken to be the initial half-width of the mixing layer, defined as the value of y at which u = 0.99). It should be emphasized that the velocity u_i appearing in Eqs. (6) and (7) is the total coherent velocity (mean plus coherent fluctuation), and the full non-linear equations of motion are employed.

Closure of the governing equations requires specification of the fine scale Reynolds stress tensor $<u''_i u''_j>$. In this study, two different approaches have been investigated, specifically, a Reynolds stress equation model and a simple eddy viscosity model. The Reynolds stress equation model adopted is Saffman's "relaxation stress model" (Saffman 1977). The turbulence model equations introduce the scalar quantities e and ω, which are intended to be similar (though not necessarily identical) to the turbulence kinetic energy and root-mean-square fluctuating vorticity of the eddies responsible for local turbulent transport. These quantities are governed by

$$\frac{De}{Dt} = \alpha'' e[2S_{ij}^2]^{\frac{1}{2}} - e\omega + \frac{\partial}{\partial x_j}[(A'' \frac{e}{\omega} + Re^{-1})\frac{\partial e}{\partial x_j}] \tag{8}$$

$$\frac{D\omega^2}{Dt} = \alpha'\omega^2[\eta' u_{i,j}^2 + 2(1-\eta')S_{ij}^2]^{\frac{1}{2}} - \beta'\omega^3 + \frac{\partial}{\partial x_j}[(A' \frac{e}{\omega} + Re^{-1})\frac{\partial \omega^2}{\partial x_j}] \tag{9}$$

where the rate-of-strain tensor S_{ij} is

$$S_{ij} = \frac{1}{2}(\frac{\partial u_i}{\partial x_j} + \frac{\partial u_j}{\partial x_i}) \tag{10}$$

and $u_{i,j} = \partial u_i/\partial x_j$. The system is completed by postulating the following relaxation equations for the fine-scale Reynolds stresses

$$\frac{D}{Dt}<u''_i u''_j> = -\theta\omega[<u''_i u''_j> - <u''_i u''_j>_E] + \frac{\partial}{\partial x_k}[(A''' \frac{e}{\omega} + Re^{-1}) \frac{\partial}{\partial x_k}<u''_i u''_j>] \tag{11}$$

where $\langle u''_i u''_j \rangle_E$ is an "equilibrium" Reynolds stress given by

$$\langle u_i u_j \rangle_E = \frac{2}{3} e\delta_{ij} - 2A \frac{e}{\omega} S_{ij} + B \frac{e}{\omega^2} S^2_{k\ell}\delta_{ij}$$

$$+ CS_{ik}S_{jk} + D \frac{e}{\omega^2} (S_{jk}\Omega_{ik} + S_{ik}\Omega_{jk}) \tag{12}$$

$$+ F \frac{e}{\omega^2} \Omega^2_{k\ell}\delta_{ij} + G \frac{e}{\omega^2} \Omega_{ik}\Omega_{jk}$$

where $\Omega_{ij} = \frac{\partial u_i}{\partial x_j} - \frac{\partial u_j}{\partial x_i}$ and δ_{ij} is the Kronecker delta. The model equations (8) to (12) contain fourteen constants, whose values were obtained (Saffman 1977, Knight and Saffman 1977), by consideration of a number of simple turbulent shear flows (e.g., decay of homogeneous turbulence and the turbulent wall layer). The values are

$$\begin{array}{llll} \alpha' = .32, & \alpha'' = .40, & \beta' = 1.67 & \\ \eta' = .25, & \theta = 2.0, & A' = .143 & \\ A'' = .143, & A''' = .286, & A = .286, & \\ B = .82, & C = 0, & D = .082, & F = .02, \\ G = 0 & & & \end{array} \tag{13}$$

It should be emphasized that the above values of the constants were determined in part by consideration of the conventional mean properties of certain turbulent shear flows (e.g., the turbulent wall layer) without specific decomposition of the coherent and incoherent (fine scale) motion. This type of calibration is, in fact, common to nearly all turbulence models to date. As a consequence, some of the constants may in fact reflect the contributions of *both* the coherent and fine scale motion. Additional theoretical and experimental research is needed to provide a more accurate determination of such constants for calculations in which the coherent and fine scale motions are both explicitly treated.

The second model employed for the fine scale turbulent Reynolds stresses is a simple eddy viscosity model,

$$-\langle u''_i u''_j \rangle = 2\,\varepsilon\, S_{ij} - \frac{1}{3} \langle q^2 \rangle\, \delta_{ij} \tag{14}$$

where $\langle q^2 \rangle = \langle u''_i u''_i \rangle$, S_{ij} is the rate-of-strain tensor Eq. (10), and ε is a simple Prandtl-type eddy viscosity taken to be

$$\varepsilon(t) = \sqrt{\frac{\pi}{2}}\, \frac{\theta(t)}{\sigma} \tag{15}$$

and $\theta(t)$ is the momentum thickness

$$\theta(t) = \frac{1}{4}\int_{-\infty}^{\infty} (1-U^2(y,t))dy \tag{16}$$

and σ is a constant. There is, of course, no rigorous argument for the use of Eq. (15) for the fine scale eddy viscosity, nor even for the assumption of an eddy viscosity formulation for the fine scale Reynolds stresses, although the results of Cantwell (1976) lend qualitative support to the latter concept (i.e., the regions of positive values of $\frac{\partial u}{\partial y}$ in Fig. 5.10b of Cantwell appear to correlate with the regions of positive values of the Reynolds stress $-\langle u''v''\rangle$ in Fig. 5.26b and vice-versa). In the authors' opinion, the primary argument for the above formulation is its extreme simplicity, in particular the fact that there is only one constant to be determined. As indicated in Section IIIB, a value for σ has been obtained for a particular experimental investigation of the turbulent mixing layer.

D. Initial and Boundary Conditions

The initial coherent fluctuations are assumed to arise from an inviscid instability in the flow. The initial profiles of the velocity u_i, pressure p, fine scale Reynolds stresses $\langle u_i''u_j''\rangle$ and scalar turbulence quantities e and ω are therefore taken to be the similarity solution of the governing equations wherein there is no streamwise (i.e., x) variation. In addition, a perturbation is added to the velocity and pressure fields, consisting of two contributions, namely 1) the most unstable eigenfunction (harmonic) based on an exact numerical inviscid (Rayleigh) stability analysis of the initial velocity profile, and 2) the first subharmonic [i.e., the eigenfunction whose wavelength is twice that of the disturbance in 1)]. The relative phase of the contributions is chosen to yield an initial velocity field which is symmetric upon reflection through an origin located at the center of the computational region. As a consequence, the pair of eddies which initially develop have the same "strength" as indicated later. For the Reynolds stress equation model, the initial profiles of the fine scale turbulence quantities e, ω, and $\langle u_i''u_j''\rangle$ are not perturbed, under the assumption that the turbulence structure does not immediately respond to an abrupt perturbation to the flow. For the simple eddy viscosity model, the nature of Eq. (14) implies that the initial fine-scale Reynolds stresses are perturbed through S_{ij}. The dimensional wavelength of the harmonic disturbance λ is 11.64 L_o for the Reynolds stress model and 7.1 L_o for the simple eddy viscosity model due to the different initial velocity profiles. The initial profiles in the former case were obtained numerically, while for the eddy viscosity model the initial velocity profile is $U = \text{erf}\,(1.8215\,y)$.

Periodic boundary conditions are employed for all flow variables in the streamwise direction x over the length of the computational domain, which is equal to the wavelength of the subharmonic (i.e., 2λ). In the y-direction, freestream boundary conditions are applied to $y \to \pm \infty$ with u approaching its freestream value (i.e., $u \to \pm 1$ as $y \to \pm \infty$) and all fine scale Reynolds stresses vanishing.

An important consideration is the strength of the initial perturbation. In temporally-developing mixing layer, the prescription of periodic boundary conditions over a finite streamwise domain constrains the spectrum of the developing coherent scales. In the present case, for example, the streamwise extent of the computational domain is equal to the subharmonic wavelength (i.e., 2λ), and consequently no flow structure can develop with wavelength greater than 2λ. The present investigation is therefore limited to the consideration of a single interaction between two eddies, which may therefore be compared to a conditionally-averaged eddy interaction "event" in the spatially-developing case. As the mixing layer grows in time, it eventually becomes stable to all disturbances of wavelength less than or equal to the streamwise extent of the computational domain. Consequently, in these calculations the coherent eddy interactions have a finite period of growth, and therefore the energy of the coherent motion at any instant depends on both the strength and form of the initial perturbation. In comparing with experimental results, there is no precise method for determining the "proper" initial perturbation strength, and therefore a parametric study needs to be performed. It is also instructive to investigate various initial strengths to determine their effect on the evolution of the coherent structures.

For the purpose of categorizing the initial perturbation strength, it is helpful to separately define the mean kinetic energy defect and kinetic energy of the coherent fluctuations, noting that the total coherent motion is the combination of the mean and coherent fluctuation. The mean kinetic energy defect is

$$E_{mean}^{(t)} = \frac{1}{2} \int_{-\infty}^{\infty} (U^2-1)\,dy \qquad (17)$$

where U(y,t) is defined in Eq. (4). The initial value of E_{mean} is -.5635 for the Reynolds stress model and -.4359 for the eddy viscosity model. The kinetic energy of the coherent fluctuations is

$$E_{coh}^{(t)} = \frac{1}{L} \int_{-\infty}^{\infty} \int_{0}^{L} \frac{1}{2}(u'^2 + v'^2)\,dx\,dy \qquad (18)$$

A convenient measure of the strength of the initial perturbation is the ratio $E_{coh}/|E_{mean}|$. A variety of initial strengths were investigated during the course of this investigation. For all results presented herein, the initial E_{coh} is 0.1% of the initial E_{mean}, with the initial perturbation energy equally divided between harmonic and subharmonic.

E. Numerical Method

The numerical method is described in detail in Knight (1979) and is therefore summarized here. The governing equations are solved in streamfunction-vorticity form using an explicit finite-difference method. The numerical treatment of all terms is spatially and temporally second-order accurate, with the exception of some of the diffusion terms which are essentially first-order accurate in time.

The mesh distributions for the two cases are presented in Table 1. The mesh spacing in the y-direction is exceedingly fine. In particular, Δy is equal to 0.113 and 0.10 for the Reynolds stress and eddy viscosity models, respectively, implying that initially, when the mixing region is the smallest, there are typically twenty mesh points across the mixing region. The upper and lower boundaries of the computational domain, located at $y = \pm 10.4$ for the Reynolds stress model and $y = \pm 9.2$ for the eddy viscosity model, remain sufficiently beyond the mixing region for the boundary conditions to be valid for the time period of interest. The additional computer memory required for the Reynolds stress calculations imposed a reduction in the number of mesh points in the streamwise direction compared to the eddy viscosity model. A truncation error study was conducted for the case of the growth of a single eddy (e.g., $L = \lambda$) with initial E_{coh} equal to 0.1% of E_{mean} for $\Delta x/\lambda = .0625$ and .0313. The computed results were essentially identical, indicating that $\Delta x/\lambda = .0625$ is sufficient for the Reynolds stress model calculations.

Table 1. Summary of Mesh Distribution

Case	IL	$\Delta x/\lambda$	JL	$\Delta y/\lambda$	Δt
Reynolds Stress Model	33	.0625	185	.0097	0.1
Eddy Viscosity Model	65	.0313	185	.0141	0.05

Note: IL = Number of mesh points in x-direction
JL = Number of mesh points in y-direction
Δx = mesh spacing in x-direction
Δy = mesh spacing in y-direction
Δt = time step

All calculations were performed on an IBM 370/168. The typical computer time required for a simulation using the Reynolds stress model is 1.25 hours (for $t_{final} = 46.5$), and 0.75 hours (for $t_{final} = 30.4$) for the eddy viscosity model. The memory requirements are 574 K bytes for the Reynolds stress model (using a 33 × 185 grid) and 384 K bytes for the eddy viscosity model (using a 65 × 185 grid).

F. Energy Budget of Coherent Fluctuations

The energetics of the eddy interaction are of particular importance. The governing equation for the kinetic energy of the coherent fluctuations is (Liu and Merkine 1976)

$$\frac{d}{dt} E_{coh} = \underbrace{\int_{-\infty}^{\infty} - \overline{u'v'} \frac{\partial U}{\partial y} dy}_{\text{TERM 1}}$$

$$+ \underbrace{\int_{-\infty}^{\infty} \overline{[\langle u_i''u_j''\rangle - \langle \overline{u_i''u_j''}\rangle] \frac{\partial u_i'}{\partial x_j}} dy}_{\text{TERM 2}} \tag{19}$$

$$\underbrace{-Re^{-1} \int_{-\infty}^{\infty} \overline{\frac{\partial u_i'}{\partial x_j} \frac{\partial u_i'}{\partial x_j}} dy}_{\text{TERM 3}}$$

where the overbar operation is defined in Eq. (4) and the Einstein summation convention assumed. The first term on the right side (TERM 1) represents the energy flux between the mean flow and coherent fluctuations associated with the mean coherent Reynolds stress $-\overline{u'v'}$. The second term (TERM 2) represents an energy flux between the coherent fluctuations and fine scale turbulence. The third term (TERM 3) represents viscous dissipation of the coherent fluctuations and is negligible in all cases discussed herein.

III. RESULTS

A. Reynolds Stress Equation Model

In Fig. 2a-d, a sequence of vorticity contour plots are displayed at four successive times (t = 5.7, 11.4, 17.1, and 22.8, where t is non-dimensionalized by L_o/U_∞) together with contour plots of -u'v' which is the local Reynolds stress associated with the coherent fluctuations. The flow near the top of each figure is from left to right, and in the opposite direction near the bottom (see Fig. 1a). The vorticity ζ is defined by $\zeta = \frac{\partial v}{\partial x} - \frac{\partial u}{\partial y}$ and is non-dimensionalized by U_∞/L_o. The coherent Reynolds stress -u'v' is made dimensionless by U_∞^2. The vorticity contours are displayed at nine different values of the vorticity equally spaced between the minimum and maximum of ζ (the minimum and maximum values themselves are not plotted). The contours of -u'v' are shown at ten equally spaced values.

The sequence of vorticity contours indicate that the two eddies grow, interact and eventually amalgamate into a single larger structure in qualitative agreement with the observations of Brown and Roshko (1974), Winant and Browand (1974), Dimotakis and Brown (1976) and others. In particular, the eddies are subjected to a significant elongation in the flow direction during amalgamation ($t = 17.1$) similar to the observations of Winant and Browand (1974). These results are also similar to the calculations of Mansour et al. (1978) and Riley and Metcalf (1980). Other calculations have reproduced a behavior reminiscent of the "tearing" interaction observed by Dimotakis and Brown (1976). This phenomenon depends on the initial relative phase of the harmonic and subharmonic perturbations (Patnaik et al. 1976).

The sequence of $-u'v'$ contours (the Reynolds shear stress associated with the coherent fluctuations) in Fig. 2 provides information on the momentum transport by the coherent fluctuating motion. In the early stages of eddy growth ($t = 5.7$ and 11.4), regions of positive values of $-u'v'$ occur in the vicinity of the leading and trailing edges of the eddy pair. The positive values of $-u'v'$ near the right side indicate the transfer of streamwise momentum from the "top" to the "bottom" of the mixing layer (i.e., $u' > 0$ and $v' < 0$), while the positive values of $-u'v'$ at the left side reflect streamwise momentum transfer from the "bottom" to the "top". The fluctuating coherent motion is extracting energy from the mean flow due to the predominantly positive sign of $-u'v'$ (see Eq. (19); the sign of $\partial U/\partial y$ is positive for the entire interaction as discussed later). As the eddies amalgamate ($t = 17.1$ and 22.8), the regions of positive $-u'v'$ shift, and at the approximate time of coalescence ($t = 22.8$), broad regions of positive $-u'v'$ occur on the left and right sides of the amalgamated eddy. This sequence of figures suggests that the entrainment pattern consists of fluid being swept around the leading and trailing edges of the eddy pair (Knight 1979). Following the eddy pairing, large regions of negative $-u'v'$ develop (see Sect. III.B) implying a reverse energy flux from the coherent fluctuating motion to the mean flow (Knight 1979). The $-u'v'$ contours at $t = 22.8$ (where a line joining the eddy "centers" is approximately perpendicular to the mean flow) were qualitatively compared with the phase-averaged measurements of Hussain and Zaman (1979) in a perturbed axi-symmetric mixing layer (i.e., the near field of a circular jet) at the phase corresponding to their "radial configuration" (Phase 2 and 3 in Figs. 19b and 19f of Hussain and Zaman with due consideration of the difference in definition of the direction of the mean flow between the present case (Fig. 1a) and Hussain and Zaman (Fig. 1)). Although both displayed regions of positive and negative values of $-u'v'$, the results of Hussain and Zaman show positive values of $-u'v'$ in a vertical "strip" of approximate width $\lambda/4$ on the right side of the vertical line through the eddy centers, and negative values of $-u'v'$ in a vertical strip to the left. The disagreement may be attributable to the difference in flow fields (i.e., the two-dimensional vs. the axi-symmetric mixing layer). In particular, at the "radial" configuration of Hussain and Zaman the vorticity distributions in the two eddies are substantially different (at Phases 2 and 3, the ratio of the maximum values of vorticity are approximately 2 and 1.5, respectively), and the eddy coalescence occurs at a later time.

The kinetic energy of the coherent fluctuations is indicated in Fig. 3. The value of E_{coh} increases from an initial level of 5.64×10^{-4} (0.1% of the magnitude of the initial mean kinetic energy defect) to a maximum value of 6.31×10^{-2} at $t = 32.4$. For this case, the kinetic energy of the coherent fluctuations is small compared to the kinetic energy of the fine scale motion. Defining the fine-scale kinetic energy associated with the x- and y-directions as

$$E^*(t) = \frac{1}{L}\int_{-\infty}^{\infty}\int_{0}^{L}\frac{1}{2}[\langle u''u''\rangle + \langle v''v''\rangle]dxdy$$

the maximum value of E_{coh}/E^* achieved during the calculation is 0.10.

The contributions to dE_{coh}/dt are presented in Fig. 4, where the terms are defined in Eq. (19). The results indicate a substantial energy flux from the mean flow to the fluctuating coherent motion (TERM 1) which is enhanced during the eddy pairing (see $t = 17.1$ and 22.8 in Fig. 2). Subsequent to the amalgamation, a reverse energy flux appears for $t > 30$. Surprisingly, the results indicate an energy flux from the fine scale to the fluctuating coherent motion during the entire pairing process (TERM 2). Similar results were found for a variety of initial perturbation strengths and wavelengths. Further discussion is provided in Knight (1979). It should be noted that there exists a substantially larger energy flux from the mean motion to the fine scale, and consequently the rate of change of E^* is always positive.

The mean streamwise velocity $U(y,t)$ remained self-similar during the entire calculation. The computed mean growth rate was linear for $t \lesssim 24$. Defining the non-dimensional velocity-profile maximum slope thickness δ_ω by

$$\delta_\omega = 2/\frac{\partial U}{\partial y}\bigg|_{max} \qquad (20)$$

and using the Galilean transformation, the equivalent spatial growth rate for $t \lesssim 24$ is

$$\frac{d\delta_\omega}{dx} = 0.24\,\frac{(U_1-U_2)}{(U_1+U_2)} \qquad (21)$$

The experimental best-fit indicated by Brown and Roshko (1974) is

$$\frac{d\delta_\omega}{dx} = 0.18\,\frac{(U_1-U_2)}{(U_1+U_2)} \qquad (22)$$

although the numerical coefficient is dependent on the state of the boundary layer at the end of the splitter plate (Batt 1975) with values ranging from 0.16 (Liepmann and Laufer 1947) to 0.24 (Wygnanski and Fiedler 1970). The computed growth rate Eq.

(21) is essentially identical to the similarity growth rate of the governing equations, indicating that in the present calculation the growth rate during the eddy interaction is dominated by the fine scale Reynolds stresses. Subsequent to the amalgamation ($t \gtrsim 30$), the growth rate decreases reflecting the reverse energy flux from the coherent to the mean motion. Additional calculations indicated a dependence of the growth rate during amalgamation on the strength of the perturbation, with stronger perturbations producing higher growth rates. Nonetheless, all calculations displayed a decrease in growth rate subsequent to the amalgamation.

B. Eddy Viscosity Model

The determination of the value of σ in the eddy viscosity model is based on the experimental observation that roughly one-half of the total Reynolds shear stress is attributable to the "incoherent" or phase-random fluctuations (Cantwell 1976, Kovasznay 1977, Latigo 1979). From Eqs. (14) and (15) and the periodicity of the flow in the streamwise direction, it follows that

$$-\overline{\langle u''v''\rangle} = \varepsilon\frac{\partial U}{\partial y} \tag{23}$$

The computed results (see Fig. 8) indicate that for the present case the velocity $U(y,t)$ remains self-similar throughout the eddy interaction satisfying

$$U = \mathrm{erf}(y/\sqrt{2\pi}\,\theta(t)) \tag{24}$$

where $\theta(t)$ is the momentum thickness defined by Eq. (16). Therefore Eqs. (23) and (24) yield

$$\max[-\overline{\langle u''v''\rangle}] = \frac{1}{\sqrt{\pi}\,\sigma} \tag{25}$$

where the maximum value occurs at $y = 0$. Typical experimental values for the maximum of the conventional time-mean Reynolds stress (non-dimensionalized by $U_\infty^2 = (U_1-U_2)^2/4$) range from 0.037 (Wygnanski and Fiedler 1970) to 0.052 (Spencer and Jones 1971). Based on the above argument that roughly *one-half* of the total Reynolds shear stress is due to the fine scale fluctuations, the above data imply a range for σ between approximately 21 and 30. As discussed below, the results of the eddy viscosity model calculations are compared with the phase-averaged measurements of Latigo (1979), for which the maximum value of the conventional time mean Reynolds stress was 0.04 to 0.044. In particular, the maximum value of the mean *fine-scale* Reynolds stress at the particular phase in the eddy interaction discussed below is 0.025, implying from Eq. (25) a value of σ approximately equal to 22. Therefore, for comparison with the experimental results of Latigo (1979), the single constant σ in the eddy viscosity model is set equal to 22 to insure matching of the *peak* value of the *fine scale* Reynolds shear stress at a particular phase in the eddy interaction.

Before comparing with the results of Latigo (1979), it is interesting to consider the evolution of the eddy interaction using the eddy viscosity model. In Fig. 5a-e, a sequence of vorticity contour plots are shown at five successive times (t = 5.7, 11.4, 17.1, 22.8, and 28.5) together with contour plots of -u'v'. The vorticity ζ and -u'v' are again made dimensionless by U_∞/L_o and U_∞^2, respectively. Due to the difference in shape of the initial velocity profiles for the Reynolds stress and eddy viscosity models, the initial values of the momentum thickness $\theta(t)$, non-dimensionalized by the respective values of L_o, are 0.2818 and 0.219, respectively.

The sequence of vorticity contours display the growth, interaction and coalescence of the large scale eddies. The contours do not exhibit the sharp gradient in vorticity near the edges of the mixing layer evident in Figs. 2a-d of the Reynolds stress model calculations. The more diffuse nature of the vorticity contours in the vicinity of the layer edges in Figs. 5a-e is attributable to the fact that the eddy viscosity Eq. (15) is applied for $|y| \leq 6.45\ \theta(t)$ (i.e., for a vertical extent much larger than the layer width). In the Reynolds stress model, the fine scale Reynolds stresses rapidly decrease to zero at the edge of the layer.

The sequence of -u'v' contours in Figs. 5a-e display a similar evolution as their counterparts in Figs. 2a-d, and the same description applies. During the growth of the eddies (t = 5.7 and 11.4), regions of positive -u'v' are observed near the leading and trailing edges of the eddy pair. At amalgamation (t = 22.8), a doubly periodic structure is clearly evident which is qualitatively similar to the Reynolds stress model results at the same stage. Subsequent to the pairing, the values of -u'v' are on the whole strongly negative, reflecting the reverse energy flux from the coherent fluctuations to the mean flow.

The kinetic energy of the coherent fluctuations is indicated in Fig. 6. The value of E_{coh} increases from an initial level of 4.36×10^{-4} (0.1% of the initial mean kinetic energy defect) to a maximum value of 4.9×10^{-1} at t = 24. The relative growth of the fluctuating coherent motion is substantially larger than for the Reynolds stress model. In particular, the ratio of the maximum value of E_{coh} to its initial value is 1124 in the present case and 112 for the Reynolds stress model. This difference is due to the fact that the fine scale Reynolds stresses are weaker in the present case since the value of σ in the eddy viscosity model is determined by the condition that the maximum of $-\overline{\langle u''v''\rangle}$ is approximately *one-half* the conventional peak Reynolds stress. In contrast to the Reynolds stress model calculations, the growth rate during the eddy amalgamation is roughly equally attributable to the fine scale and coherent motion (see below), and the coherent fluctuations are relatively strong.

The contributions to dE_{coh}/dt are presented in Fig. 7. The results display a strong energy flux from the mean to the coherent motion (TERM 1) during the eddy interaction in qualitative agreement with Fig. 4. The peak value of this energy flux again occurs at approximately the moment of eddy amalgamation. At subsequent times, the energy flux changes sign as seen previously. The energy flux between the coherent fluctuations and fine scale motion (TERM 2) is negative definite due to the nature of the fine scale Reynolds stress in Eqs. (14) to (16). In particular, this term is equal to

$$-2 \int_{-\infty}^{\infty} \varepsilon \, \overline{S'_{ij} S'_{ij}} \, dy \tag{26}$$

where S'_{ij} is the rate-of-strain tensor of the fluctuating coherent motion,

$$S'_{ij} = \frac{1}{2} \left(\frac{\partial u'_i}{\partial x_j} + \frac{\partial u'_j}{\partial x_i} \right) \tag{27}$$

The mean velocity profile is shown in Figure 8, where the ordinate is $y/\theta(t)$. The velocity is seen to remain closely self-similar throughout the entire interaction and satisfies Eq. (24).

The mean growth rate is presented in Fig. 9, where the momentum thickness is shown as a function of time. The conventional growth rate is indicated by the straight line, representing the standard similarity solution

$$\frac{d\theta}{dt} = \frac{1}{\sqrt{2\pi}\,\sigma_0} \tag{28}$$

where $\sigma_0 = 11$ is a typical value of the conventional mean reference spreading parameter (Mansour et al. 1978). By means of the Galilean transformation, the analogous conventional mean growth rate in the spatially-developing mixing layer is

$$\frac{d\theta}{dx'} = \frac{1}{\sqrt{2\pi}\,\sigma_0} \frac{(U_1-U_2)}{(U_1+U_2)} \tag{29}$$

The calculated growth rate is initially low, reflecting the relatively weak nature of the initial coherent perturbation. As the eddies interact, the growth rate increases substantially, and for $12 \lesssim t \lesssim 24$ the calculated growth rate exceeds the conventional mean. Subsequent to the eddy pairing, a strong reverse momentum flux appears as discussed earlier (Fig. 5e) and the growth rate is substantially reduced. It should be emphasized that, in the spatially-developing mixing layer, the conventional growth rate is a result of numerous eddy interactions (pairing, tearing, etc.) The present calculation considers the evolution of a single eddy pair, and for this reason cannot be used to predict the conventional mean growth rate.

Prior to comparison with the results of Latigo (1979), it is important to describe the nature of his experiment. A plane homogeneous turbulent mixing layer with velocity ratio $U_2/U_1 = 0.12$ was perturbed by an acoustic disturbance emanating from a spanwise array of loudspeakers at the trailing edge of the splitter plate. The disturbance consisted to a repeated set of "bursts", with each burst consisting of four separate pulses. The pulse width corresponded to the half-period of the most unstable disturbance in the near wake of the splitter plate. The disturbance was spanwise uniform. Detailed conventional mean profiles of the velocity and Reynolds stresses indicated close agreement with measurements in the unperturbed ("natural") turbulent mixing layer, indicating that the effect of the perturbation was to "organize" the large structures without evidently significantly altering the characteristics of the flow. Conditional phase-averaged measurements were obtained at a variety of streamwise locations, representing various stages in the interaction of the several eddies. In particular, phase-averaged measurements of the mean coherent Reynolds stresses $\overline{u'^2}$, $\overline{v'^2}$, $\overline{u'v'}$ and fine scale Reynolds stresses $\overline{\langle u''u''\rangle}$, $\overline{\langle v''v''\rangle}$, $\overline{\langle u''v''\rangle}$ were presented at $Re_x = (U_1-U_2)x/\nu = 1.18 \times 10^6$ corresponding to $Re_\theta = (U_1-U_2)\theta/\nu = 3.0 \times 10^4$ (Latigo 1979, Fig. 9.4a,b at x = 30.625 in). This location corresponded to a stage of amalgamation of an eddy pair.

In order to compare with the results of Latigo, it is necessary to determine the particular time in the calculation corresponding to the stage in eddy pairing discussed above. Since there was no flow visualization in the experiment, it was decided to choose the time at which the *peak* computed value of $-\overline{u'v'}$ matched the corresponding peak measured value.

In Fig. 10, the computed and measured values of $+\overline{u'v'}$ are shown as a function of y/θ. The time is 12.7 in this and all succeeding figures, and corresponds to an intermediate stage in the calculated eddy pairing (Fig. 5). The solid line is the computed results, and the symbols are the data of Latigo. As indicated above, only the *peak* values of $+\overline{u'v'}$ are matched. Nevertheless, the agreement over the entire profile is excellent. Although results are presented here for an initial E_{coh} of 0.1%, additional calculations have been performed using initial E_{coh} from 0.025% to 0.25% with essentially identical results, the effect of different initial E_{coh} being a change in the time at which the comparison with experiment is made.

In Fig. 11, the calculated and measured values of $+\overline{\langle u''v''\rangle}$ are presented. As indicated earlier, the *peak* values are matched (this determined the constant σ). Nevertheless, there is good agreement over the entire profile.

The computed and measured values of $\overline{u'^2}$ are presented in Fig. 12. It should be emphasized that there is no matching of any point in the profile. The computed peak value is 14% above the experimental value, and the profiles are in good agreement except in the region of the secondary "humps" at $y \approx \pm 4\theta$.

IV. CONCLUSIONS

The interaction of large scale eddies in the temporally-developing turbulent mixing layer have been investigated by imposing a weak perturbation consisting of a harmonic and subharmonic disturbance on the initial similarity profiles. Two different models for the fine scale Reynolds stresses were utilized, namely a Reynolds stress equation and a simple eddy viscosity model. In both cases, the large eddies were observed to grow, interact and amalgamate in qualitative agreement with experiment. The period of eddy growth and interaction was characterized by transport of momentum across the mixing layer by the coherent fluctuations in the neighborhood of the leading and trailing edges of the eddy pair. During this period, a strong flux of energy from the mean to the coherent fluctuating motion occurred. Subsequent to the pairing, the energy flux reversed and the amalgamated eddy decayed. During the entire interaction the mean velocity profile remained self-similar. Detailed quantitative comparison was made with the experimental results of Latigo (1979) at a particular stage in the eddy pairing. The computed profiles of the coherent and fine scale Reynolds shear stress and the longitudinal coherent fluctuation energy were in close agreement with the experimental results.

ACKNOWLEDGEMENTS

This work was supported by the National Science Foundation under Grant No. ENG78-05454 and the Emil Buehler Foundation.

REFERENCES

Ashurst, W., 1977, "Numerical Simulation of Turbulent Mixing Layers Via Vortex Dynamics," Symposium on Turbulent Shear Flows, Penn. State Univ., pp. 11.43-11.51.

Batt, R., 1975, "Some Measurements on the Effect of Tripping the Two-Dimensional Shear Layer," AIAA J., Vol. 13, pp. 245-247.

Breidenthal, R., 1978, "A Chemically Reacting, Turbulent Shear Layer," Ph.D. Thesis, Calif. Inst. of Tech.; also, AIAA J., Vol. 17, 1979, pp. 310-311.

Brown, G., and Roshko, A., 1974, "On Density Effects and Large Structure in Turbulent Mixing Layers," J. Fluid Mech., 64, pp. 775-816.

Browand, F., and Weidman, P., 1976, "Large Scales in the Developing Mixing Layer," J. Fluid Mech., Vol. 76, pp. 127-144.

Cantwell, B., 1976, "A Flying Hot Wire Study of the Turuulent Near Wake of a Circular Cylinder at a Reynolds Number of 140,000," Ph.D. Thesis, Calif. Inst. of Tech., Pasadena, Calif.

Couët, B., Buneman, O., and Leonard, A., 1979, "Three-Dimensional Simulation of the Free Shear Layer Using the Vortex-in-Cell Method," Second Symposium on Turbulent Shear Flows, Imperial College, London.

Delcourt, B., and Brown, G.L., 1979, "The Evolution and Emerging Structure of a Vortex Sheet in an Inviscid and Viscous Fluid Modelled by a Point Vortex Method," Second Symposium on Turbulent Shear Flows, Imperial College, London.

Dimotakis, P., and Brown, G., 1976, "The Mixing Layer at High Reynolds Number. Large Scale Dynamics and Entrainment," J. Fluid Mech., 78, pp. 535-560.

Dryden, H., 1948, "Recent Advances in the Mechanics of Boundary Layer Flow," Advances in Applied Mechanics, Vol. 1, Academic Press, New York, 1-40.

Gatski, T., and Liu, J., 1980, "On the Interactions Between Large-Scale Structure and Fine-Grained Turbulence in a Free Shear Flow. III. A Numerical Solution," Proc. Roy. Soc. London, Ser. A., to appear.

Hanjalic, K., and Launder, B., 1972, "A Reynolds Stress Model of Turbulence and Its Application to Thin Shear Flows," J. Fluid Mech., Vol. 52, pp. 609-638.

Higuchi, H., 1976, "An Experimental Investigation on Axi-Symmetric Turbulent Wakes with Zero Momentum Defect," Ph.D. Thesis, Calif. Inst. of Tech.

Hussain, A.K.M.F., and Reynolds, W.C., 1970, "The Mechanics of an Organized Wave in Turbulent Shear Flow," J. Fluid Mech., Vol. 41, pp. 241-258.

Hussain, A.K.M.F., 1977, Cardiovascular Flow Dynamics and Measurements, N.H.C. Hwang and N. Norman, ed., Univ. Park Press, Baltimore, Md.

Hussain, A.K.M.F., 1979, "Investigations of Coherent Structures in Free Turbulent Shear Flows," Sixth Biennial Symposium on Turbulence, Dept. of Chemical Engineering, Univ. of Missouri-Rolla.

Hussain, A.K.M.F., and Zaman, K.B.M.Q., 1979, "Vortex Pairing in a Circular Jet Under Controlled Excitation: Part 2," Report FM-2, Dept. of Mech. Engr., Univ. of Houston, Houston, Texas.

Kendall, J., 1970, "The Turbulent Boundary Layer Over a Wall With Progressive Surface Waves," J. Fluid Mech., Vol. 41, pp. 259-281.

Kline, S.J., et al., 1967, "The Structure of Turbulent Boundary Layers," J. Fluid Mech., 30, pp. 741-773.

Knight, D., and Saffman, P.G., 1977, "Turbulence Model Predictions for Flows With Significant Mean Streamline Curvature," Report AM 77-1, Dept. of Applied Math., Calif. Inst. of Tech., Pasadena, Calif.

Knight, D., 1979, "Numerical Investigation of Large Scale Structures in the Turbulent Mixing Layer," Proceedings of Sixth Biennial Symposium on Turbulence, Dept. of Chemical Engineering, Univ. of Missouri-Rolla.

Konrad, J., 1976, "An Experimental Investigation of Mixing in Two-Dimensional Turbulent Shear Flows with Applications to Diffusion-Limited Chemical Reactions," Project Squid Technical Report CIT-8-PU.

Kovasznay, L.S.G., 1977, "The Role of Large Scale Coherent Structures in Turbulent Shear Flows," Proceedings of the Fifth Biennial Symposium on Turbulence, Dept. of Chemical Engineering, University of Missouri-Rolla.

Latigo, B., 1979, "Large-Scale Structure Interactions in a Two-Dimensional Turbulent Mixing Layer," Ph.D. Thesis, Dept. of Aerospace Engineering, Univ. of S. Calif., Los Angeles, Calif.

Laufer, J., 1975, "New Trends in Experimental Turbulence Research," in Annual Review of Fluid Mechanics, Vol. 7, Annual Reviews, Inc., Palo Alto, Calif.

Liepmann, H., and Laufer, J., 1947, "Investigations of Free Turbulent Mixing," NACA Report 1257.

Liepmann, H., 1962, "Free Turbulent Shear Flows," Mechanics of Turbulence, Int. Symp. Natl. Science Research Center, Marseille, pp. 211-227.

Liu, J.T.C. and Merkine, L., 1976, "On the Interactions Between Large-Scale Structure and Fine-Grained Turbulence in a Free Shear Flow; I. The Development of Temporal Interactions in the Mean," Proc. Roy. Soc. London, A, Vol. 352, pp. 213-247.

Mansour, N., Ferziger, J., and Reynolds, W., 1978, "Large-Eddy Simulation of a Turbulent Mixing Layer," Report TF-11, Dept. of Mechanical Engineering, Stanford University, Stanford, Calif.

Oster, D., 1980, "The Effect of an Active Disturbance on the Development of the Two-Dimensional Turbulent Mixing Region," Ph.D. Thesis, Tel-Aviv Univ.

Papailiou, D., and Lykoudis, P., 1974, "Turbulent Vortex Streets and the Mechanism of the Turbulent Wake," J. Fluid Mech., 62, pp. 11-31.

Patnaik, P., Sherman, F., and Corcos, G., 1976, "A Numerical Simulation of Kelvin-Helmholtz Waves of Finite-Amplitude," J. Fluid Mech., Vol. 73, pp. 215-240.

Reynolds, W.C., 1976, "Computation of Turbulent Flows," in Annual Review of Fluid Mechanics, Vol. 8, Annual Reviews, Inc., Palo Alto, Calif.

Riley, J., and Metcalf, R., 1980, "Direct Numerical Simulation of a Perturbed, Turbulent Mixing Layer," AIAA, Paper 80-0274, AIAA 18th Aerospace Sciences Meeting, Pasadena, Calif.

Roshko, A., 1976, "Structure of Turbulent Shear Flows: A New Look," AIAA J., Vol. 14, pp. 1349-1357.

Saffman, P.G., 1977, "Results of a Two Equation Model for Turbulent Flows and Development of a Relaxation Stress Model for Application to Straining and Rotating Flows," Project Squid Workshop on Turbulence in Internal Flows, Hemisphere Publishing, pp. 191-231.

Spencer, B., and Jones, B., 1971, "Statistical Investigation of Pressure and Velocity Fields in the Turbulent Two-Stream Mixing Layer," AIAA Paper 71-613, AIAA 4th Fluid and Plasma Dynamics Conference, Palo Alto, Calif.

Winant, C., and Browand, F., 1974, "Vortex Pairing: The Mechanism of Turbulent Mixing-Layer Growth at Moderate Reynolds Number," J. Fluid Mech., 63, pp. 237-255.

Wygnanski, I., and Fiedler, H., 1970, "The Two-Dimensional Mixing Region," J. Fluid Mech., Vol. 41, pp. 327-361.

Wygnanski, I., Oster, D., and Fiedler, H., 1979, "The Forced Plane Turbulent Mixing Layer: A Challenge for the Predictor," Proceedings of the Second Symposium on Turbulent Shear Flows, London.

Yule, A., Chigier, N., and Thompson, N., 1977, "Coherent Structures in Combustion," Symposium on Turbulent Shear Flows, Penn. State Univ.

Zaman, K.B.M.Q., and Hussain, A.K.M.F., 1979, "The Mechanics of Vortex Pairing in a Circular Jet Under Controlled Excitation," Second Symposium on Turbulent Shear Flows, London.

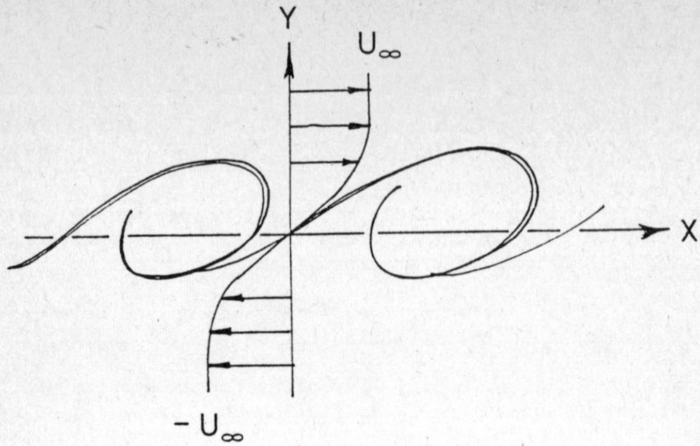

Fig. 1a. Temporally-Developing Mixing Layer

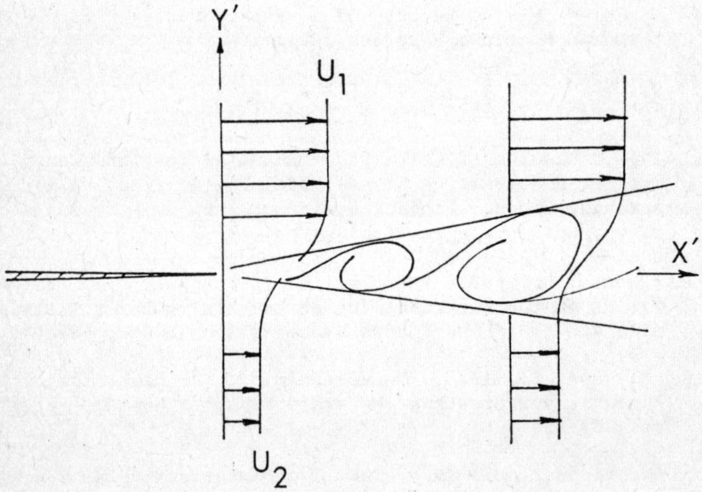

Fig. 1b. Spatially-Developing Mixing Layer

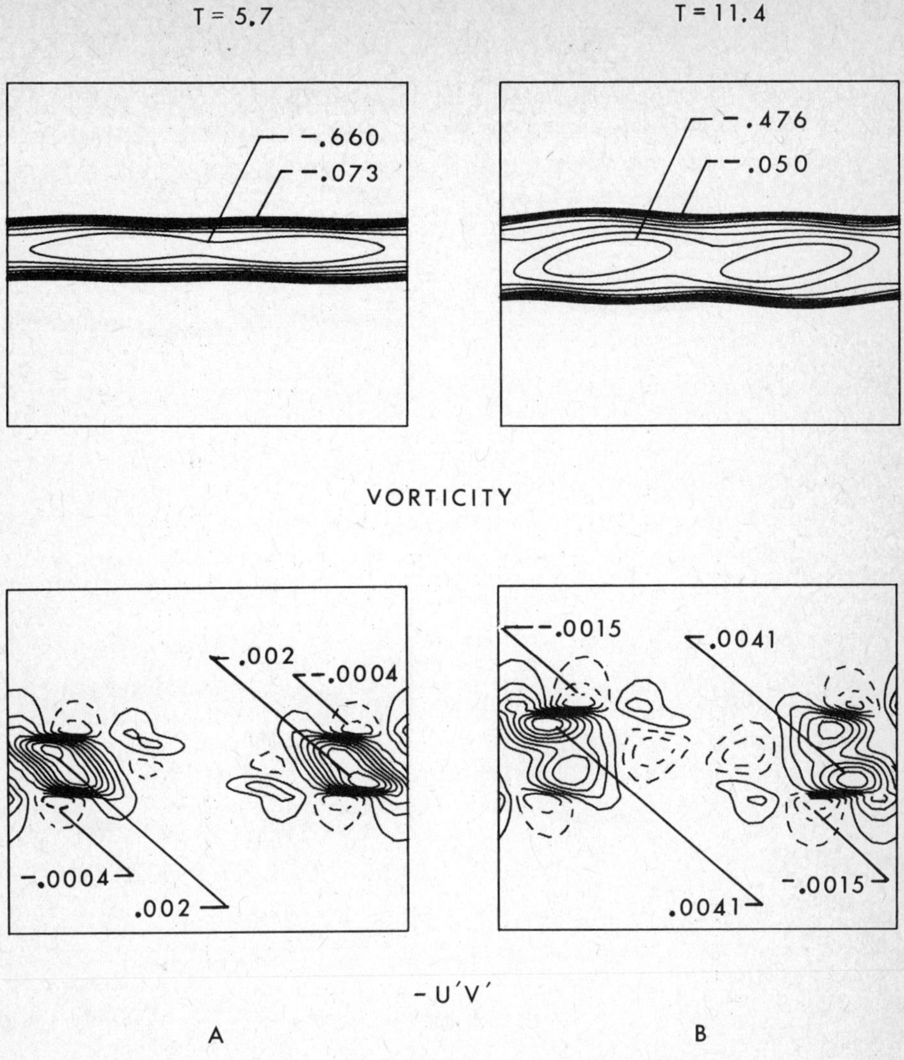

Fig. 2. Vorticity and -u'v' Contours (Reynolds Stress Equation Model) Solid Lines Represent Positive Values

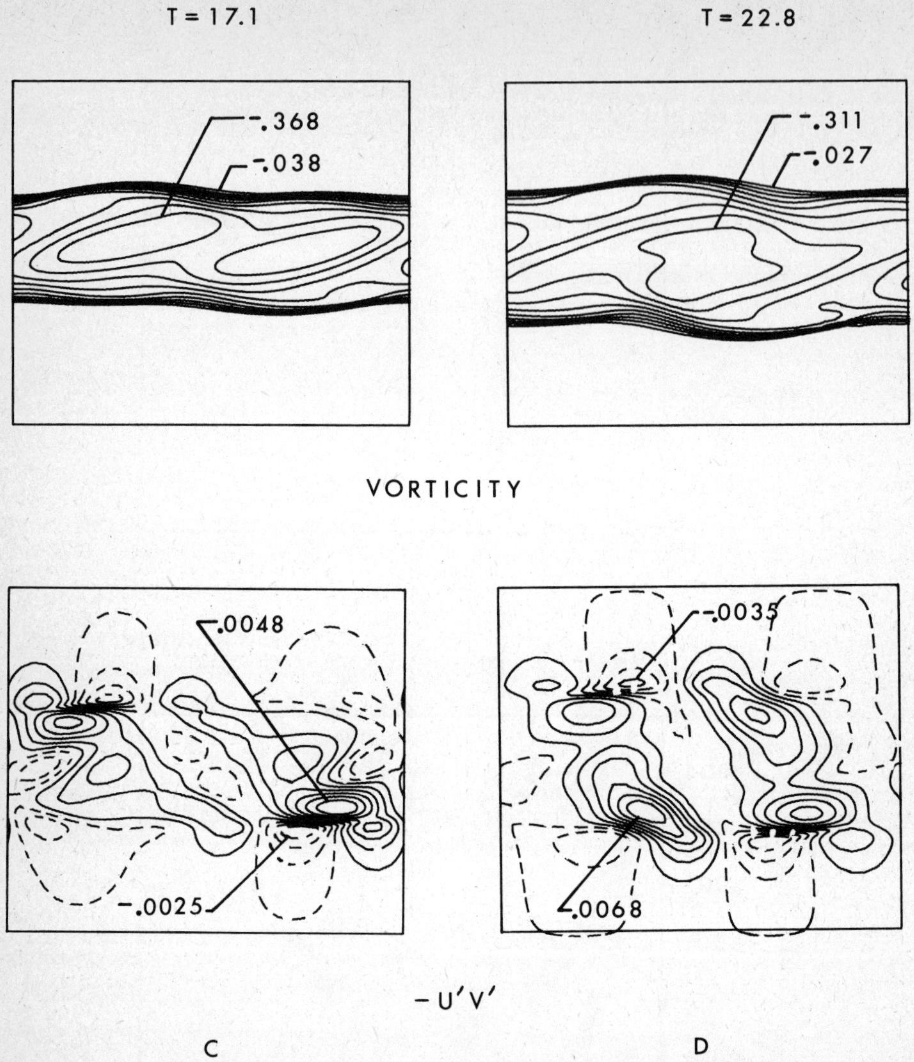

Fig. 2. Continued

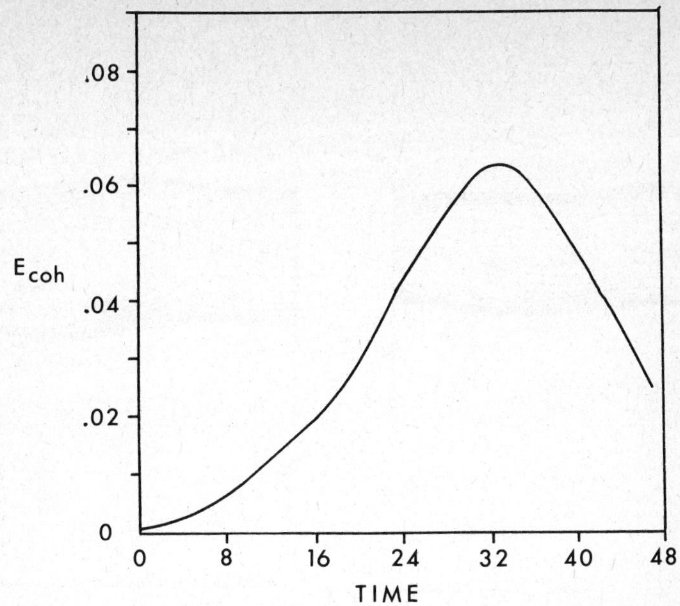

Fig. 3. Kinetic Energy of Coherent Fluctuating Motion (Reynolds Stress Equation Model)

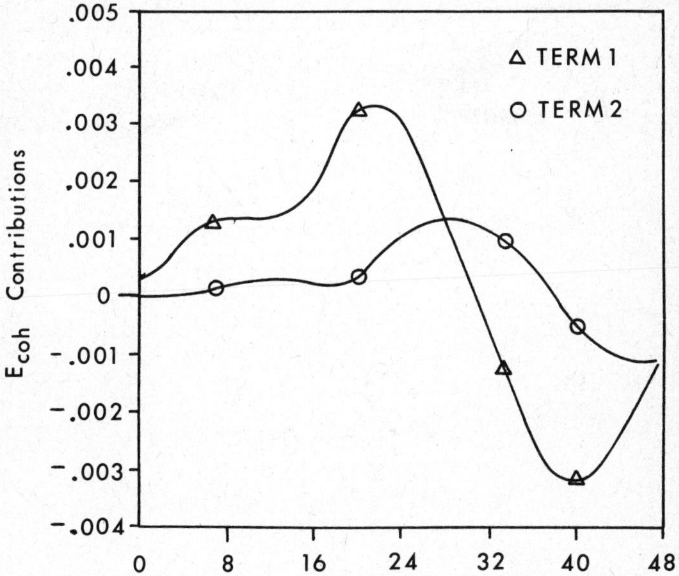

Figure 4. Terms in Rate of Change of E_{coh} (Reynolds Stress Equation Model)

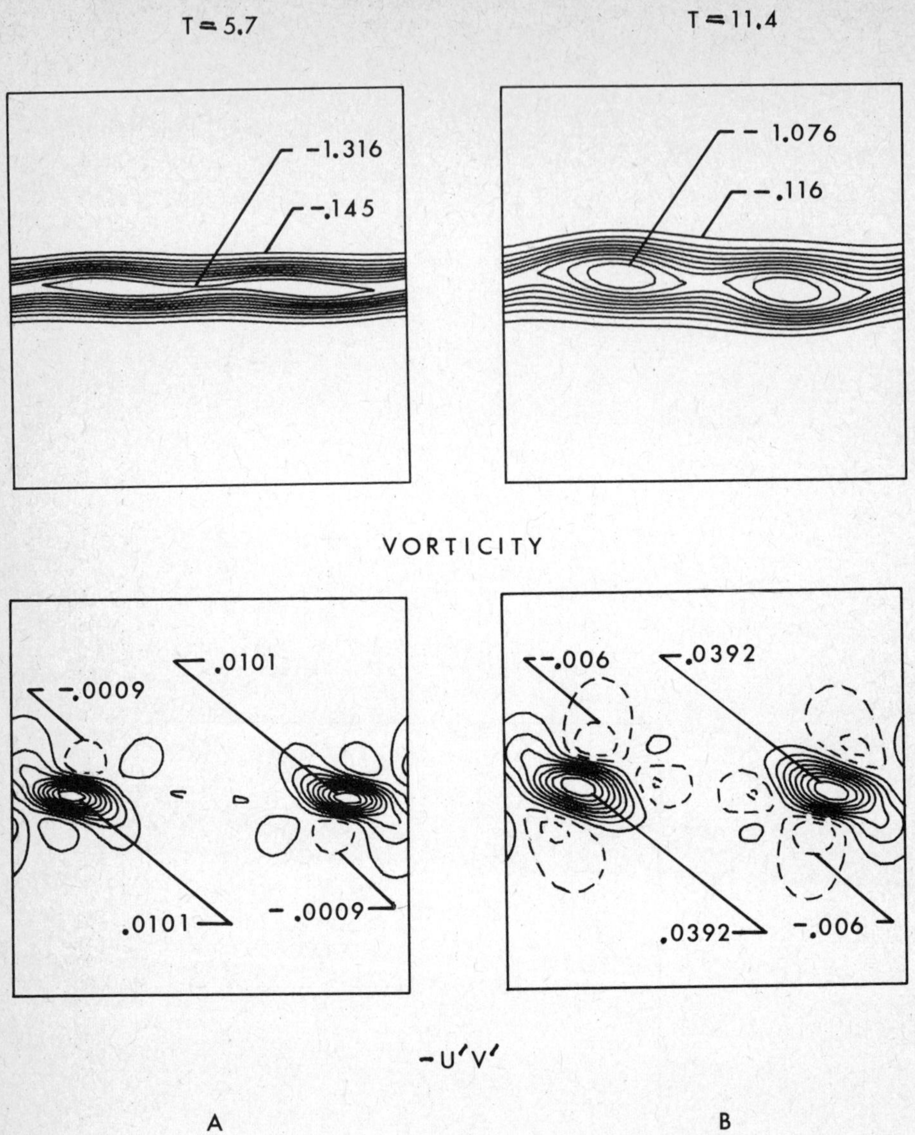

Fig. 5. Vorticity and -u'v' Contours (Eddy Viscosity Model)

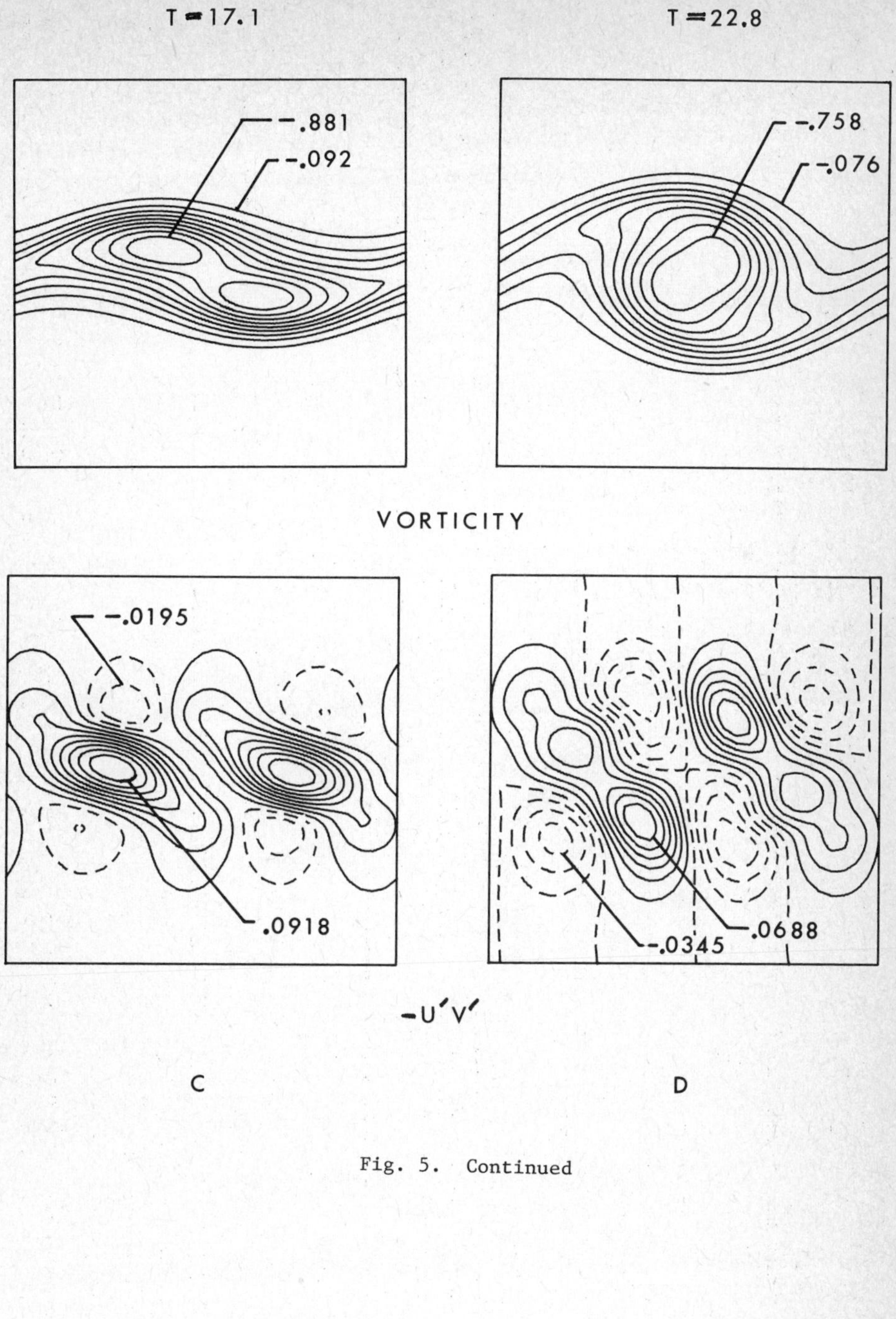

Fig. 5. Continued

T = 28.5

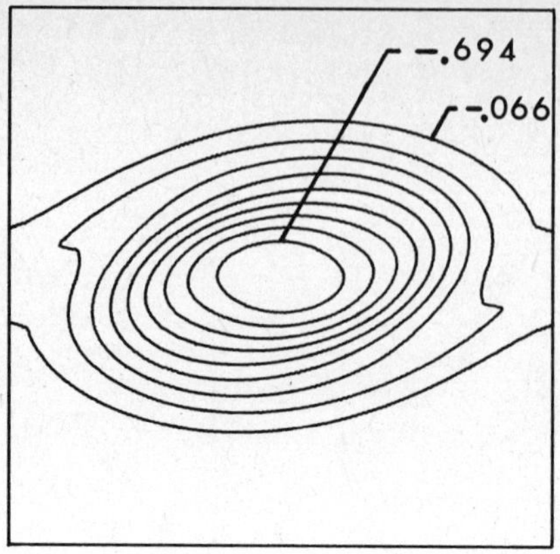

VORTICITY

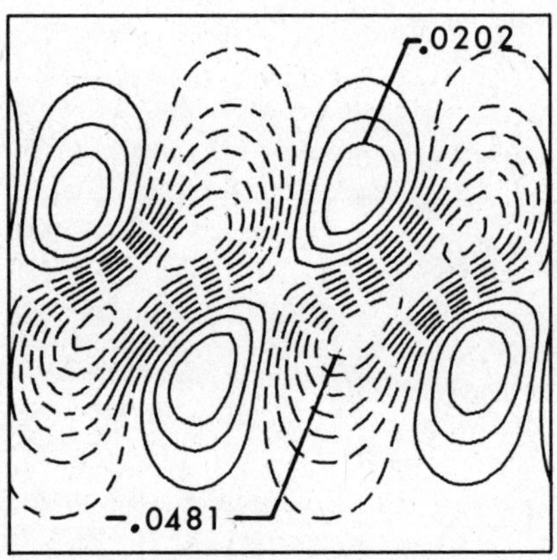

$-U'V'$

E

Fig. 5. Continued

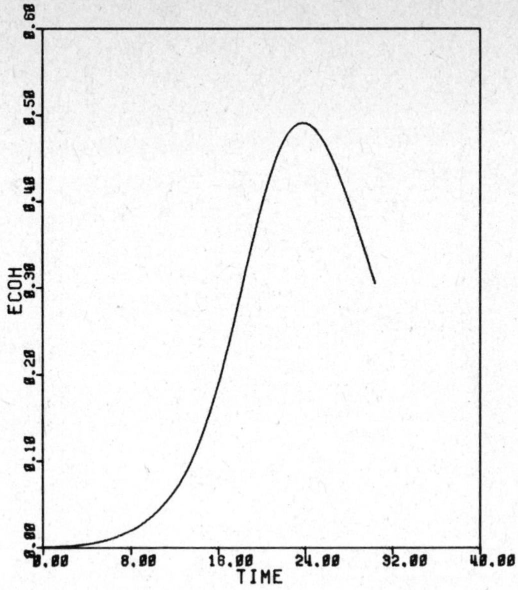

Fig. 6. Kinetic Energy of Coherent Fluctuating Motion (Eddy Viscosity Model)

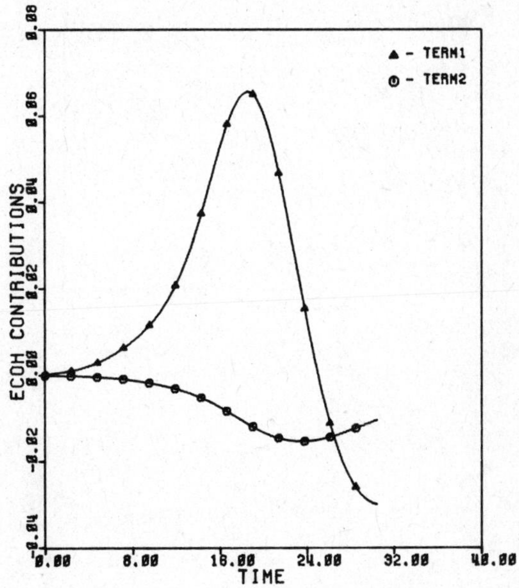

Fig. 7. Terms in Rate of Change of E_{coh} (Eddy Viscosity Model)

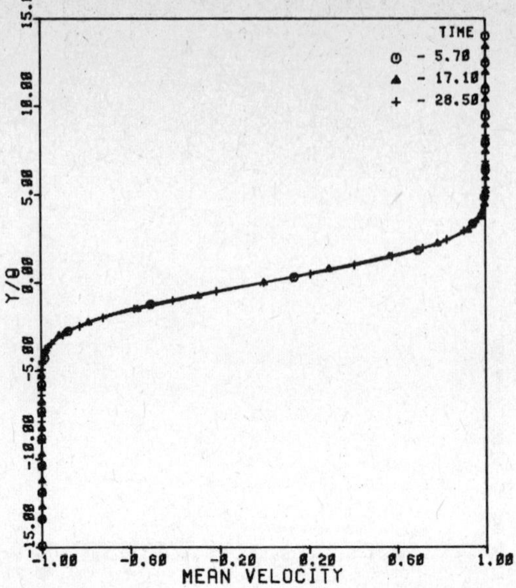

Fig. 8. Mean Velocity of Selected Times (Eddy Viscosity Model)

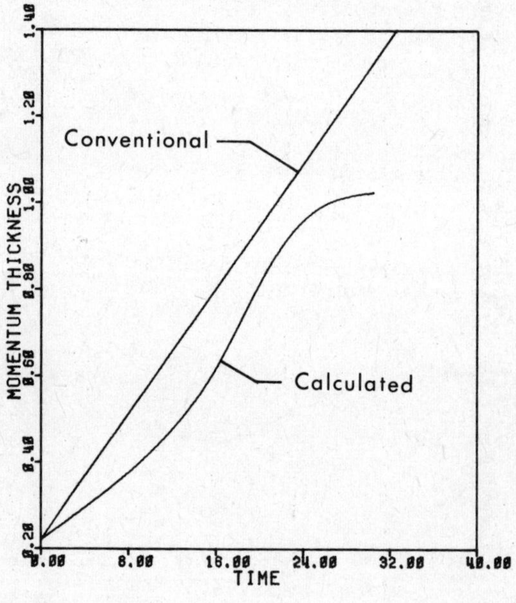

Fig. 9. Momentum Thickness (Eddy Viscosity Model)

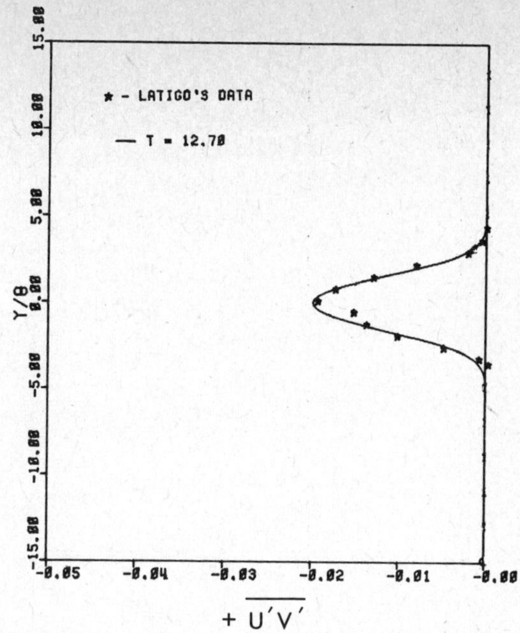

Fig. 10. Coherent Reynolds Stress $+\overline{u'v'}$ at a Stage in Eddy Pairing

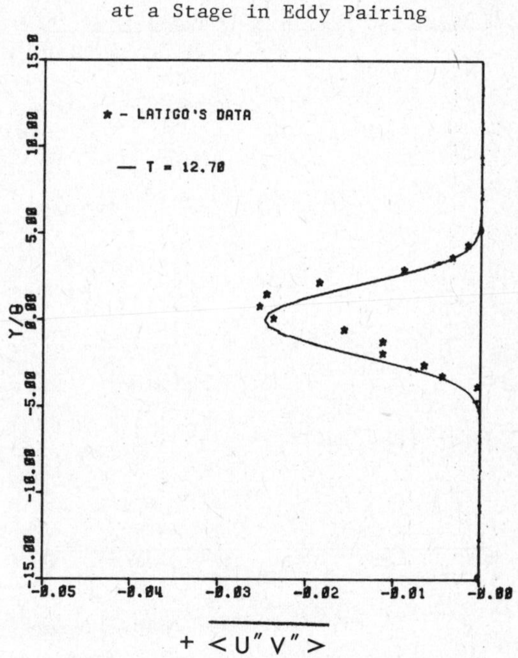

Fig. 11. Fine Scale Reynolds Stress $+\overline{<u''v''>}$ at a Stage in Eddy Pairing

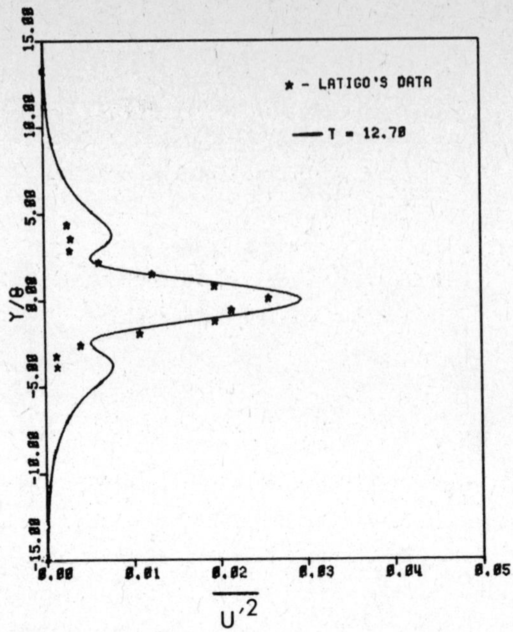

Fig. 12. Coherent Longitudinal Reynolds Stress $+\overline{u'^2}$ at a Stage in Eddy Pairing

A MODEL FOR PERIODIC STRUCTURES IN TURBULENT BOUNDARY LAYERS

A.C.M. Beljaars

Royal Netherlands Meteorological Institute
De Bilt, The Netherlands

K. Krishna Prasad
Eindhoven University of Technology
Eindhoven, The Netherlands

Abstract

Many experimental studies emphasize the importance of periodic recognizable flow patterns for the transport process in turbulent flow. In this paper a model is formulated for the large scale part of the turbulent motion. The experimental observation that the structures in the outer region run in phase with the bursting cycle in the wall layer forms the basis of the model. The wall layer, where viscous stresses are important and the outer region where the inviscid approximation holds, are treated separately. The small scale part of the turbulent motion, which is assumed to be important in localized regions only (bursts regions), couples the wall region and the outer region.

The mean wall shear stress calculated with this model agrees reasonably well with the empirical formulae for the friction coefficient even for the more complex case of the transpired boundary layer. The main conclusion of the model calculations is that the transport of momentum can be very well explained in terms of turbulent structures. The model clearly illustrates how momentum is transported in three stages: (i) Thin elongated layers near the wall slow down as the result of viscous forces. (ii) The retarded fluid is ejected in localized regions or bursts. (iii) The large scale motion in the outer region takes over the transport.

In this paper special attention will be given to the function of the longitudinal vortices in the wall layer. In turns out that they hardly influence the turbulent exchange, but that they are very important for the creation of locally unstable regions. It is believed that the strength of the longitudinal vortices is influenced by the large scale structures in the outer region. By this mechanism the large scales in the outer region can influence the burst frequency.

In a discussion some ideas are presented about what this can mean for special flow phenomena as: drag reduction by polymer solutions or along compliant walls and rapid shear stress change along curved walls.

1. Introduction

The large amount of experimental data published during the past 15 years clearly show the importance of recognizable flow patterns in turbulent shear flows. These patterns, often called structures, are periodic and retain their coherence over relatively long distances in the dominant flow direction. A well known example of these structures is that of vortices in a mixing layer (Brown & Roshko, 1974).

The wavy interface between turbulent and non-turbulent fluid is a clear manifestation of these structures in boundary layers. Blackwelder &Kovasznay (1972) attribute this to the large scale vortices with their axes in the spanwise direction. Their identity is preserved over a distance of 20 boundary layer thicknesses.

The wall region of a turbulent boundary layer, where viscosity plays a role, also exhibits periodic flow patterns. The flow visualization studies of Kline et al. (1967) and Corino & Brodkey (1969) establish a sequence of events in the wall layer. (i) Streaks of low momentum fluid with a length of the order of boundary layer thickness (δ) and a width of about $30\nu/u_*$ (where ν is the kinematic viscosity and u_* the friction velocity) show a periodicity in the streamwise as well as spanwise directions. (ii) A horizontal shear layer is formed on top of these streaks at $y^+ = yu_*/\nu \simeq 40$; the local velocity profile shows an inflexion point. (iii) The streaks are terminated by the ejection of low momentum fluid into the outer region. (iv) The ejected fluid is replaced by high momentum fluid from the border of the buffer region (sweep). The last two parts of the cycle - the ejection and sweep - cover only a small fraction of the total time but the major part of the Reynolds shear stress comes from them. Thus the periodicity in the wall layer is seen to be important for the transport process.

Two other features are important in this connection. (i) The period of the wall layer cycle scales on outer layer parameters (Laufer & Badri Narayanan, 1971). (ii) Brown & Thomas (1977) show that the structures in the outer region run in phase with the periodic phenomena in the wall layer.

These observations have been used to construct a simple deterministic model for the large scale part of turbulent motion. A detailed description of the model and results obtained from it is available elsewhere (Beljaars, 1979, Beljaars, Krishna Prasad & de Vries, 1980). We will limit ourselves to a brief presentation of these aspects here. The purpose of the present paper is to discuss the possible role of such modelling in understanding and predicting turbulent flow properties.

2. The Model

An idealized phenomenological picture of the turbulent boundary layer will be derived from the experimental features discussed above. There are three distinct elements in the model: (i) wall layer where viscosity is important; (ii) outer region where viscosity can be ignored; and (iii) interactions between the two. A crucial assumption is that the structures in both the wall layer and the outer region move downstream with a constant convection velocity. This permits the use of a coordinate reference frame that moves downstream with this velocity resulting in a more or less stationary picture for the structures in both the regions.

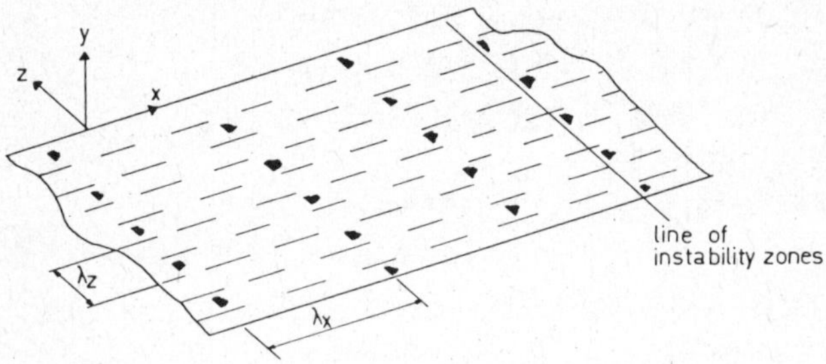

Fig. 1 The moving instability zones in the wall layer.

The burst events in the wall region are represented by moving instability zones with spacing λ_x and λ_z in the x and z directions respectively (see Fig. 1). The dimensions of these zones in both the directions are small compared to λ_x. These zones are assumed to serve the functions of ejection and sweep events in the flow visualization studies. Such a representation results in the zones providing periodic and intermittent records at a fixed observation point in the laboratory frame. The high momentum fluid from the edge of the wall layer is retared by the development of a viscous boundary layer which is terminated by a new instability zone.

The periodicity in the z-direction is assumed to arise out of counter-rotating longitudinal vortices in the wall region between two lines of instability zones. These counter-rotating vortices are not only responsible for the periodicity in the z-direction, but also produce a streaky structure. The latter situation is illustrated

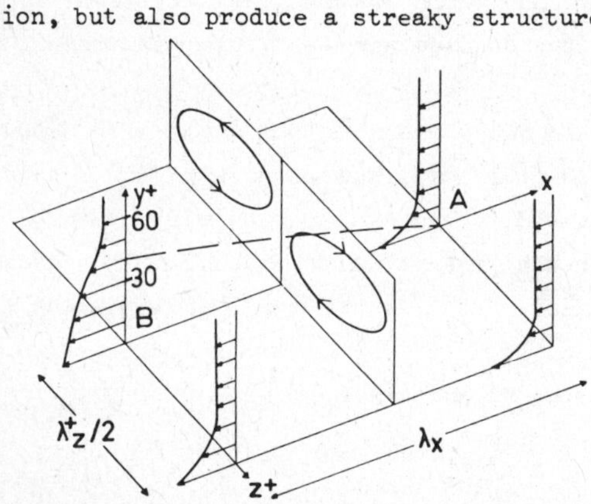

Fig. 2 Developing wall layer between two instability zones as seen by an observer in the moving coordinate system.

in the convected frame representation of Fig. 2. The counter-rotating vortices produce a streak with wall velocity at z-locations where the secondary motion is outwards; on the other hand the boundary layer development is suppressed at z-locations where the secondary motion is wallward. These correspond to low and high speed streaks respectively in the laboratory frame.

The structures in the outer region are assumed to be two-dimensional. This is supported by experiments of Brown & Thomas (1977), Falco (1977), and Kovasznay et al. (1970) who show that these structures are vortex-like and have dimensions of the order of boundary layer thickness in the spanwise direction. Thus one structure in the outer region covers many spanwise periods in the wall region. The period between these structures is also assumed to be λ_x.

Laufer and Badri Narayanan (1971) show that the dimensionless mean period, $\overline{T}_b u_\infty/\delta$, is 5 for the bursting cycle and 2.5 for the large scale structures in the outer region. This difference can be reconciled with our assumption by noting that the low speed streaks are shifted by $\frac{1}{2}\,\lambda_z$ in successive cycles (see Fig. 3). This

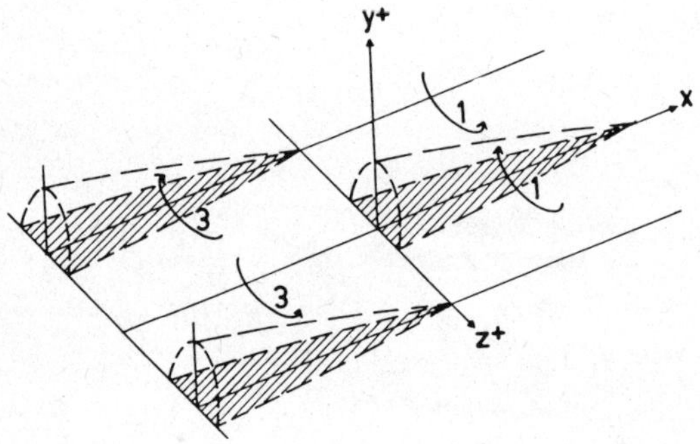

Fig. 3 The relative position of low-speed streaks (the edge is marked by a dashed line) in two successive cycles. The arrows 1 and 3 indicate the direction of rotation of the longitudinal vortices.

was also observed by Offen and Kline (1975). Thus a high speed streak will alternate with a low speed streak to an observer in the laboratory frame. Since a burst occurs only at the end of a low speed streak, a doubled period will be measured in the wall region.

The burst events bring the ejected fluid from the wall region to the outer region. The trajectory of this ejected fluid will have an almost fixed position with respect to the large scale structures and is responsible for bringing vorticity from the wall region to the outer-layer in the present model. The outer-layer will influence the inner layer by the periodicity of pressure and by providing high momentum fluid to replace the ejected fluid. All these different phenomena represent the interaction mechanisms between the inner and outer layers.

A complete picture of the boundary layer model is shown in Fig. 4.

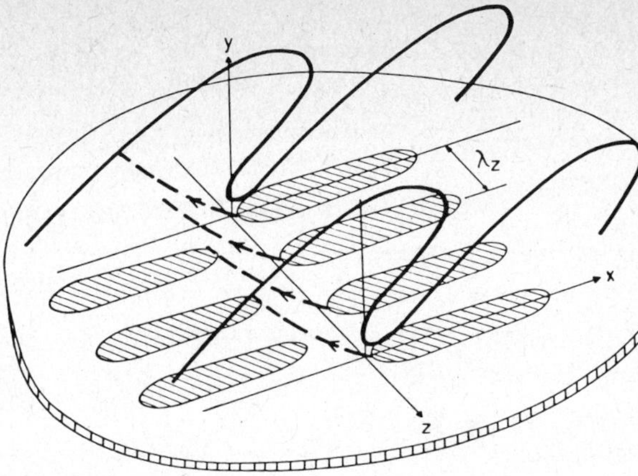

Fig. 4 A composite picture of the proposed structures in a turbulent boundary layer. The dashed lines are the trajectories of the ejected fluid parcels. The shaded areas represent the low-speed streaks.

The mathematical description of the model above starts with the hypothesis that two scales are sufficient to describe the boundary layer: λ_x of the large scale structures and $\lambda'(\ll\lambda_x)$ of the fluctuations in the burst events. The Navier-Stokes equations in the moving frame of reference are averaged according to the rule

$$\tilde{q}_c = \frac{1}{l_i} \int_{x + l_i/2}^{x + l_i/2} q_c dx, \quad q'' = q_c - \tilde{q}_c \tag{1}$$

where subscript c stands for quantities in the moving frame and $\lambda' \ll l_i \ll \lambda_x$. This latter condition is adequately satisfied in a turbulent boundary layer. The averaged equations for the large-scale motion contain small-scale stress terms as unknowns, which are assumed to be zero everywhere except in the burst region (dashed line in Fig. 4).

Approximate equations are derived for the wall layer and the outer region by carrying out an order-of-magnitude analysis on the two-dimensional version of the averaged equations. Because of inadequate information on the longitudinal vortices, the three-dimensional effect is incorporated in parametric form. Such a procedure is not expected to alter the two layer structure of the boundary layer as obtained from the order-of-magnitude analysis.

The resulting equations for the wall region and the outer layer are given below.

Wall region:

$$\tilde{u}_c \frac{\partial \tilde{u}_c}{\partial x_c} + \tilde{v}_c \frac{\partial \tilde{u}_c}{\partial y_c} + \tilde{w}_c \frac{\partial \tilde{u}_c}{\partial z_c} = \frac{1}{\rho} \frac{\partial \tilde{p}_c}{\partial x_c} + \nu \frac{\partial^2 \tilde{u}_c}{\partial y_c^{\,2}} + \nu \frac{\partial^2 \tilde{u}_c}{\partial z_c^{\,2}} + \frac{\partial (\widetilde{u''v''})}{\partial y_c} \tag{2}$$

$$\frac{\partial \tilde{p}_c}{\partial y_c} = 0 \tag{3}$$

$$\frac{\partial \tilde{u}_c}{\partial x_c} + \frac{\partial \tilde{v}_c}{\partial y_c} + \frac{\partial \tilde{w}_c}{\partial z_c} = 0 \tag{4}$$

Outer region:

$$\frac{\partial \tilde{u}_c}{\partial t_c} + \tilde{u}_c \frac{\partial \tilde{u}_c}{\partial x_c} + \tilde{v}_c \frac{\partial \tilde{u}_c}{\partial y_c} = \frac{1}{\rho} \frac{\partial \tilde{p}_c}{\partial x_c} + \frac{\partial (\widetilde{u''v''})}{\partial y_c} \tag{5}$$

$$\frac{\partial \tilde{v}_c}{\partial t_c} + \tilde{u}_c \frac{\partial \tilde{v}_c}{\partial x_c} + \tilde{v}_c \frac{\partial \tilde{v}_c}{\partial y_c} = \frac{1}{\rho} \frac{\partial \tilde{p}_c}{\partial y_c} + \frac{\partial (\widetilde{-v''^2})}{\partial y_c} \tag{6}$$

$$\frac{\partial \tilde{u}_c}{\partial x_c} + \frac{\partial \tilde{v}_c}{\partial y_c} = 0 \tag{7}$$

In the above system, the wall region is treated as a stationary problem while the time dependent terms in the outer region are retained to permit the growth of structures to keep in step with the boundary layer growth in the flow direction. The small scale stress terms in the equations are to be retained only in regions C and D (see Fig. 5).

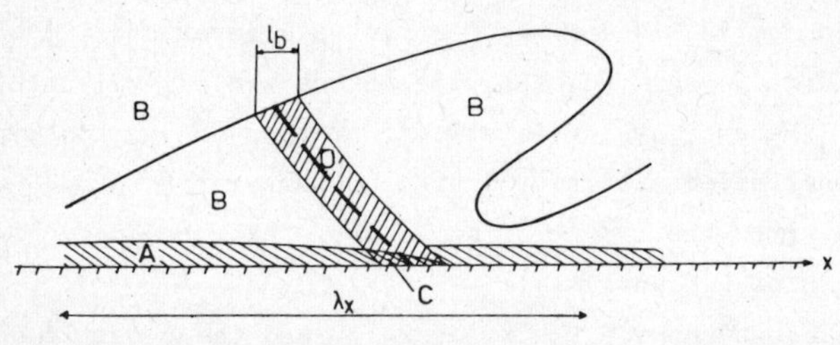

Fig. 5 The location of the different regions in the boundary layer. A and C form the wall layer, B and D the outer region. The small-scale turbulent stress is important in the regions C and D only.

The final element of the model, the interaction mechanism between the wall and outer regions, really comprises of 3 parts. The first part involves the calculation of the trajectory of the fluid parcel released by the instability zones. This is done by using a Lagrangian description with a damping mechanism proportional to the difference in velocities of the ejected fluid and the surrounding fluid. The relevant equations are

$$\frac{d^2X_{bc}}{dt^2}(t,t_o) = -C_x \frac{dX_{bc}}{dt}(t,t_o) - \tilde{u}_c(X_{bc}, Y_{bc}, t) \tag{8}$$

$$\frac{d^2Y_{bc}}{dt^2}(t,t_o) = -C_y \frac{dY_{bc}}{dt}(t,t_o) - \tilde{v}_c(X_{bc}, Y_{bc}, t) \tag{9}$$

where X_{bc}, $Y_{bc}(t,t_o)$ denote the coordinates at time t of the fluid parcel that left the wall region at time t_o. The vorticity added to the large scale outer field is given by

$$\omega_c = \int_{t_o}^{t} \frac{\partial^2(-v''^2)}{\partial y_c \partial x_c} - \frac{\partial^2(u''v'')}{\partial y_c^2}\, dt \tag{10}$$

where $u'' = (u_{bc}-\tilde{u}_c)$, $v'' = (v_{bc}-\tilde{v}_c)$, $u_{bc} = dX_{bc}/dt$ and $v_{bc} = dY_{bc}/dt$

The other two parts of the interaction mechanism, namely the pressure imposed by the outer layer and the arrival of the high momentum fluid at the wall layer, are not at present easily amenable to mathematical description. This problem has been circumvented in the present study by treating the equation sets (2)-(4) and (5)-(7) separately with boundary-conditions for them derived by experimental information when necessary.

The boundary-conditions for equations (2)-(9) are given below:

Wall layer:

$$\begin{aligned}
&\tilde{u}_c = U_c,\ \tilde{v}_c = 0 && \text{for } y_c = 0\\
&\tilde{u}_c = u_{cr}(x_c) && \text{for } y_c \to \infty\\
&\tilde{u}_c = U_c - u_o(y_c) && \text{for } x_c = 0
\end{aligned} \tag{11}$$

Outer region:

$$\begin{aligned}
&\tilde{u}_c = \tilde{u}_c(x_c, y_c), && \tilde{v}_c = \tilde{v}_c(x_c, y_c) \text{ at } t = 0\\
&\tilde{u}_c = u_\infty - U_c && \text{for } y_c \to \infty
\end{aligned}$$

TABLE 1

Parameter specification for system (2)-(13)

Parameter	Value/Equation/Hypothesis	Reference
Wall layer		
Convection velocity	$U_c = 0.8u_\infty$	Willmarth & Wooldridge (1962)
Mean-burst interval	$\overline{T}_b u_\infty/\delta = 2.5$	Laufer & Badri Narayanan (1971)
Initial velocity profile	Blasius profile with a free stream velocity $u_0^+(\infty) = 16$ and a wall friction of $1.7\ \rho u_*^2$	Einstein & Li (1956) and Eckelmann (1974)
$\tilde{w}_c$-profile	$\tilde{w}_c^+ = \tilde{w}_c/u_* = \emptyset(y^+) \sin 2\pi/\lambda_z$	Stuart (1965)
	$\emptyset(y^+) = \beta\gamma\ [-\exp(-\gamma y^+) + 4 \exp(-2\gamma y^+) - 3 \exp(-3\gamma y^+)]$	
	$\lambda_z^+ = 100$	Kline et al. (1967)
	β and γ are allowed to vary parametrically.	
u_{cr}	Allowed to vary parametrically	
Outer layer		
Initial distribution of point vortices	Start with a vorticity distribution corresponding to 1/7 power law velocity profile discretized to concentrate vorticity in four layers	Beljaars (1979)
Interaction mechanism		
Surface area of ejection jet	Continuity arguments	
u_{bco}	Based on the idea that the contribution of the ejected jet to the shear stress when smeared out over one period in the x-direction is equal to the wall shear stress.	Beljaars (1979)
v_{bco}	Based on the observation that u'v' has a peak of about $20u_*^2$.	
X_{bco}	Taken as a constant.	
C_x and C_y	Obtained from the behaviour of equations (8) and (9) close to the wall.	

$$\tilde{v}_c = 0 \qquad \text{for } y_c \to 0 \tag{12}$$

Interaction equation:

$$X_{bc}\ (t_o,t_o) = X_{bco},\ (dX_{bc}/dt)_{t_o,t_o} = u_{bco}$$
$$Y_{bc}\ (t_o,t_o) = Y_{bco},\ (dY_{bc}/dt)_{t_o,t_o} = v_{bco} \tag{13}$$

A number of parameters have to be specified in order to obtain numerical results from the set of equations (2)-(13). Most of these have been selected from available experimental information. Some, on which sufficient experimental information is not available, have been handled parametrically to establish the extent of their influence. Table 1 summarizes this information as was used to obtain the results presented in the next section.

The wall layer equations were solved by the "Keller Box Method" (Cebeci & Smith, 1974); the outer layer equations were solved by the discrete vortex approximation procedure (Acton, 1976), the ejection jet trajectory was computed by a simple marching procedure.

The detailed arguments for the choice of parameters in Table 1 as well as the methods of obtaining numerical solutions are available in Beljaars (1979) and Beljaars et al. (1980).

3. Results

In this section a few results are presented to illustrate the role of structures in determining the transport processes in a turbulent boundary layer. An extensive discussion of the behaviour of the model is available in Beljaars (1979) and Beljaars et al. (1980). During the course of the investigation, it was sought to obtain answers to three principal questions:

(i) Can a model of this type explain skin friction?

(ii) What are the mechanisms responsible for the continuous regeneration of the bursts?

(iii) What are the mechanisms that contribute to the transport of momentum in the outer layer?

All results pertain to the problem of flow over a smooth flat plate.

Fig. 6 shows the mean wall shear stress obtained from the wall layer calculations as a function of the Reynolds number based on the momentum loss thickness. Because of the scaling on ρu_*^2, the comparison is in fact with the empirical friction law (Hinze, 1975)

$$\frac{1}{2}\, c_f = \left(\frac{u_*}{u_\infty}\right)^2 = 0.012\ \mathrm{Re}^{-0.25} \tag{14}$$

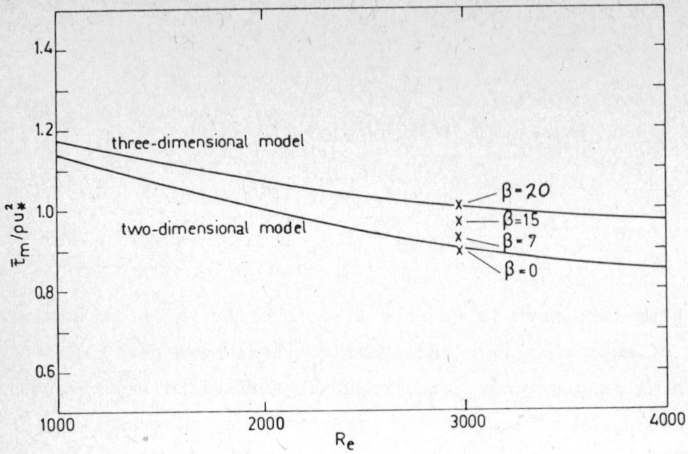

Fig. 6 The influence of the counter-rotating vortices on the predicted mean wall shear stress. $u_\infty \overline{T}_b/\delta = 2{,}5$; $(du_o^+/dy^+)_{y=o} = 1{,}7$; $\gamma = 0{,}06$.

The agreement is within 10%. The figure also shows the effect of the longitudinal vortices on skin friction. β, the strength of these vortices, when zero reduces the model to a two-dimensional problem. It is seen that the longitudinal vortices have only a marginal influence on the skin friction.

Turning now to the second question above, we start by noting that the bursts are treated as instability zones. It is next hypothesized that the occurrence of the bursts should be preceded by the formation of inflexional velocity profiles. The velocity profiles obtained from the two-dimensional as well as three-dimensional models were examined for the presence of points of inflexion. The two-dimensional profiles did not show any points of inflexion, while the three-dimensional calculations did produce such profiles, but displaced by $z^+ = 50$ in the spanwise direction (see Fig. 7).

In order to quantify this effect, the stability of the calculated velocity profiles was estimated using a correlation provided by Shen (1964) between the shape factor and the critical Reynolds number. The critical Reynolds numbers for the two-dimensional profiles were about two orders of magnitude higher than the prevailing local Reynolds number. Imposition of measured pressure gradients of Willmarth (1975) and Blackwelder & Kaplan (1972) did not alter the situation very much. On the other hand, the use of the three-dimensional model changed the situation dramatically as can be seen from Fig. 8. The effect of the strength of the longitudinal vortices, i.e. β, on $t/\overline{T}_b$ where $R_\Delta = R_{cr}$ is tabulated in Table 2, showing the rather strong influence this parameter has on stability.

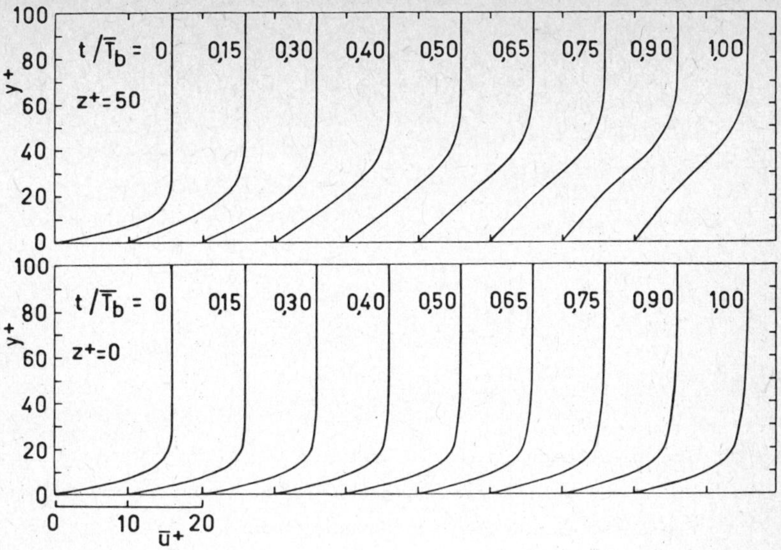

Fig. 7 Time dependent profiles, calculated with the three-dimensional model.

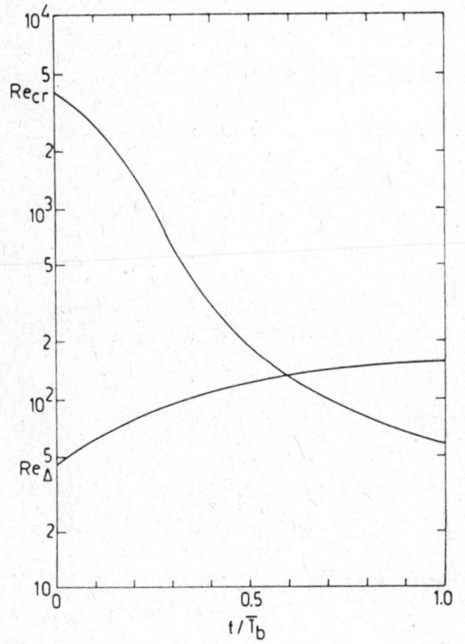

Fig. 8 The critical Reynolds number compared with the actual Reynolds number for the three-dimensional model as a function of the stage of development of the wall layer at $z^+ = 50$.

Table 2 The value of $t/\overline{T}_b$, where $Re_\Delta = Re_{cr}$ and the mean wall shear stress for different values of β; (Re=3000).

β	$(t/\overline{T}_b)_{Re_\Delta = Re_{cr}}$	$\overline{\tau}_m/\rho u_*^2$
0	1400	0,904
10	1,40	0,933
15	0,81	0,968
20	0,61	1,008
25	0,48	1,065
30	0,39	1,121

A question of some importance in this connection is the synchronization mechanism of the structures in the wall layer and those in the outer layer. Nychas et al. (1973) suggest that the burst events are initiated by sudden acceleration at the end of the cycle. Imposition of such pressure gradients on the wall layer hardly influences the stability properties of the wall layer as can be seen from the results in Fig. 9.

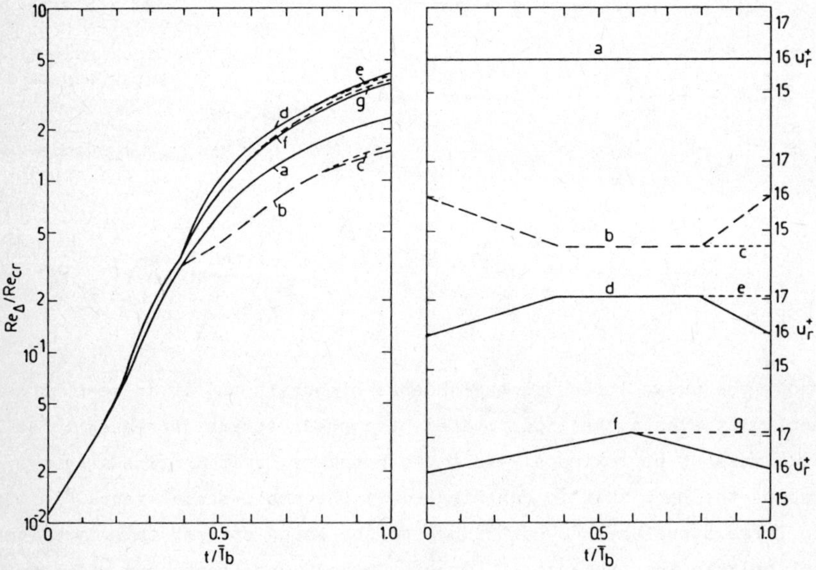

Fig. 9 The actual Reynolds number Re_Δ divided by the critical Reynolds number Re_{cr} as a function of the stage of development $t/\overline{T}_b$ (left figure). The different curves correspond to different upper boundary conditions u_r^+ as indicated in the right hand part of the figure.

The outer layer equations were solved for three specific cases:

(i) without the ejection events in a single structure:

(ii) with the ejection events in a single structure; and

(iii) with two successive structures interacting, but in the absence of burst events.

The computation produced streamline plots. Examination of these plots showed that while case (i) produced only weak vorticity concentrations, case (ii) clearly showed a build up of new vorticity concentrations with the passage of time. For case (iii), there was a tendency of one structure to absorb the other and the process strongly resembles vortex pairing observed in mixing layers.

A more important result that could be derived from these calculations is the shear stress profile in the outer layer. These profiles are shown in Fig. 10.

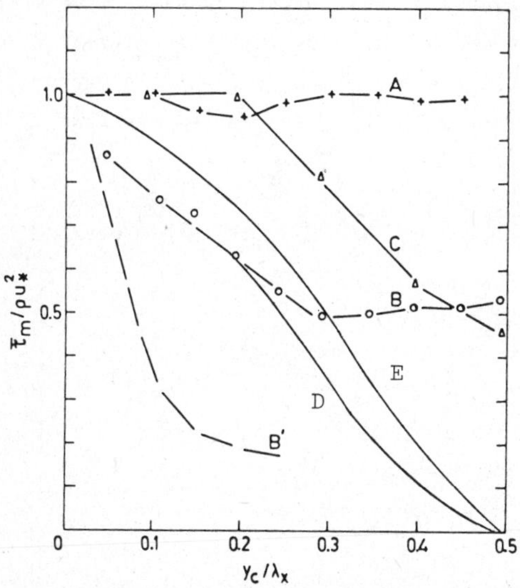

Fig. 10 Shear stress profiles calculated for three different cases: outer layer calculation without burst events (A), with burst events (B) and with vortex pairing (C). The dashed line indicates the contribution to the shear stress of the "ejection jets" only. y_c/λ_x = 0,5 corresponds to the edge of the boundary layer.

Curve A shows the shear stress in the absence of bursts and it is seen to be constant. When burst events are incorporated, the shear stress decreases to half the wall shear stress at approximately half the boundary layer thickness (curve B). Curve B' shows the part that is contributed by the small-scale events. In other words, the large structures do contribute to the shear stress; this contribution is significant only in the presence of bursts. Curve C represents the effect of vortex amalgamation and this process contributes to shear stress only in the outer half of the boundary layer.

A composite shear stress profile was constructed from the profiles B and C by subtracting the common constant τ_w to avoid the double counting. Such an additive composition while strictly not justifiable, is plausible because the ejection jet interaction mechanism assumes constants relevant to a region near the wall while the structure interaction mechanism is shown to be relevant only far away from the wall. It is remarkable that the shear stress profile, D, bears a striking resemblance to curve E, obtained from Coles' Law of the Wake (Hinze, 1975). Considering the rather crude assumptions employed in arriving at the numerical results, the quantitative agreement could be thought of as acceptable.

4. Discussion

The results presented in the previous section lend support to the thesis that transport phenomena in turbulent shear flows can be explained in the framework of a "structural approach". In particular, the results also imply that background turbulence, that comes down the energy cascade, has if at all any a secondary role in determining the turbulent transport.

We shall in the rest of this section examine several theoretical attempts as well as conjectures based upon experimental observations to explain the behaviour of the turbulent boundary layer in the light of the present work.

The first thing to note is that the present model as far as the calculation procedure goes indentifies only two regions - a wall region where viscosity plays a dominant role and an outer nonviscous, but rotational region. However, the results in the outer region show a further subdivision into two regions. In other words the turbulent boundary layer is composed of different layers with distinct properties. This feature is a recurring theme employed in correlating mean velocity profiles in turbulent boundary layers over the years (see for a more recent example, Ramaprian and Shivaprasad, 1977) and in constructing prediction methods (see for example the discussion on the length scale equation by Bradshaw, 1973). This situation is also consistent with the theories of turbulent boundary layers based on matched asymptotic analysis (Tennekes & Lumley, 1974; Mellor, 1972, Fendell, 1972). In these theories the logarithmic velocity profile comes out as the overlap region of inner and outer layers. In the present model the coupling between the inner and outer regions is achieved by the burst events or more specifically by the ejected fluid lumps that travel from the inner to the outer region. We therefore believe that the logarithmic region coincides with the region where the burst events dominate the transport process.

We wish to turn our attention to a series of theories that have been proposed for the wall region. The principal features of these theories are summarized in Table 3. Barring the last entry in the table, the authors present a more detailed discussion on these theories in the references cited earlier. We will concentrate here on the work of Orszag (1978). The model was developed to test a proposal for explaining drag reduction by compliant walls (Bushnell et al., 1977). The model involved the solution of the two-dimensional Navier-Stokes equations with a background turbulence model in the wall region with prescribed inflow and outflow boundary conditions, and imposed inflow velocity and large-scale pressure

TABLE 3

SUMMARY OF THEORIES FOR THE WALL LAYER

	Reference	Model description	Remarks
1.	Einstein & Li (1956) Hanratty (1956)	"Surface renewal" theory. A linear diffusion equation	Modest predictive capability. Has proved very popular in many engineering applications.
2.	Sternberg (1962) Schubert & Corcos (1967)	Linear model. A basic flow $\overline{u}(y)$ is assumed and the behaviour of u' subject to suitable initial and boundary conditions.	Yields spectrum of u', phase relationships between u' & v'; assigns a passive role for the wall layer; does not predict Reynolds stresses.
3.	Landahl (1965, 1967, 1975) Bark (1975)	Two levels of decomposition of the velocity field. Model specifically allows for localized burst regions.	Successful in prediction of the decay of pressure fluctuations in the streamwise direction. Arguments for burst regeneration do not seem to work in practice (Beljaars, 1979).
4.	Orszag (1978)	See Text.	See Text.

pulse at 'infinity'.

The mean flow profiles thus obtained are examined for their stability properties by solving the relevant Orr-Sommerfeld equations. These results are used to calculate the amplification ratios; when this ratio exceeds a prescribed threshold level, a burst is presumed to have occurred.

The report available to the authors was too sketchy and it was difficult to arrive at a definitive assessment of the work. However, it seems worthwhile to record one result obtained in the study. The pressure pulse imposed at 'infinity' produced an inflexional velocity profile only in the presence of an inflow velocity at that boundary. This is in conformity with our finding that pressure fluctuations produced by the outer structures are unable to provide the necessary conditions for two-dimensional velocity profiles to become unstable.

This brings us to the question of the three-dimensionality of the wall layer. We have tacitly assumed the existence of longitudinal counter-rotating vortices in the wall layer. These structures have a relatively small influence on the mean skin friction, but are shown to have a profound influence on the stability properties of the wall layer and hence on the burst regeneration process. These longitudinal counter-rotating vortices have never been directly observed in the flow visualization experiments. Their presence has been inferred from the observations of Offen and Kline (1974, 1975) who show that the low speed streaks in two successive cycles shifted over $\lambda_z/2$ with respect to each other in the z-direction. Blackwelder and Eckelmann (1979) demonstrate the existence of these structures in a more convincing manner by applying a quadrant probability analysis technique to signals obtained from a set of flush-mounted wall probes. There were of course difficulties in locating these vortices in the normal direction as well as in determining their extent in the streamwise direction. However, some recent large eddy simulation tend to dispute such a conceptual model (Kim and Moin, 1979). It is not clear whether the numerical procedures employed in this investigation would be able to resolve such relatively weak rotatory motions that are envisaged in the present model.

A suggestion that has been often made is that the longitudinal vortices are caused by a Taylor-Görtler type of instability. Witting (1958) has shown that the stream-line curvature caused by a very small amplitude Tollmien-Schlichting wave is adequate for the onset of Taylor-Görtler instability. The postulated cycle of events in the wall layer comprising of a deceleration in the beginning and an acceleration at the end of the cycle can certainly produce curved steamlines. Brown and Thomas (1977) show by means of order-of-magnitude estimates that the outer layer causes sufficient stream-line curvature near the wall for the occurrence of Taylor-Görtler instability.

Additional work seems essential to identify the origin of these vortices in the wall region. Two lines of investigation are possible. The first one requires the estimation of pressure perturbation imposed by the outer region on the wall layer.

The stability of the two-dimensional velocity profiles can then be investigated in a manner similar to that of Witting. Such an analysis would provide evidence for the existence or otherwise of these vortices; but if the evidence is favourable it can give the wavelength λ_z for the most unstable mode.

The second approach is to carry out a three-dimensional boundary layer calculation of the type usually done for corner flows (Mager, 1964). Should our speculations prove correct, longitudinal vortices should come out of such calculations and our assumption on $\tilde{w}_c$ can then be dropped. We believe that calculations of this type or those suggested in the previous paragraph are likely to produce more reliable information about the existence or otherwise of the longitudinal vortices than the direct numerical simulation approach based on the full Navier-Stokes equations.

The next set of observations in this section concerns the conceptual models developed on the basis of experimental information. Offen & Kline (1975) interpret the burst event as a flow separation (although no actual backflow is observed) induced by the pressure gradient of a vortex coming from upstream. In the present work, the burst event is treated as a local instability. This latter seems more plausible especially in the light of the observation that the burst event is preceded by an inflexional velocity profile. Similar attitudes seem to be prevalent in the theories of Landahl and Orszag discussed earlier. It must be conceded that the origin of bursts is still shrouded in mystery and as of now is inaccessible to a mechanistic description. The situation appears to be similar as far as the production of intense shear stress in the burst regions is concerned.

However, the assumptions made in deriving the present model and the results obtained therefrom seem consistent with the quadrant analysis of uv signals by Brodkey et al. (1974). This analysis shows that most of the contribution to the Reynolds stress in the wall region comes from sweep events, and that in regions away from the wall it comes from the ejections*. Large eddy simulation results of Kim and Moin also show this.

Falco (1977) postulates what he calls "typical eddies" as being responsible for the turbulent stress throughout the boundary layer. The size of these eddies is said to be about 100 to 200 ν/u_*, which is much smaller than the boundary layer thickness. According to Falco these eddies are formed at several distances from the wall although it is not clear how they could be recognized from his smoke pictures. It seems more plausible that the eddies originate in the wall layer and could be considered 'bloomed' out ejection jets. This certainly will explain their scaling behaviour. If this were true, such eddies can have very little ability to transport momentum when they arrive at the edge of the boundary layer. The ejection jets as modelled in this work lose their importance at about 0.2δ from the wall.

---

* of course we need to qualify 'away from the wall' to mean the inner part of the outer layer.

Brown & Thomas (1977) observed a constant phase relation between the inner and outer layer as well as a convection velocity of 0.8 u_∞. Their picture of structures bears a considerable similarity with the present model. Even the deep incursions of nonturbulent fluid into the boundary layer at an angle were noticed in our outer layer calculations by following a marker line in the computational field. Brown & Thomas state that the angle is 18°. These inclined structures have been identified in duct flows, but at an angle of 13° by Rajagopalan & Antonia (1979). The numerical simulation studies cited earlier also showed the existence of such structures. In our calculations, no attempt was made to estimate the angle of these structures.

We now turn to that hardy perennial of all such discussions: does the wall layer control the processes in the outer layer or does it simply respond to the 'happenings' in the outer layer? (see the arguments of Sternberg and Kline at the Stanford Conference, 1968). The argument has recently been joined by Praturi & Brodkey (1978) who used a stereoscopic visualization technique for the turbulent boundary layer. While many of the observed features in their study are consistent with the present work, their interpretation is not supported by our results. The "high speed front" in the outer region is held to be the principal cause of all other processes in the boundary layer: the engulfment of the nonturbulent fluid and the burst activity. The wall layer is said to respond to the outer layer and not influence it; the longitudinal vortices are seen as the result of spanwise periodicity and not the origin of it. The present results do not justify these simple cause and effect relationships. It has been clearly shown here that: the burst events can only occur when longitudinal vortices are present; and the burst events have a significant influence on the outer layer structures and their capability to transport momentum.

It is appropriate here to make a few remarks on another approach which is gaining ground in recent years - the large eddy simulation. The early work in this field was carried out by Deardorff (1970) and Schumann (1975). The idea is to decompose the flow field variables into a large-scale field (the filtered field) and a residual field (the subgrid field) in a manner very much similar to what has been done in the present work. The smoothed equations of motion, which are 3-dimensional and time dependent, have the small scale stress terms as unknowns. The equations are closed by using an eddy viscosity model for the unknown stress terms. The calculations so far have been only on channel flows and the most advanced ones of this class are, as far as the authors are aware, due to Kim and Moin (1979).

There are some significant differences between the present work and the large eddy simulations. Generally stated, our attitude has been to make use of the available experimental information to construct a model and see whether it is possible to throw some light on the dynamics of the problem. The large eddy simulation should be viewed, at least in our opinion, as an alternative to experiments to derive data on structural aspects of turbulent flow. Because of the enormous amount of information produced by the computer, there will remain a problem of "pattern recognition" that has to be resolved by the worker concerned. However, the flexibility of a computer is not available to an experimentor.

Apart from the general aspects mentioned above, there are many questions of a detailed nature. The treatment of small scale stresses and handling of the wall layer differ in the two approaches. A local closure assumption for the subgrid field may not be appropriate for depicting burst events which have non-local effects; they originate in the wall layer and exert an important influence on the structures in the outer region. The direct effect of the small scale stress terms, in our opinion, is restricted to small regions of the flow. Carrying such terms over the entire flow field seems to us to be an enormous effort without adequate compensation. Of course it should be stated that we skirt around the closure problem by using experimental data.

The cyclic behaviour of the wall layer between the viscous phase and the burst events would prove difficult to reproduce. It seems to demand a time-dependent version of the law of the wall. Kim & Moin recognize this problem. However, their computations which go all the way down to the wall, without recourse to such refinements mentioned above, has produced remarkable results. They clearly show the formation of streaky structures in the wall region.

The large eddy simulation is certainly attractive. Given the present state of development of either methodology, it is obviously inappropriate to comment on the effectiveness of either approach in producing a greater physical insight into the turbulence problem. The choice at this moment should remain one of personal taste.

It seems unsatisfactory to end any serious discussion of turbulence without taking note of its statistical nature. It is possible to include certain amount of statistics in the present model by the application of probability density functions to the different parameters like λ_z, T_b, U_c, etc.. No serious efforts were made to incorporate these effects in the present work. A major problem in this connection is the lack of adequate experimental information on these aspects. Some limited computations were made to demonstrate the possible effects of such variations on the wall layer results. The distribution function for T_b is said to be log-normal (Kim et al., 1971) with a rather large spread. The calculations with this distribution function showed that the mean wall shear stress changed by only a few percent. Similar results were obtained when the other variables in the model were changed parametrically.

If we grant that the structures in the inner and outer layers run in phase with one another, a stochastic variation in T_b will result in different size and phase relationships among the outer structures. Our results in the outer layer demonstrate that such differences are essential for the amalgamation process of the outer structures. It is our thesis that while the stochastically varying parameters will give the periodic structures a random aspect, the physics governing the individual structures remain unchanged. However, this randomness, when it occurs in scales of successive structures in particular, poses a formidable problem for the detection of these structures in experiments. Averages over structures without reference to the scales associated with them will result in considerable blurring of information. Blackwelder & Kaplan (1976) discuss a method to account for the randomness in the convection velocity to minimize the effect of blurring.

5. The Model and More complex flows

So far the discussion has centred around the flow over a flat plate at zero incidence. A model could be considered useful only if it is able to provide insight into a wider class of flows. An obvious difficulty in the application of the model as it stands now to more complex flows is inadequate experimental information. It should be remembered that we are seeking information of a specialized kind like the burst interval time, convection velocity, etc. - quantities that are not easy to measure.

We discuss in this section the possibilities of applying the model to a few flows. The choice of these flows is purely a personal one, rather than being a representative selection of possible turbulent flows. With the current status of the model, it is to be expected that the discussion here would be somewhat speculative. It is presented in the belief that the model holds sufficient promise to warrant further work. We hope it will also stimulate experiments to measure "structure parameters" in more complex flows.

If we were asked to choose the most vital parameter for the implementation of the model from the rather long list in Table 1, we would place our money on the mean burst interval time. There is a distinct possibility for calculating this along the lines suggested by Orszag (1978). Even if this were to come to pass, we shall still be requiring experimental information from complex flows to validate the methodology under diverse conditions.

At the present moment there is evidence to believe that the concept "the dimensionless mean burst interval time is a constant, i.e. $\overline{T}_b u_\infty/\delta = 2.5$" is more generally valid than we have indicated. This value seems appropriate for boundary layers under pressure gradients (Kline et al., 1967) and for transpired boundary layers (Senda, 1979). Some calculations were performed for the latter case on the wall layer by setting $\tilde{v}_c = v_w$ at the wall where v_w is the blowing velocity. The calculated friction coefficients are shown in Fig. 11 in comparison with the following empirical friction law for the boundary layer with blowing (Senda et al., 1979).

$$\frac{C_{fb}}{C_f} = \frac{\ln(1+B)}{B}^{\frac{5}{4}} (1+B)^{\frac{1}{4}} \tag{15}$$

where $B = 2v_w/(C_{fb}U_\infty)$ and C_f is given by Eq. (14). The model behaves surprisingly

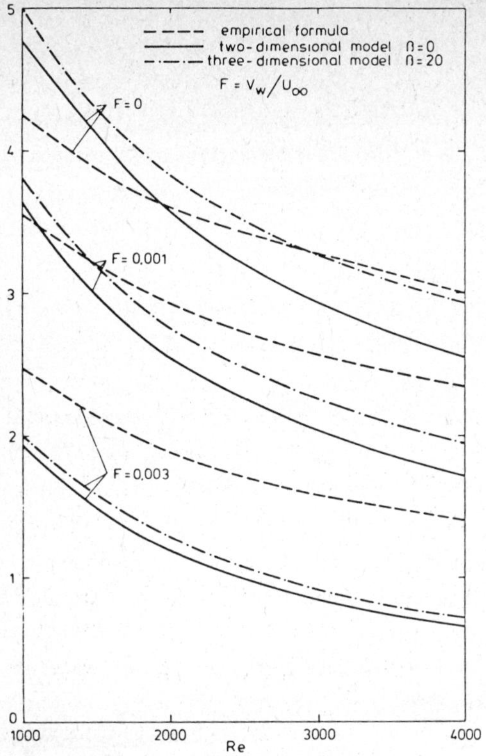

Fig. 11 Friction coefficients calculated with the two-dimensional (——) and with the three-dimensional (—•—•) wall layer model, compared with the empirical formula (---), for different blowing rates ($F = v_w/u_\infty$).

well, which suggests that the transport mechanisms are not altered by small blowing rates. Of course more extensive testing of the model would be required before making any definitive assertions. The considerable body of data that exists on transpired boundary layers would come in handy for a clear assessment of the model.

The next problem we consider is the very interesting phenomenon of drag reduction in pipe flows due to small additions of polymers. A detailed review on the problem is provided by Virk (1975). There are many aspects of the problem, but for the purposes here it is sufficient to mention that drag reduction mainly comes about by the alterations of structures in the wall region caused by the polymers. Two changes were noticed by Fortuna and Hanratty (1972): the correlation length in the flow direction of the turbulent fluctuations increases with the addition of polymers; drag reduction is accompanied by an increase in the size of longitudinal vortices. The former result is interpreted to mean that drag reduction is achieved through the postponement of burst events - an increase in $\overline{T}_b$. This was verified by Mizushina & Usui (1977) by measuring $\overline{T}_b$. The results showed that $\overline{T}_b$ does not correlate with the bulk parameters (equivalent of outer layer parameters in a boundary layer) as in the case for Newtonian flow, but correlates according to

$$\frac{\overline{T}_b\, u_*^2}{\nu} = 0.112\ \mathrm{Re}^{0.93}. \qquad (16)$$

The correlation clearly establishes that for a given flow rate, the mean burst interval time shows an increase with reducing skin friction - a consequence of polymer addition.

These are interesting results that could be used for the study of the present model. According to the hypothesis advanced earlier, the longitudinal vortices are responsible for burst regeneration. We could adopt the attitude taken in the present work, introduce appropriate rheological properties of the fluid and determine the onset of instability on the basis of examining the velocity profiles for inflexion points. The real difficulty of such an approach lies in the fact that the fluid exhibits anisotropic behaviour. It appears therefore that a three-dimensional formulation of the wall layer problem is essential if we wish to get at the correct physics of the problem. If we press on and do the outer layer calculations (Note that we are talking in terms of boundary layers and the experiments are on pipe flows.), the model would be under some severe test. At any rate, according to Virk, the mechanism of drag reduction remains obscure and it appears very worthwhile pursuing the line of thinking sketched above.

Drag reduction in external flows is sought to be achieved through the use of so-called compliant walls. The review of Bushnell et al. (1977) suggests that neither experiments nor the theories are conclusive about the drag reducing properties of such walls. The idea is based upon the fact that the pressure fluctuations caused by the outer layer are intimately connected with the formation of bursts. It is believed that the negative results from experiments are due to improper matching of the dynamics of the wall motion and the dynamics of the flow. Orszag (1978) has studied the problem according to the model discussed in the previous section. The study suggests that there exists certain class of wall motions that do postpone burst formation and hence lead to reduced drag. The differences between Orszag's approach and the present approach has already been delineated; suffice it to say that this is another candidate problem for the model testing.

The last problem we will be considering in this short list of complex flows that could serve as a test for the model is the boundary layer flow along curved walls. The rather surprisingly large influence that one notices on boundary layer properties with even mild curvatures has attracted considerable attention in the past 10 years. An extensive review on the subject has been published by Bradshaw (1973). The most important effect is that the streamline curvature changes the Reynolds stresses of turbulent flow by as much as ten times compared to changes in viscous stresses; generally speaking there is an increase along a concave wall and a decrease along a convex wall.

If it is accepted that the longitudinal vortices in the wall region of a turbulent boundary layer arises due to a Taylor-Görtler type of instability, then it is reasonable to expect that streamline curvature should have a profound influence on them. A concave wall will enhance the Taylor-Görtler instability leading to stronger longitudinal vortices, resulting in lower $\overline{T}_b$. An opposite situation will prevail for the flow over a convex wall. The flow visualization studies of Johnston (1972) in a rotating duct lend support to this line of argument: the characteristic streaky structure and bursting phenomena could be almost suppressed on the stable side of the flow at certain characteristic rotation numbers despite the presence of vigorous activity on the unstable side of the flow.

The most recent work on this class of flows is due to Ramapriyan and Shivaprasad (1977, 1978). Their mean flow measurements suggest that the conventional flat plate log law is valid for these flows; but the extent of the region where the log law is valid is reduced by convex curvature. Clearly these results could provide a valuable check for the outer layer formulations of the present model.

Ramapriyan & Shivaprasad (1978) also reported measurements of turbulent structure in such flows. According to these measurements, the production and outward diffusion of the turbulent kinetic energy is enhanced over a concave wall and reduced over a convex wall; in the latter case it appears that the production of turbulent energy is almost stopped. Also the curtosis of the probability density function for the u-fluctuations is increased at the concave wall and decreased at the convex wall. These results support the conjectures made earlier about the bursting activity in flows with streamline curvature. It should prove interesting to verify these conjectures by actual calculations.

6. Concluding Observations

In the foregoing pages we have presented:

(a) a conceptual model for the treatment of wall bounded turbulent shear flows from a structural viewpoint using observations from flow visualization and conditional averaging experiments;

(b) results from limited calculations performed with the model;

(c) a discussion of the model and computational results with reference to other theories and conjectures on phenomena associated with turbulent boundary layers; and

(d) a few speculative remarks about the usefulness of the model vis-a-vis the behaviour of other complex flows.

The present theory is distinct from the earlier theories in that it explicitly accounts for the existence of a wall layer and an outer layer. Such an approach is preferable since there is no need to make an a priori decision as to which layer is the controlling one. Computations with the model - if it's any good - ought to bring out this aspect without too much ambiguity. We believe that the modelling efforts of the present type, comprising as they do of several distinct elements, are better equipped to provide physical insight into the behaviour of turbulent boundary layers under different conditions. A second role for the models in this class would be that they could serve as important tools in interpreting experimental data. Because of this connection with experiments, it might in the ultimate analysis provide significant inputs to the prediction of turbulent flows in general.

References

Acton, E., (1976). The modelling of large eddies in a two-dimensional shear layer, J. Fluid Mech., 76, 561.

Bark, F.H., (1975). On the wave structure of the wall region of a turbulent boundary layer, J. Fluid Mech., 70, 229.

Beljaars, A.C.M., (1979). A model for turbulent exchange in boundary layers, Ph.D. thesis, Eindhoven University of Technology, Eindhoven, Netherlands.

Beljaars, A.C.M., Krishna Prasad, K. and Vries, D.A. de, (1980). A structural model for turbulent exchange in boundary layers, submitted to J. Fluid Mech.

Blackwelder, R.F. and Eckelmann, H., (1979). Streamwise vortices associated with the bursting phenomenon, J. Fluid Mech., 94, 577.

Blackwelder, R.F., and Kaplan, R.E., (1972). The intermittent structure of the wall region of a turbulent boundary layer, Univ. S. Calif. Rep. USCAE, 1-22.

Blackwelder, R.F., and Kaplan, R.E., (1976). On the wall structure of turbulent boundary layers, J. Fluid Mech., 76, 89.

Blackwelder, R.F., and Kovasznay, L.S.G., (1972). Time scales and correlation in a turbulent boundary layer, Phys. Fluids, 15, 1545.

Bradshaw, P., (1973). Effects of streamline curvature on turbulent flow. AGARDograph No. 169.

Bradshaw, P., (1973). The strategy of calculation methods for complex turbulent flows, Imperial College Aero report 73-05.

Brodkey, R.S., Wallace, J.M., and Eckelmann, H., (1974). Some Properties of Truncated Turbulence Signals in Bounded Shear Flows. J. Fluid Mech., 63, 209.

Brown, G.L., and Roshko, A., (1974). On density effects in turbulent mixing layers, J. Fluid Mech., 64, 775.

Brown, G.L., and Thomas, A.S.W., (1977). Large structures in a turbulent boundary layer, Phys. Fluids, 20, S243.

Bushnell, O.M., J.N. Hefner and R.L. Ash, (1977). Compliant wall drag reduction for turbulent boundary layers. Phys. Fluids, 20, S31.

Cebeci, T., and Smith, A.M.O., (1974). Analysis of turbulent boundary layers, Academic Press, New York.

Corino, E.R., and Brodkey, R.S., (1969). A visual investigation in the wall region of turbulent flow, J. Fluid, Mech., 37, 1.

Deardorff, J.W., (1970). A numerical study of three-dimensional turbulent channel flow at large Reynolds number, J. Fluid Mech., 41, 453.

Eckelmann, H., (1974). The structure of the viscous sublayer and the adjacent wall region in a turbulent channel flow, J. Fluid Mech., 65, 439.

Einstein, H.A., and Li, H., (1956). The viscous sublayer along a smooth boundary, J. Engng. Mech. Div. Am. Soc. Civ. Engrs., 82 (EM2), 945.

Falco, R.E., (1977). Coherent motion in the outer region of turbulent boundary layers, Phys. Fluids, 20, S124.

Fendell, F.E., (1972). Singular perturbation and turbulent shear flow near walls, J. Astronautical Sc., 20, 129.

Fortuna, G. and Hanratty, T.J., (1972). The influence of drag-reducing polymers on turbulence in the viscous sublayer, J. Fluid Mech., 53, 575.

Hanratty, T.J., (1956). Turbulent exchange of mass and momentum with a boundary, A.I.Ch.E.Je, 2, 359.

Hinze, J.O., (1975). Turbulence, McGrawhill, New York, 2nd ed.

Johnston, J.P., (1972). The Suppression of shear-layer turbulence in rotating systems. ARARD Conf. Proc. 93.

Kim, H.T., Kline, S.J., and Reynolds, W.C., (1971). The production of turbulence near a smooth wall in a turbulent boundary layer, J. Fluid Mech., 50, 133.

Kim, J. & Moin, P., (1979). Large eddy simulation of turbulent channel flow - Illiac IV Calculation, AGARD Symposium on Turbulent Boundary Layer-Experiment, Theory, and Modelling.

Kline, S.J., (1968). Discussion, Proc. of AFOSR-IFP-Stanford Conf. on computation of Turbulent Boundary Layers. Ed: S.J. Kline et al., Vol. 1 p. 527.

Kline, S.J., Reynolds, W.C., Schraub, F.A. and Rundstadler, P.W., (1967). The structure of turbulent boundary layers. J. Fluid Mech., 30, 741.

Kovasznay, L.S.G., Kibens, V., and Blackwelder, R.F., (1970). Large scale motion in the intermittent region of a turbulent boundary layer, J. Fluid Mech., 41, 283.

Landahl, M.T., (1965). A wave-guide model for turbulent shear flow, NASA, CR-317.

Landahl, M.T., (1967). A wave-guide model for turbulent shear flow, J. Fluid Mech., 29, 441.

Landahl, M.T., (1975). Wave breakdown and turbulence, SIAM, J. Appl. Mech., 28, 735.

Laufer, J., and Badri Narayanan, M.A., (1971). The mean period of the production mechanism in a boundary layer, Phys. Fluids, 14, 182.

Mager, A., (1964). Three-dimensional laminar boundary layers in "Theory of Laminar flows", Ed. F.K. Moore, Princeton University Press.

Mellor, G.L., (1972). The large Reynolds number asymptotic theory of turbulent boundary layers. Int. J. Engng. Sci., 10, 851.

Mizushina, T., and Usui, H., (1977). Reduction of eddy diffusion for momentum and heat in viscoleastic fluid flow in a circular tube. Phys. Fluids, 20, S100.

Nychas, S.G., Hershey, H.C., and Brodkey, R.S., (1973). A visual study of turbulent flow, J. Fluid Mech., 61, 513.

Offen, G.R., and Kline, S.J., (1974). Combined dye-streak and hydrogen bubble visual observations of a turbulent boundary layer, J. Fluid Mech., 62, 223.

Offen, G.R., and Kline, S.J., (1975). A proposed model of the bursting process in turbulent boundary layers, J. Fluid Mech., 70, 209.

Orszag, S.A., (1978). Prediction of compliant wall drag reduction- Part II, Cambridge Hydrodynamics Report 11.

Praturi, A.K., and Brodkey, R.S., (1978). A stereoscopic visual study of coherent structures in turbulent shear flow, J. Fluid Mech., 89, 251.

Rajagopalan, S., and Antonia, R.A., (1979). Some properties of the large structures in a fully developed turbulent duct flow, Phys. Fluids, 22, 614.

Ramapriyan, B.R. and B.G. Shivaprasad, (1977). Mean flow measurements in turbulent boundary layers along mildly curved surfaces. AIAA Journal, 15, 189.

Ramapriyan, B.R., and B.G. Shivaprasad, (1978). The structure of turbulent boundary layers along mildly curved surfaces, J. Fluid Mech., 76, 561.

Schumann, U. (1975). Subgrid scale model for finite difference simulations of turbulent flows in plane channels and annuli, J. Comp. Phys., 18, 376.

Senda, M., Susuki, K. and Sats, T., (1979). Turbulence structure related to the heat transfer in a turbulent boundary layer with injection, 2nd symposium on Turbulent Shear Flows, London, 1979.

Shen, S.F., (1964). Stability of laminar flows, in: Theory of laminar flows, edited by F.K. Moore, Princeton Univ. Press, New Jersey.

Shubert, G., and Corcos, G.M., (1967). The dynamics of turbulence near a wall according to a linear model, J. Fluid Mech., 29, 113.

Sternberg, J., (1962). A theory for the viscous sublayer of a turbulent flow, J. Fluid Mech., 13, 241.

Sternberg, J., (1968). Discussion, Proc. of AFOSR-IFP-Stanford Conf. on computations of turbulent boundary layers. Ed: S.J. Kline et al., Vol. 1, p. 411.

Stuart, J.T., (1965). The production of intense shear layers by vortex stretching and convection, AGARD Rep. 514.

Tennekes, H., and Lumley, J.L., (1974). A first course in turbulence, MIT Press, Cambridge.

Virk, P.S., (1975). Drag reduction fundamentals, AICHE J., 21, 625.

Willmarth, W.W., (1975). Structure of turbulence in boundary layers, Adv. App. Mech., vol. 15, p. 159. Academic Press.

Willmarth, W.W., and Woodridge, G.E., (1962). Measurements of the fluctuating pressure at the wall beneath a thick turbulent boundary layer, J. Fluid Mech., 14, 187.

Witting, H., (1958). Über den Einfluss der Strömlinien-Krümmung auf die stabilität Laminarer Strömingen, Arch. Rat. Mech. Anal., 2, No. 3, 243.

TURBULENT STRUCTURES IN WALL-BONDED SHEAR FLOWS OBSERVED VIA THREE-DIMENSIONAL NUMERICAL SIMULATIONS

A. Leonard
Ames Research Center, NASA
Moffett Field, California 94035, U.S.A.

I. INTRODUCTION

In the past few years, numerical techniques in fluid dynamics have matured to the point that meaningful simulations of three-dimensional transitional or turbulent shear flows bounded by a wall are possible. At the same time, availability of advanced computers, over and above a CDC 7600-class machine, has allowed roughly a doubling of the spatial resolution in each direction, greatly enhancing the quality of the results. We are in a position now to expand our set of objectives for flow simulations in the next decade to the following:

1. Increase our understanding of the physics of turbulence
2. Provide statistical and other information useful to turbulence modelers
3. Provide a testbed for the development of improved numerical methods and guidance for improved modeling of fine-scale effects.

Objectives (1) and (2) are, of course, in common with the goals of most experimental (and purely theoretical) investigations of turbulent flows. The numerical simulation, however, provides a different way to look at a turbulent flow; the results of computer simulations are complementary to those of laboratory experiments. In the computer experiment, the corresponding laboratory time of the simulation is relatively small, but complete flow-field information is available throughout that time interval. The third objective is related to the desire to improve future simulations and to make it possible to investigate more complex flows. Advanced computers, as they become available, will also help in this regard.

In this paper we report on three recent simulations, using the ILLIAC IV computer, that begin to meet the above objectives:

1. Study of the evolution of a wave-like disturbance in a laminar boundary layer corresponding to vibrating-ribbon experiments.
2. Study of the evolution of a spot-like disturbance in a laminar boundary layer.
3. Investigation of turbulent channel flow.

In particular, we will concentrate on the turbulent structures observed in these flow simulations and comment on their role (1) in the dynamical process, for example, the growth and maintenance of turbulence, and (2) as contributors to turbulent statistics of interest to turbulence modelers. As will be shown, these persistent flow structures include streamwise and vertical vorticity distributions near the wall, low-speed and high-speed streaks, and local regions of intense vertical velocity.

The problem of representing the large range of physical scales in a computer simulation will also be discussed, along with some prospects for future improvement.

Quite different numerical techniques are employed in the three simulations. The first and third simulations listed above use an Eulerian grid with pressure and velocity as dependent variables and periodic boundary conditions in two directions, whereas the second simulation uses a Lagrangian vortex method in a semi-infinite domain. The first two simulations studied the evolution in time of an initial disturbance and have no explicit model to account for "subgrid-scale" turbulence — fine-scale structures not resolvable by the computational mesh. Consequently, in these simulations the evolution time during which meaningful results were obtained was limited, due to the importance of subgrid effects at later times. In the study of turbulent channel flow, however, the objective was to simulate a fully developed turbulent flow having a range of scales much larger than the computational mesh would allow. Thus, the use of a subgrid model was mandatory. In Section II, further details of the simulations are given along with major findings regarding the observed turbulent structures. The problem of accounting for turbulence scales not resolved or represented by the computational mesh — subgrid or supergrid modeling or both — is discussed in Section III, and some possibilities for improvement are suggested. In Section IV, a summary of the results and conclusions are presented.

II. RESULTS OF THREE NUMERICAL SIMULATIONS

A. Flat-Plate Transition

In a series of computer experiments Wray and Hussaini [1] (hereinafter referred to as WH) studied the instability of a flat-plate boundary layer in three dimensions by solving numerically the incompressible Navier-Stokes equations. In another three-dimensional numerical study, Orszag and Kells [2] investigated transition of plane Poiseuille and plane Couette flow. The ILLIAC computer code of WH is capable of simulating a wide range of boundary-layer stability problems. In this particular investigation they simulated the vibrating-ribbon experiments of Kovasznay et al. [3] by a careful choice of initial conditions for the three-dimensional perturbation to the Blasius profile. Two oblique (three-dimensional) eigensolutions of the Orr-Sommerfeld equation were added to a two-dimensional eigensolution to form the initial perturbation. To correspond with the laboratory experiment the streamwise and spanwise wavelengths of the perturbation waves were 25.2 δ^* and 30 δ^*, respectively, where δ^* is the initial displacement thickness of the laminar boundary layer. A schematic of the numerical simulation of WH is shown in Fig. 1. Wray and Hussaini used Fourier expansions (spectral method) in the streamwise (x) and spanwise (z) directions with 64 points each and second-order finite differences in the normal (y) direction with 128 points on a stretched mesh. Fourier transformation of the momentum and mass

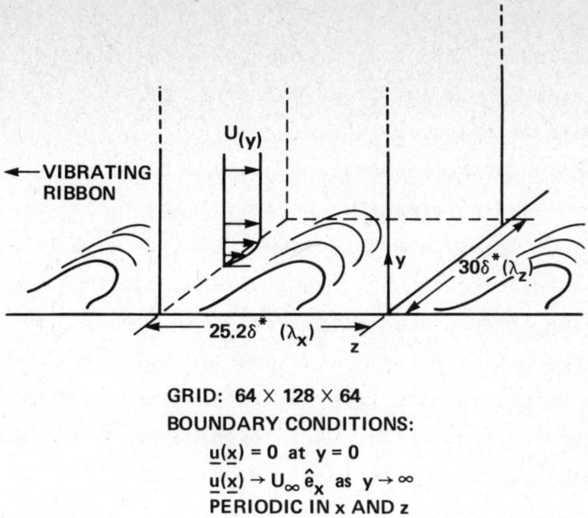

FIG. 1 Simulation of a wave-like disturbance in a laminar boundary layer.

conservation equations in x and z and some manipulations to minimize storage requirements yield the following system of equations for every (k_x, k_z):

$$\frac{\partial \hat{u}}{\partial t} + ik_x(\widehat{u^2} - \widehat{v^2}) + \frac{\partial \widehat{uv}}{\partial y} + ik_z\widehat{uw} + ik_x\hat{q} = -\nu k^2\hat{u} + \nu\,\frac{\partial^2\hat{u}}{\partial y^2}\;; \tag{1}$$

$$\frac{\partial \hat{v}}{\partial t} + ik_x\widehat{uv} + ik_z\widehat{vw} + \frac{\partial \hat{q}}{\partial y} = -\nu k^2\hat{v} + \nu\,\frac{\partial^2\hat{v}}{\partial y^2}\;; \tag{2}$$

$$\frac{\partial \hat{w}}{\partial t} + ik_x\widehat{uw} + \frac{\partial \widehat{vw}}{\partial y} + ik_z(\widehat{w^2} - \widehat{v^2}) + ik_z\hat{q} = -\nu k^2\hat{w} + \nu\,\frac{\partial^2\hat{w}}{\partial y^2}\;; \tag{3}$$

$$ik_x\hat{u} + \frac{\partial \hat{v}}{\partial y} + ik_z\hat{w} = 0\;. \tag{4}$$

Equations (1)-(4) give the time evolution of the Fourier transformed velocity components $\hat{u}$, $\hat{v}$, $\hat{w}$ and a pseudopressure $\hat{q} = \hat{p} + \widehat{v^2}$. Here $k^2 = k_x{}^2 + k_z{}^2$. In computing the nonlinear terms above, WH zero any aliasing contributions by the 2/3 rule, yielding a calculation that is equivalent to a Galerkin spectral method with 22 wave numbers in the x and z directions.

The use of periodic boundary conditions in the streamwise direction leads to an evolution of the disturbance in time rather than space and to a slight departure from the Blasius profile. However, this assumption allows WH to apply 64 grid points (22 wave numbers) to each wavelength of the disturbance, thereby achieving high resolution of streamwise variations in the flow field. In the laboratory experiment, the evolution of the disturbance occurs in space over a distance of about 10 wavelengths. Thus, in the laboratory a particular structure at one streamwise location is a neighbor to the same structure located 1 wavelength downstream, but in a slightly advanced state of evolution. Similarly, an earlier state of that structure is situated 1 wavelength upstream. As found by WH, however, the difference between the spatial and temporal developments in the linear part of the evolution (exponential growth) is quite small as measured by the amplitudes and phases of the eigenmodes of the two problems. Apparently the nonlinear evolution is very similar as well, judging by the excellent agreement between the two experiments. In Fig. 2 are shown contours of $\partial u/\partial y$ for both the laboratory experiment and the computer simulation at the one-spike stage — the spike being the kink in the high amplitude $\partial u/\partial y$ contours [3]. The contours are taken in the x-y plane containing the maximum initial perturbation — the peak plane. At this point the time of evolution in the simulation is $340\ \delta^*/U_\infty$. The agreement is quite good as is the comparison of the corresponding normal velocity contours given in Fig. 3. The maximum vertical velocity is roughly $0.07\ U_\infty$ at this stage. The boundary-layer thickness, δ, is approximately 3 times the displacement thickness.

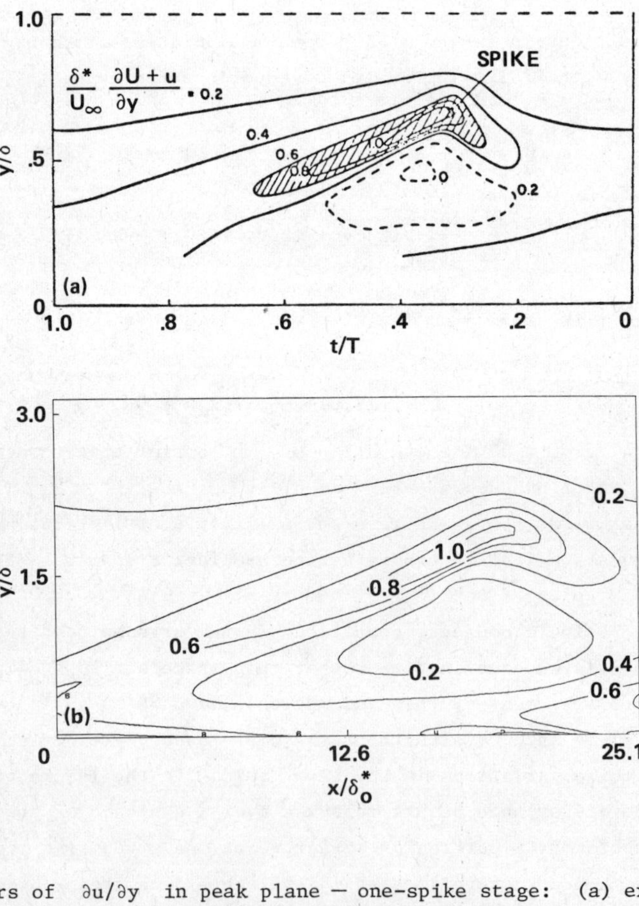

FIG. 2 Contours of $\partial u/\partial y$ in peak plane — one-spike stage: (a) experiment [3]; (b) numerical simulation of WH.

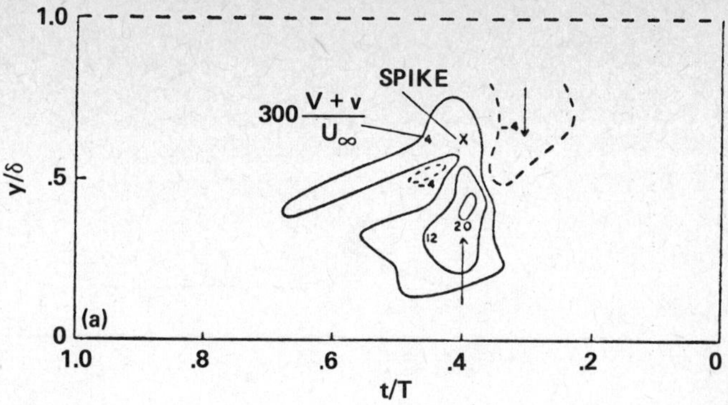

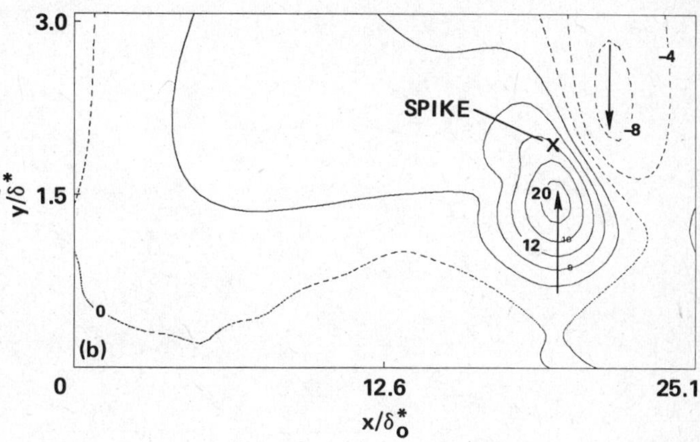

FIG. 3 Vertical velocity in peak plane — one-spike stage: (a) experiment [3]; (b) numerical simulation of WH.

As nonlinear effects become increasingly significant, the evolution of the computational disturbance reaches the two-spike stage at $t = 360\ \delta^*/U_\infty$. Contours of $\partial u/\partial y$ and normal velocity at this stage are shown in Figs. 4 and 5, respectively. Note that again there is very good agreement on the structure of the disturbance in the peak plane. The relative shift of the structures along the abscissa is irrelevant [1], but other discrepancies are evident. For example, the computation is producing peak values in v ($\approx 0.21\ U_\infty$) that are higher than the experiment. Considering the large streamwise gradient of v in the vicinity of this maximum, it is quite possible that this type of discrepancy is due to lack of resolution or to ensemble averaging or both in the laboratory experiment. On the other hand, the appearance of oscillatory fine-scale features in Fig. 4 indicates that the computation may be starting to have a resolution problem also. A three-spike stage occurs in the simulation at $t \approx 365\ \delta^*/U_\infty$. Soon after this time significant errors begin to appear due to the fact that the energy of the disturbance is trapped to remain within the computational wave numbers.

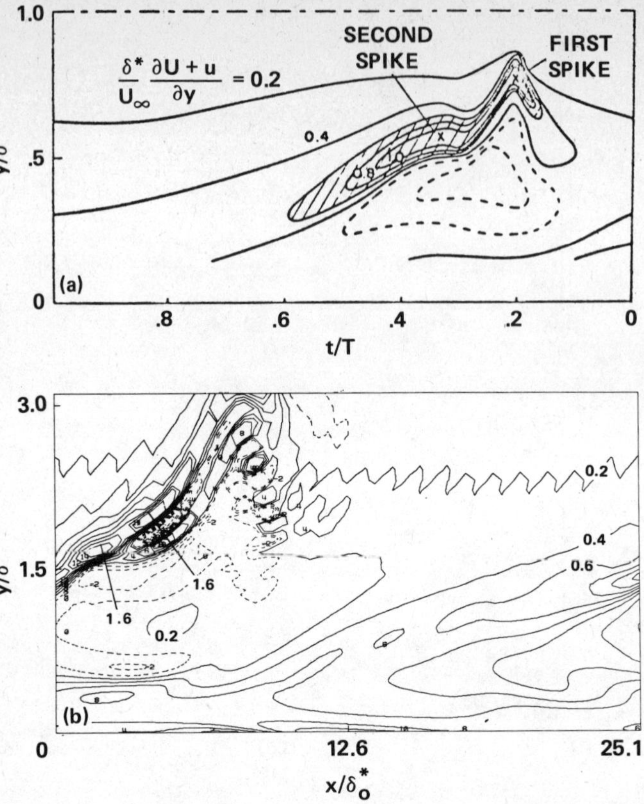

FIG. 4 Contours of $\partial u/\partial y$ in peak plane — two-spike stage: (a) experiment [3]; (b) numerical simulation of WH.

Thus, in a remarkable computational effort WH have demonstrated the capability of simulating a complex, three-dimensional transitional flow in an "honest" calculation — no subgrid modeling was used. In the particular case that was stimulated the results appear to be valid up to and including the two-spike stage when $v_{max} \approx 0.21\ U_\infty$. After that stage, presumably, scales with wavelengths $\leq 30\delta^*/22$ in the spanwise direction or $\leq 25\delta^*/22$ in the streamwise direction become important. The vertical resolution appears to be quite adequate throughout the simulation.

B. Spot-Like Disturbance in a Laminar Boundary Layer

In another numerical investigation of boundary-layer mechanics, Leonard [4] studied the growth of localized disturbances in a laminar boundary layer as they evolve into spot-like turbulent structures. As in the study of WH, these are computer experiments, but using a quite different numerical method. Vortex filaments are used to represent the three-dimensional, perturbed vorticity field within the spot. Each filament is defined by a three-dimensional space curve $\underline{x}(\xi,t)$, circulation Γ, and a

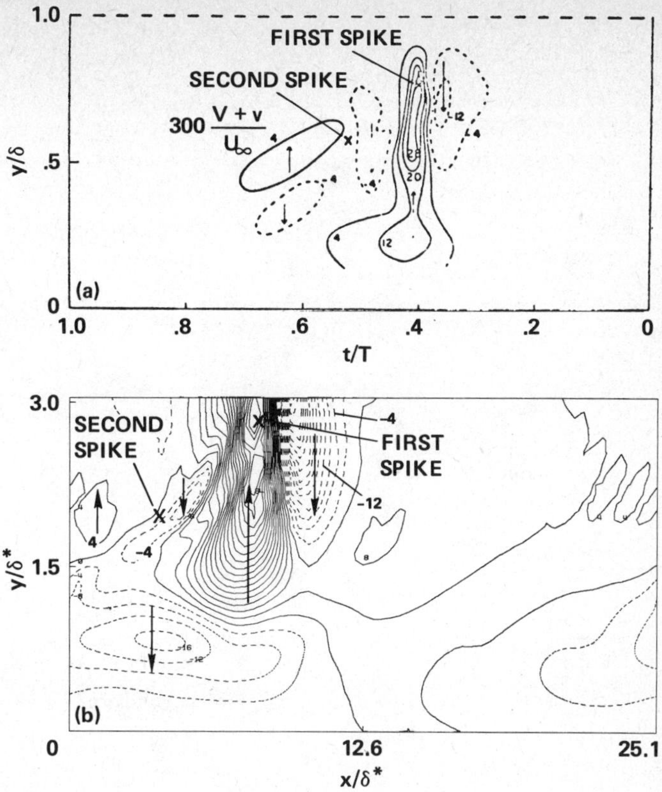

FIG. 5 Vertical velocity in peak plane — two-spike stage: (a) experiment [3]; (b) numerical simulation of WH.

core size, σ, which parameterizes an assumed Gaussian vorticity within the core. Thus, the perturbed vorticity field is given by a summation over filaments,

$$\underline{\omega}'(\underline{x},t) = \sum_i \Gamma_i \int_{C_i} \exp[-|\underline{x} - \underline{x}_i|^2/\sigma^2] \Big/ (\sqrt{\pi}\sigma)^3 (\partial\underline{x}_i/\partial\xi)d\xi \ . \tag{5}$$

The vortex filaments represent a continuum of vorticity which initially is in the form of an infinite sheet of finite thickness, that is, a boundary layer. As indicated in the schematic (Fig. 6) the disturbance in the simulation is created by a local distortion of the vortex filaments at $t = 0$. For numerical purposes each tube of vorticity in the boundary layer is decomposed into its straight, unperturbed contribution and into a loop representing its contribution to the perturbed vorticity field, as shown in Fig. 7. The velocity field is decomposed, accordingly, as

$$\underline{u}(\underline{x}) = U(y)\hat{e}_x + \underline{u}'(\underline{x}) \ , \tag{6}$$

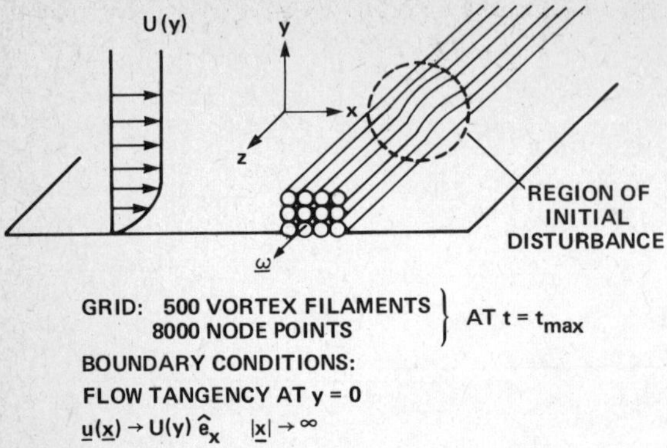

FIG. 6 Simulation of a spot-like disturbance in a laminar boundary layer.

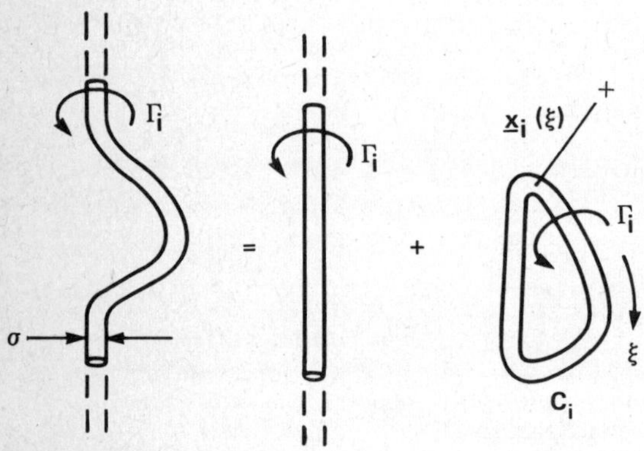

FIG. 7 Decomposition of vorticity field.

where $U(y)$ is the laminar profile and $\underline{u}'$ is given by a sum of Biot-Savart integrations over all the filament curves, C_i, plus their images. The curves are marked with a sequence of node points that are tracked in a Lagrangian reference frame, according to the following equation of motion:

$$\frac{\partial \underline{x}_i}{\partial t} = -\frac{1}{4\pi}\sum_j \Gamma_j \int_{C_j} \times \frac{(\underline{x}_i - \underline{x}_j)\times(\partial \underline{x}_j/\partial\xi')\,d\xi'}{\left[|\underline{x}_i - x_j|^2 + \alpha(\sigma_i{}^2 + \sigma_j{}^2)\right]^{3/2}}$$

$$+ \hat{e}_x\, U(y) + \text{image contributions.} \qquad (7)$$

The equation is seen to be the Biot-Savart induction law for an infinitesimal filament, but corrected in the near-field for finite core size. Using $\alpha = 0.2065$ we recover the correct speed of a ring vortex in the limit σ << ring radius. Up to numerical truncation error, the method simulates the inviscid dynamics of the vorticity field and therefore satisfies the Kelvin and Helmholtz theorems. For a more complete description of the method including simulation of viscous effects see [5].

Biot-Savart integration yields exact boundary conditions at infinity in all directions. The image contributions ensure tangency of the flow at the wall. In this initial study we ignore the generation of vorticity at the wall boundary due to the disturbance, that is, we use the inviscid boundary condition at the wall for $\underline{u}'$. This appears justified because the relatively high initial disturbance level ($0.07 - 0.10\ U_\infty$) and the initially small length scale of the perturbation $\sim O(\delta^*)$ (corresponding to laboratory experiments on turbulent spots [6,7]) leads to the rapid evolution of a sizable, strong disturbance at the end of the simulation, $t \approx 50\delta^*/U_\infty$. The viscous length scale for this time, assuming $Re_{\delta^*} = 1000$, is only $0.3\delta^*$.

Computational filaments are added automatically as the disturbance grows in size. At the end of the simulation, about 8000 points represent one half of the spot (symmetry of the flow about the midplane is enforced). The computation could continue beyond this time, but (1) some of the filaments are quite stretched at this point, indicating that remeshing is required, and (2) the cpu time per time step is quite large at this stage being proportional to the square of the number of computational points for vortex methods. Total run times are ≈2 hr on the ILLIAC.

In Fig. 8, the space curves defining the vortex filaments originally in a single layer (out of a total of six representing the boundary layer) are shown at three different times. The spreading of the disturbance and the increasing deformations suffered by the vortex lines are quite evident. In Fig. 9, vertical and streamwise vorticity contour plots at a late time are shown in horizontal (x-z) and spanwise-vertical (z-y) planes, respectively. Note the development of intense streamwise and vertical vorticity generated in the flow in agreement with the pattern of vortex lines displayed in Fig. 8. The structures are cell-like, $O(\delta^*)$ in the spanwise and vertical directions, but elongated in the streamwise direction. For example, the region of negative ω_y just above the x-axis in Fig. 9a together with its antisymmetric counterpart (not shown) just below the axis would constitute one cell approximately $4\delta^*$ in spanwise extent. Velocity contour plots are shown in the same (x-z) and (y-z) planes in Figs. 10 and 11, respectively. The presence of streaky structures in the horizontal plane is compatible with experimental observations of passive particle motion in turbulent spots [7] and turbulent boundary layers [8]. The maximum normal velocity is quite high at $\approx 0.4\ U_\infty$.

Most laboratory experiments on turbulent spots report ensemble-averaged measurements several hundred δ^* downstream of the initial disturbance. A typical streamwise velocity history plot from Cantwell et al. [7] measured in the midplane of the spot is shown in Fig. 12a. Note that compared with the laminar profile there is a velocity excess near the wall and a defect away from the wall — typical of a turbulent boundary-layer profile. The experiment of Wygnanski et al. [6] produced similar results. However, as indicated in Figs. 10 and 11, the particular numerical simulation presented above yields a velocity excess in the midplane for all heights above the boundary. On the other hand, if the velocity field from the numerical simulation is span-averaged over one cell of width $4\delta^*$ centered on the midplane we obtain the velocity plots shown in Fig. 12b. (To obtain better statistics, the results shown in Fig. 12b are based on the velocity field at a single time late in the simulation. Thus, decreasing x is used along the simulation rather than increasing time.) Note the close similarity in the behavior of these computational results averaged over one cell in the span direction at $t = 43\ \delta^*/U_\infty$ to the ensemble-averaged laboratory measurements of Fig. 12a, taken at a fixed streamwise location, at later times ($t \gtrsim 1500\ \delta^*/U_\infty$). The good comparison leads to the speculation that the internal structure of a single realization of a large turbulent spot ($\gtrsim 100\delta^*$ in streamwise extent) is dominated by the presence of many cell-like substructures such as those revealed in the numerical simulation. (For experimental support of this speculation see I. Wygnanski, these proceedings.)

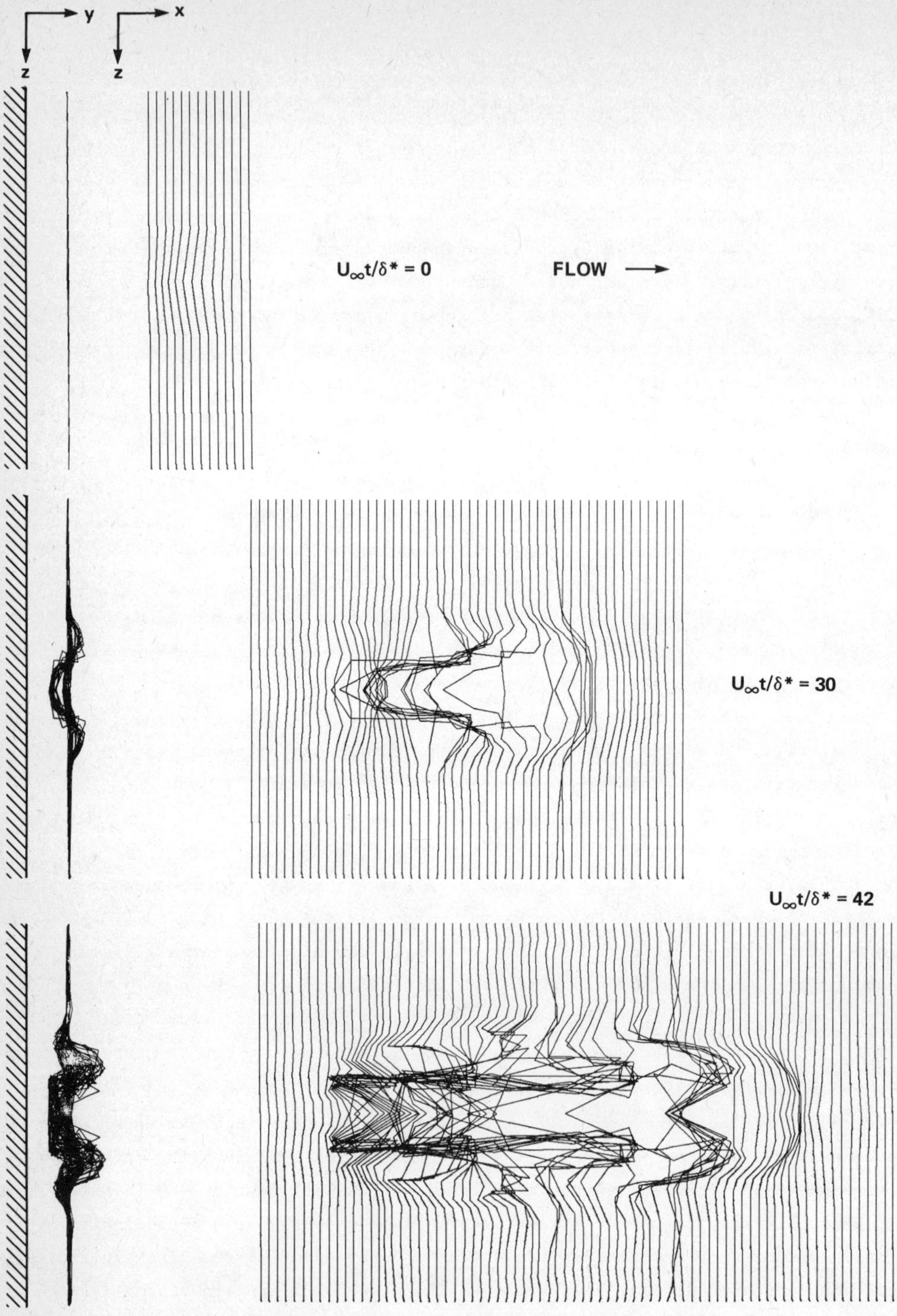

FIG. 8 Top and rear views of one plane of the computed vortex lines.

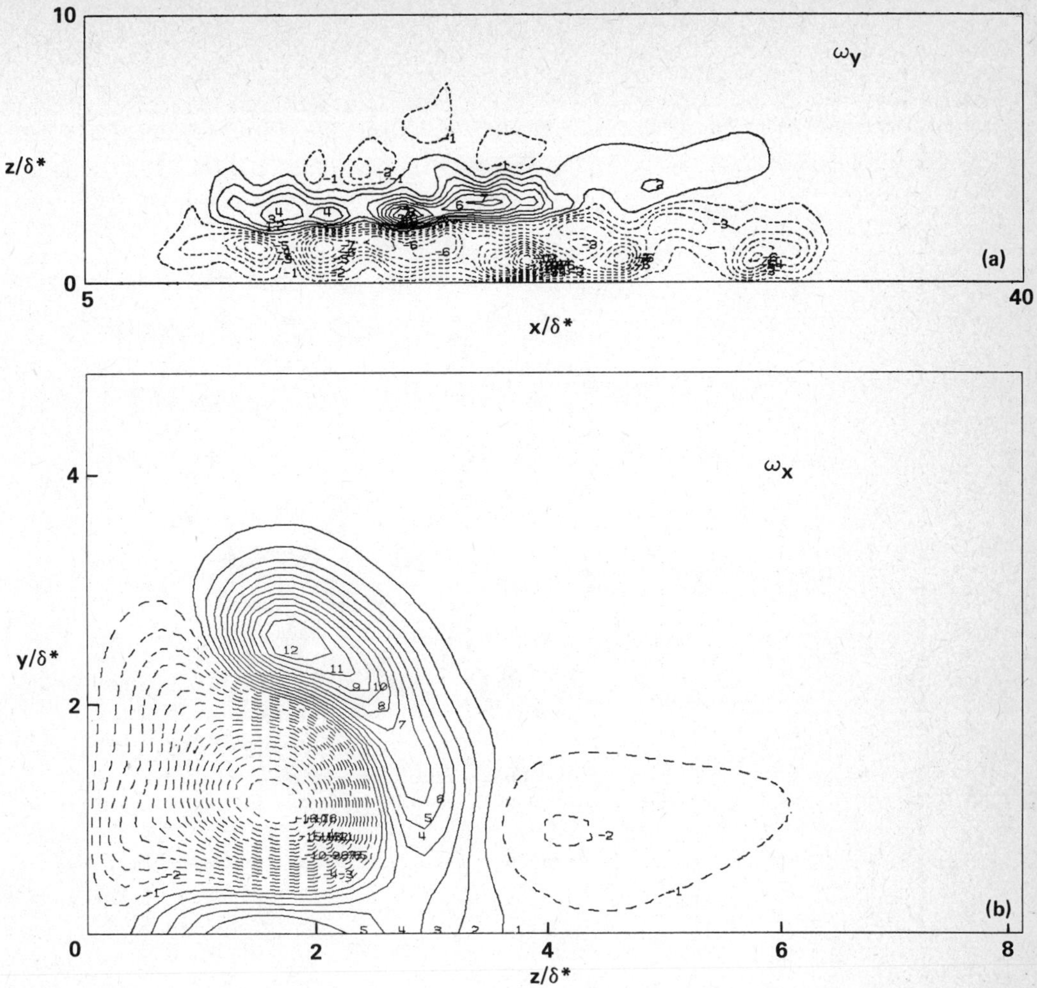

FIG. 9 Contours of vorticity for $U_\infty t/\delta^* = 41$: (a) ω_y in a horizontal plane, $y/\delta^* = 0.66$; (b) ω_x in a vertical (z-y) plane $x/\delta^* = 25$. ——, positive contours; — — —, negative contours; contour levels ±0.05, ±0.15, ±0.25 . . . of $\partial u/\partial y|_{wall}$.

Three-dimensional vortex simulations have therefore demonstrated the rather rapid, nonlinear evolution of small localized disturbances in a laminar boundary layer. The boundary layer appears to deform into a corrugated surface with significant amounts of streamwise and vertical vorticity — leading to strong negative u'v correlations and, hence, intense turbulent mixing. Velocity perturbations induced in the nearby, relatively undisturbed vorticity layer lead to sufficient deformations in the vortex lines to produce local nonlinear growth. Thus, the spot appears to grow in the spanwise direction by "transverse contamination," in agreement with recent experiments using dye interjection [9], which demonstrated that turbulent convection is not

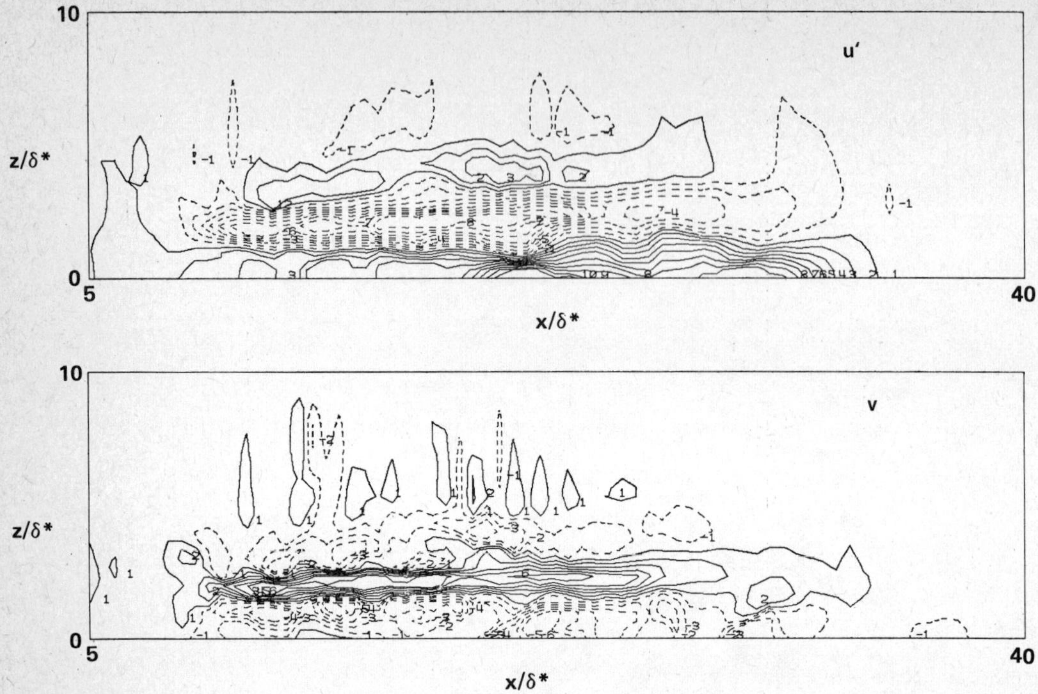

FIG. 10 Velocity contours in the horizontal plane of Fig. 9a. Contour levels ±0.02, ±0.06, ±0.10 . . . of U_∞.

responsible for the spanwise growth of a spot. Further study is needed to simulate properly larger turbulent spots or a collection of spots and the decay of turbulence rearward of the spot.

C. Turbulent Channel Flow

In our third example of a wall-bounded shear flow we discuss the numerical simulation of turbulent channel flow by Kim and Moin [10] (hereinafter referred to as KM). As shown in Fig. 13, a 64^3 grid is used to simulate flow at Re = 13,800 with periodic boundary conditions in the spanwise and streamwise directions and the no-slip condition at the channel walls. In an earlier investigation of turbulent channel flow, Moin et al. [11] used a 16×64×16 grid with the same boundary conditions. Because the laboratory flow at this Reynolds number contains a significant amount of turbulent energy at length scales smaller than the computational grid, KM used the filtered Navier-Stokes equations to determine the dynamics of the turbulence field that is resolvable on their mesh (large eddy simulation technique). Thus, a flow variable f may be decomposed as follows:

$$f = \bar{f} + f' , \tag{8}$$

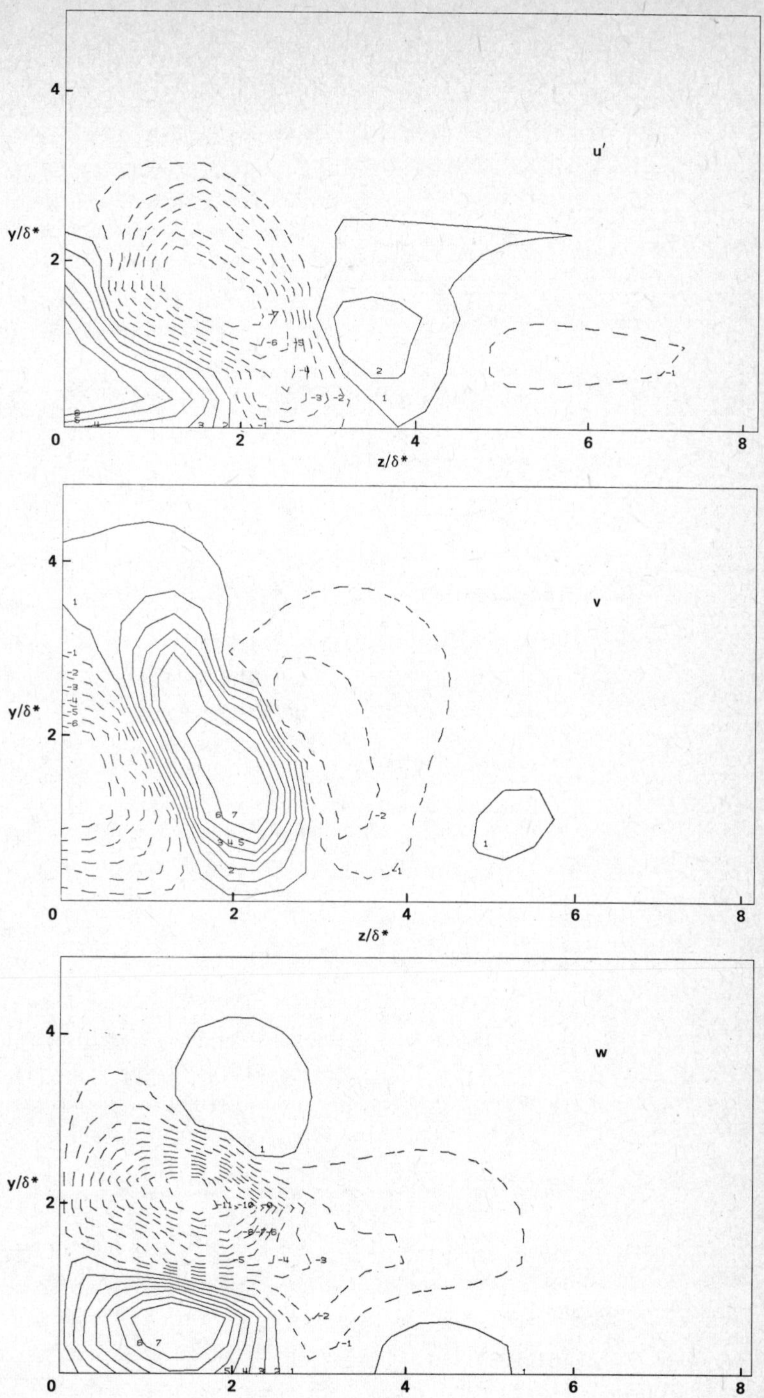

FIG. 11 Velocity contours in the vertical plane of Fig. 9b. Same contour levels as Fig. 10.

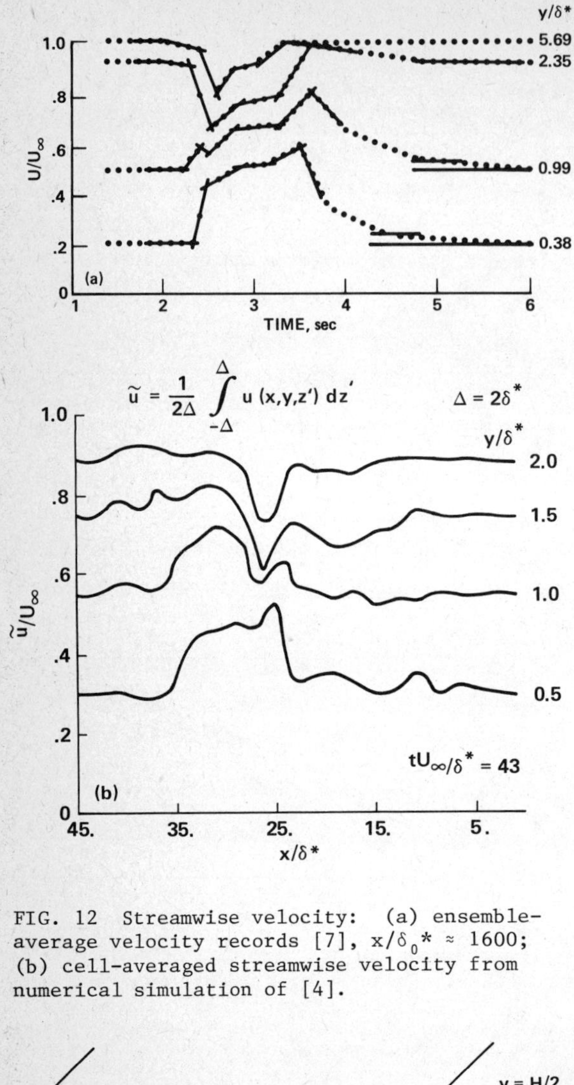

FIG. 12 Streamwise velocity: (a) ensemble-average velocity records [7], $x/\delta_0{}^* \approx 1600$; (b) cell-averaged streamwise velocity from numerical simulation of [4].

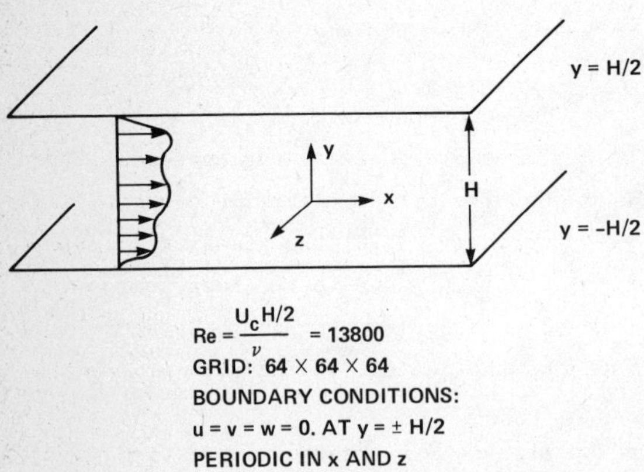

FIG. 13 Numerical simulation of turbulent channel flow.

where $\overline{f}$ is the large-scale field, defined as the result of a filtering or smoothing operation of f, and f' is the subgrid field.

After applying the filtering operation to the incompressible momentum and continuity equations, the governing equations for the filtered field may be written

$$\frac{\partial \underline{\overline{u}}}{\partial t} + \overline{\underline{\overline{\omega}} \times \underline{\overline{u}}} = -\nabla p^* + \frac{dP}{dx}\,\hat{e}_x - \nabla\cdot\underline{\tau} + \nu\nabla^2\underline{\overline{u}} \tag{9}$$

and

$$\nabla\cdot\underline{\overline{u}} = 0\ , \tag{10}$$

where dP/dx is the constant mean pressure gradient, $\underline{\overline{\omega}}$ is the large-scale vorticity field, and the subgrid-scale stress tensor, $\underline{\tau}$, (having zero trace) has components defined by

$$\tau_{ij} = R_{ij} - R_{kk}/3\ \delta_{ij} \tag{11}$$

and

$$R_{ij} = \overline{u_i'u_j'} + \overline{u_j'\overline{u}_i} + \overline{\overline{u}_j u_i'}\ . \tag{12}$$

The quantity p^* is given by

$$p^* = \overline{p}/\rho + 1/2\ \overline{\overline{u}_k\overline{u}_k} + R_{kk}/3\ . \tag{13}$$

The system of equations (9) and (10) becomes deterministic once $\underline{\tau}$ is specified. Kim and Moin use an eddy viscosity model,

$$\tau_{ij} = -2\nu_T S_{ij}\ , \tag{14}$$

where

$$S_{ij} = \frac{1}{2}\left(\frac{\partial\overline{u}_i}{\partial x_j} + \frac{\partial\overline{u}_j}{\partial x_i}\right), \tag{15}$$

and ν_T is given by the Smagorinsky model,

$$\nu_T = (C_S\Delta)^2\sqrt{S_{ij}S_{ij}}\ . \tag{16}$$

The constant C_S is taken to be 0.1 and Δ is a measure of the grid resolution and is taken to be

$$\Delta = (\Delta x \; \Delta y \; \Delta z)^{1/3}[1 - \exp(-y^+/50)] \tag{17}$$

where the exponential function damps the eddy viscosity near the wall.

The channel height is H. The lengths of computational box in the streamwise and spanwise directions were chosen to be $L_x = \pi H$ and $L_z = 2\pi H/3$, respectively, based on two-point correlation measurements in laboratory experiments and the desire to capture all the important large eddies in the flow. The nondimensional grid spacings that result are $\Delta x^+ = 63$ and $\Delta z^+ = 42$, expressed in wall units. (The channel height H is 1280 wall units.) The grid spacing in the normal direction varies from $\Delta y^+ \approx 2$ near the wall to $\Delta y^+ \approx 50$ at the middle of the channel. The need for subgrid modeling is clear. Even near the wall, where grid is finely meshed in the normal direction, the horizontal spacing is not sufficient to resolve the experimentally observed wall streaks with a mean spanwise spacing of $\lambda_z^+ \approx 100$. Kim and Moin find, however, that wall streaks do appear in their simulation, but at a larger scale, the streaks apparently accommodating themselves to the available resolution.

The agreement of KM's simulation with laboratory measurements is remarkably good. After a time of evolution of $15.4H/U_c$ (U_c is the mean velocity of the center of the channel) KM obtain the mean-velocity profile (horizontally averaged) shown, along with experimental data, in Fig. 14. Near the wall the simulated profile is somewhat low; away from the wall a distinct logarithmic region is revealed with the correct slope (Kárman constant). The computed large-scale turbulent intensity profiles and the turbulent shear-stress profile are shown in Figs. 15 and 16, respectively. The $\langle \bar{u}''^2 \rangle^{1/2}$ and $\langle \bar{w}^2 \rangle^{1/2}$ profiles are in excellent agreement with experiment at all distances from the wall. (Superscript double prime denotes deviation from horizontal average.) In the wall region ($y^+ \lesssim 90$) the resolved vertical turbulence intensity is lower than experiment, possibly due to the fact that a significant amount of $\langle v^2 \rangle^{1/2}$ resides in the subgrid scales or that the subgrid model has deficiencies near the wall, as suggested by the deficit of the mean profile in this region.

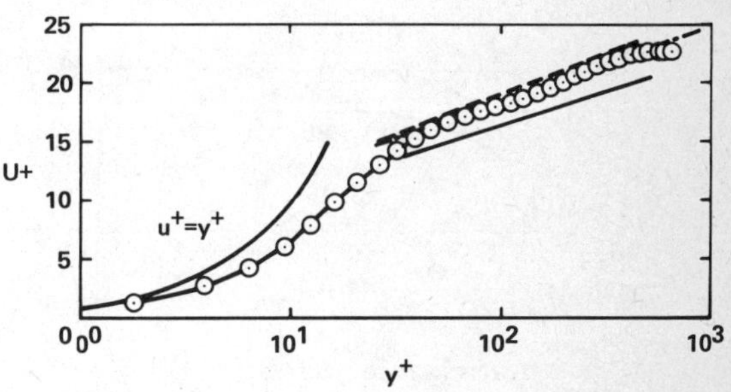

FIG. 14 Mean-velocity profiles; —⊙—, numerical simulation of [10]; experiment, ———, Re = 13800; — — —, Re = 57000; — — —, Re = 12200-61600.

Contour plots of instantaneous Reynolds stress $\bar{u}''\bar{v}$ reveal interesting phenomena. In a horizontal plane near the wall ($y^+ = 16$, Fig. 17) a number of hotspots occur where $\bar{u}''\bar{v}$ is large and negative. Kim and Moin determined that a great

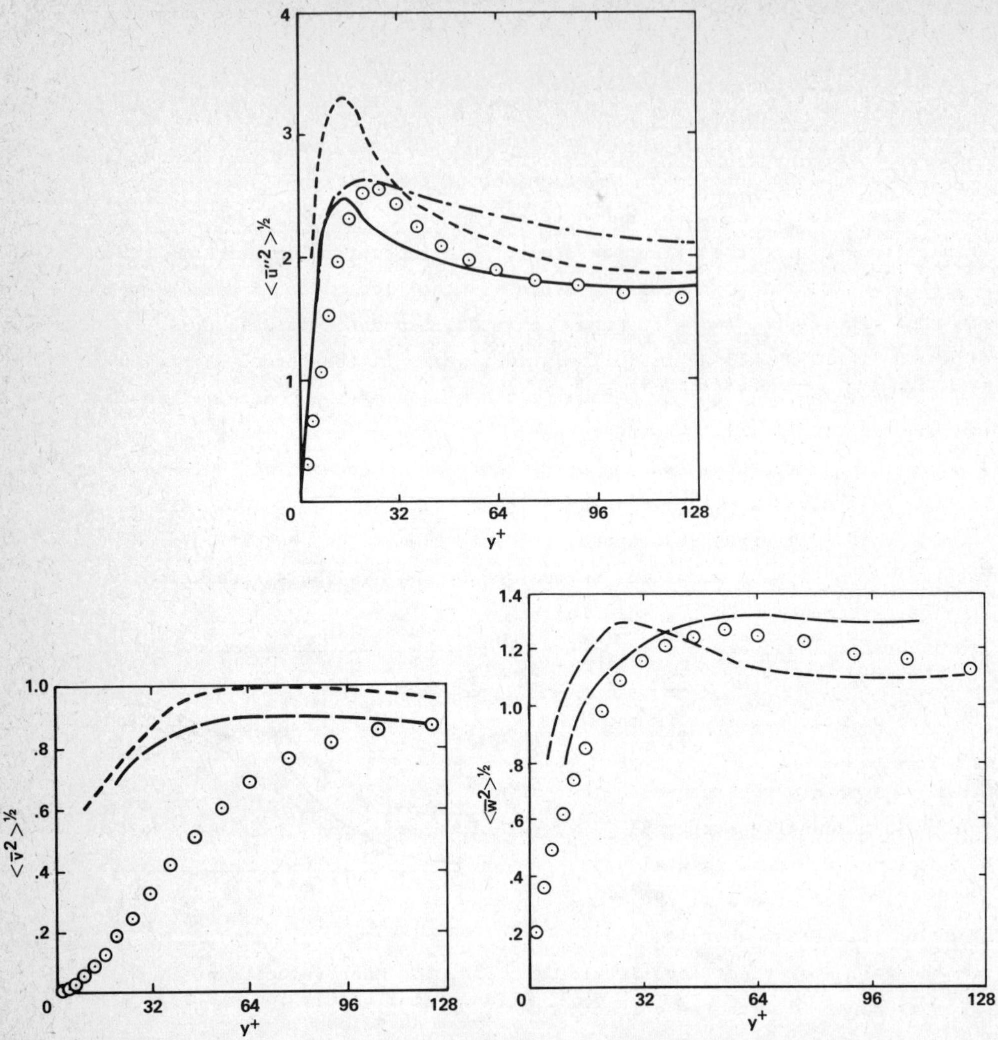

FIG. 15 Comparison of the horizontally averaged resolvable turbulence intensities normalized by u_τ with experimental data; ⊙, numerical simulation of [10]; experiment, ——, Re = 13800; — — —, Re = 15200; —— ——, Re = 25000; —— - ——, Re = 120000.

majority of these active regions is associated with $\bar{u}'' > 0$, $\bar{v} < 0$, or "sweeps." Away from the wall ($y^+ = 90$, Fig. 18) the hotspots are associated with "bursts" or $\bar{u}'' < 0$, $\bar{v} > 0$. All these findings are in agreement with experiment [12].

Kim and Moin also investigated pressure-velocity gradient correlations in the region of $y^+ < 100$. These correlations provide for energy transfer between the individual components and turbulent energy and, as such, are of great interest to turbulence modelers. The computed correlations are shown in Fig. 19. Their behavior for $y^+ \gtrsim 20$ corresponds to the expected transfer from $\langle \bar{u}''^2 \rangle$ to $\langle \bar{v}^2 \rangle$ and $\langle \bar{w}^2 \rangle$. Close to the wall, however, KM show a substantial loss of $\langle \bar{v}^2 \rangle$ and a corresponding gain of $\langle \bar{w}^2 \rangle$ through the action of these correlations. By studying the fluid motion near the wall using "snapshots" or instantaneous contour plots, KM determined that the energy transfer was due to high-speed parcels of fluid approaching the wall and then spreading laterally ("splatting" effect).

D. Comparison of the Simulations

In Table I we compare a number of features of the three simulations. The natural length and time scales for the first two studies or the evolution of a disturbance in a

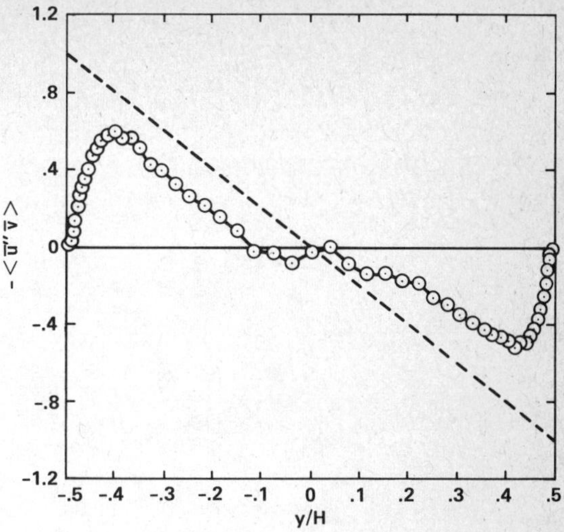

FIG. 16 Vertical profile of horizontally averaged resolvable turbulent shear stress.

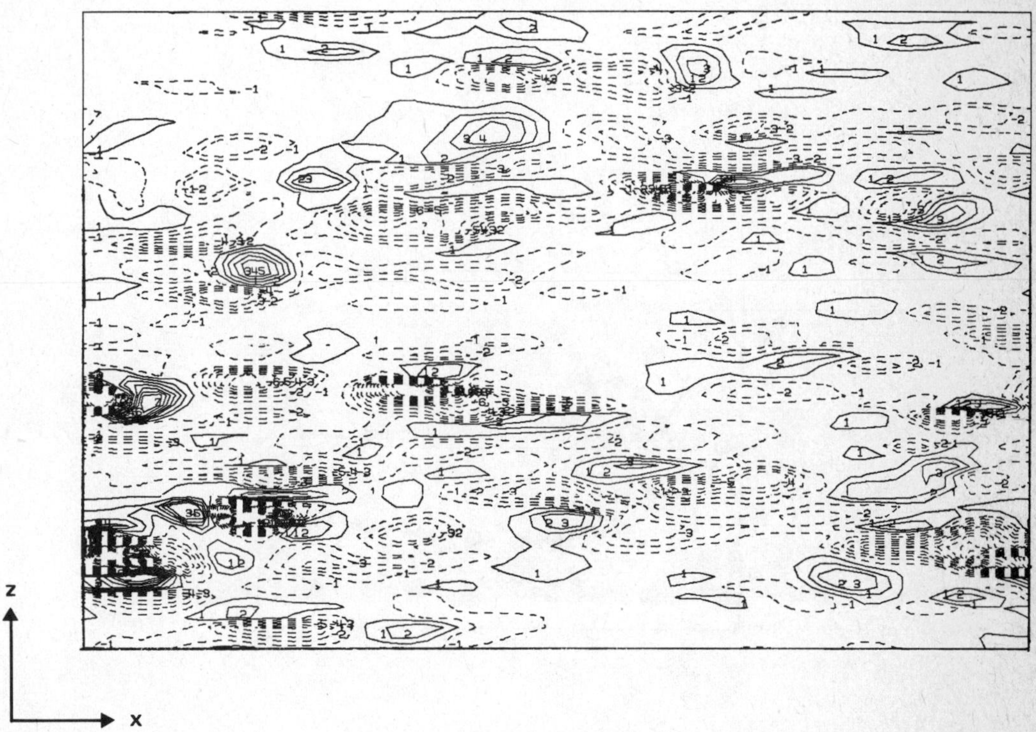

FIG. 17 Contours of $\bar{u}''\bar{v}$ in the x-z plane at $y^+ = 16$.

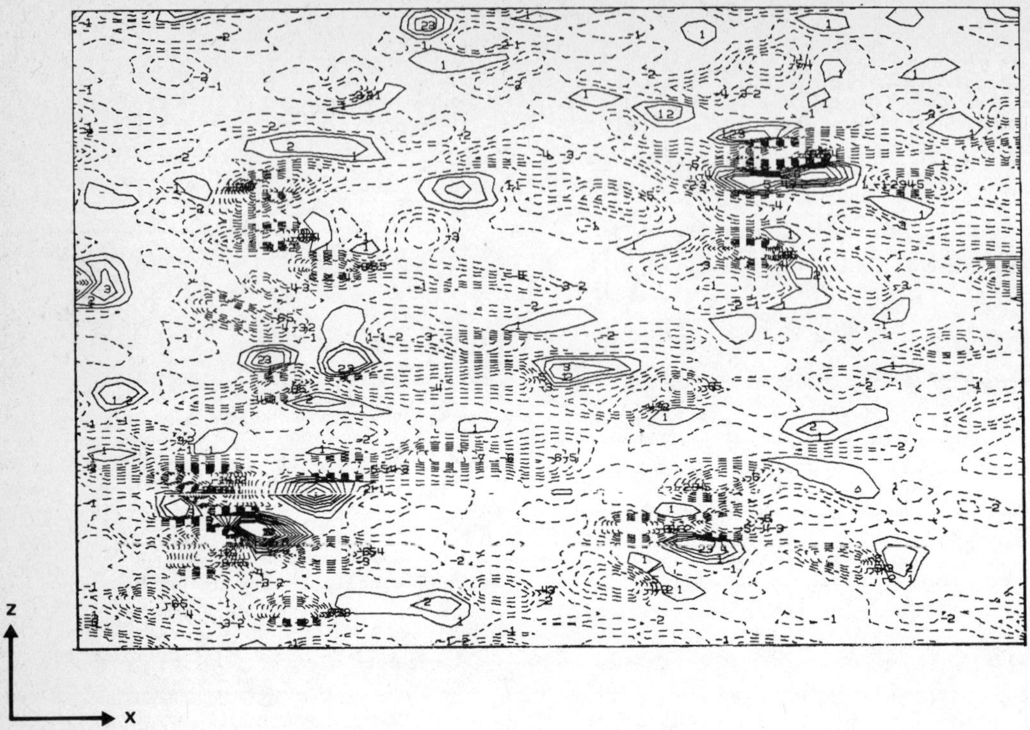

FIG. 18 Contours of $\bar{u}''\bar{v}$ in the x-z plane at $y^+ = 90$.

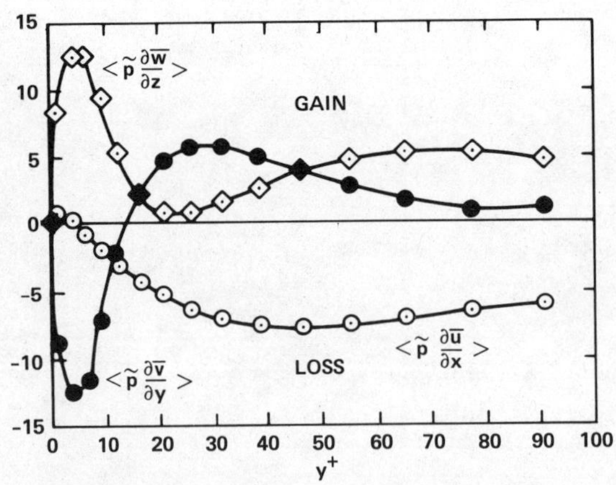

FIG. 19 Vertical profiles of horizontally averaged, pressure-velocity gradient correlations in the vicinity of the wall ($y^+ \leq 100$).

TABLE I

SIMULATION PARAMETERS

	Wave disturbance	Spot disturbance	Channel flow
Grid			
$N_x \times N_y \times N_z$	64×128×64	500 filaments 8000 points (1/2 spot)	64×64×64
Mesh spacing			
$\Delta x/\delta^*$	0.39	0.41	---
$\Delta y/\delta^*$	$0.0078 \rightarrow \infty$	0.41	---
$\Delta z/\delta^*$	0.47	0.66	---
(wall units)			
Δx^+	16^a	16^a	63
Δy^+	$0.31 \rightarrow \infty$	16	$1.8 \rightarrow 50$
Δz^+	19	26	42
Turbulence volume (wall units^3)	0.3×10^{9b}	0.4×10^{9b}	12×10^9
Time step			
$\Delta t U_\infty/\delta^*$	0.24	0.49	---
Δt^+	0.38^a	0.78^a	0.61
Number steps/run	1500	100	2000
ILLIAC time/run, hr	12	2	12

[a]Estimate using $\delta^{*+} = (u_\tau/U_\infty)Re_{\delta*} \approx 40$; $(\delta^*/U_\infty)^+ = (u_\tau^2/U_\infty^2)Re_{\delta*} \approx 1.6$.

[b]Estimated vertical extent of turbulence = 250 wall units.

laminar boundary layer are δ^* and δ^*/U_∞, respectively. In the simulation of turbulent channel flow, appropriate length and time scales for the near-wall region are ν/u_τ and ν/u_τ^2, where u_τ is the wall friction velocity, $u_\tau = \sqrt{\nu\, \partial \langle u \rangle \partial y|_{wall}}$. To compare the transition studies with the channel flow simulation, the scales of the first two studies were converted to approximate turbulent scales as follows. Assume that the laminar boundary layer completes its transition to a turbulent layer. Then the original laminar displacement thickness in wall units is given by

$$\delta^{*+} = \frac{\delta^* u_\tau}{\nu} = \frac{u_\tau}{U_\infty}\,\frac{\delta^* U_\infty}{\nu}$$

$$= \frac{u_\tau}{U_\infty}\,Re_{\delta^*}$$

$$\approx 40\ ,$$

where we have used $u_\tau/U_\infty = 0.04$ for the turbulent layer and $Re_{\delta^*} = 1000$. The time scale in wall units is found to be

$$\left(\frac{\delta^*}{U_\infty}\right)^+ = \frac{u_\tau^{\,2}}{U_\infty^{\,2}}\,Re_{\delta^*}$$

$$= 1.6\ .$$

Note that the two transitional studies have comparable resolution in the horizontal directions, but the vertical grid of WH is much finer. Both are able to resolve structures having length scales $O(\delta^*)$ in any direction, and, as shown in earlier figures, many such significant structures were found. Recall, for example, the vertical velocity contour plot (Fig. 5) at the two-spike stage of the study by WH, exhibiting large peak values ($\approx 0.21\ U_\infty$) and very large streamwise gradients. As shown in Fig. 20a, a similar structure with a peak v' of $0.42\ U_\infty$ was found in the spot simulation. The results are shown in a z-plane ($z = 0.8\ \delta^*$) containing mostly upward fluid motion (bursting ?) similar to the peak-plane behavior observed by WH. Contours of the perturbed spanwise vorticity are shown in Fig. 20b in the same x-y plane at $z = 0.8\ \delta^*$. As might be expected, the region of large $\partial v/\partial x$ is associated with a concentration of ω_z'.

The time evolution of the maximum vertical velocities is shown in Fig. 21. Also shown is the range of peak values in the

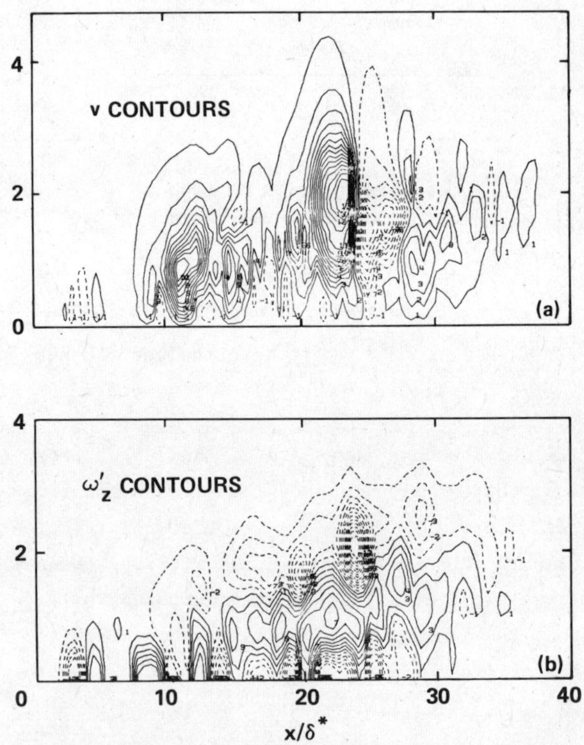

FIG. 20 Contours in an x-y plane from the numerical simulation of a spot-like disturbance: (a) vertical velocity; (b) perturbed spanwise vorticity.

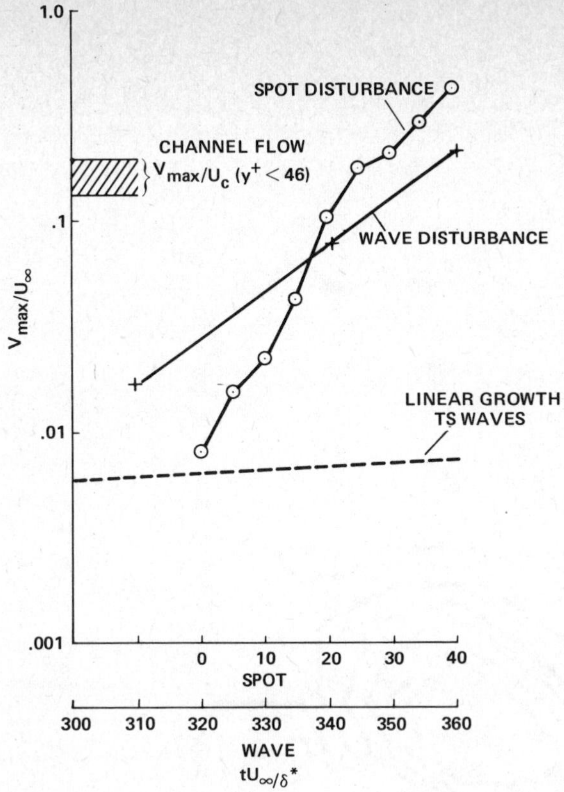

FIG. 21 Maximum vertical velocities in a plane.

channel flow simulation by KM in several y-z planes for $y^+ < 46$. Through the final stages of evolution the peak vertical velocity in the transition study of WH is experiencing exponential growth at 15 times the growth rate of the initial two-dimensional perturbation wave. Apparently this peak is associated with a small-scale ($O(\delta^*)$) secondary instability of the larger ($O(25\delta^*)$) slower-growing primary instability. The growth rate seen in the spot simulation is even larger and not exponential, presumably due to non-linear interactions between small-scale perturbations.

Note that in the channel-flow study a relatively coarse mesh in the horizontal plane was used, clearly dictated by the relatively large volume of turbulence being simulated. A number of flow structures seen in the transitional studies that were ($O(\delta^*)$) and smaller in the spanwise and streamwise directions would not be resolved adequately by this mesh. (Recall that $\delta^{*+} \approx 40$.) Nevertheless, KM see in their simulation considerable turbulent activity near the wall, as evidenced by Figs. 15-19. As was mentioned above, turbulent structures do appear (wall streaks, etc.), but at a larger scale than in experiment, accommodating themselves to the available resolution. In Figs. 22 we see this effect illustrated in a contour plot of streamwise vorticity from KM's simulation. In Fig. 22a, about one seventh of the y-z computational plane is shown. Note that near the wall the spanwise extent of the cell-like structures is only a few times the smallest available scale in the span direction ($\Delta z^+ = 42$). In Fig. 22b a blowup of one of the more active regions near the wall is shown. With one exception, the structure is similar to the one shown in Fig. 9b from the spot simulation. The vertical scales in Figs. 22a and 22b have been expanded by a factor of 4 over the horizontal scales so that the actual structure depicted in Fig. 22b is considerably flattened. For comparison, Fig. 23 shows Fig. 9b from the spot simulation using the same scale in wall units of the channel-flow results. The smaller computational domain affords the luxury of seeing the fine-scale structures depicted in the figure.

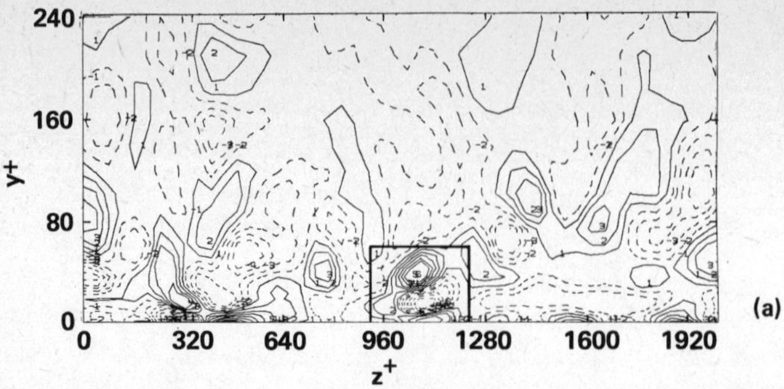

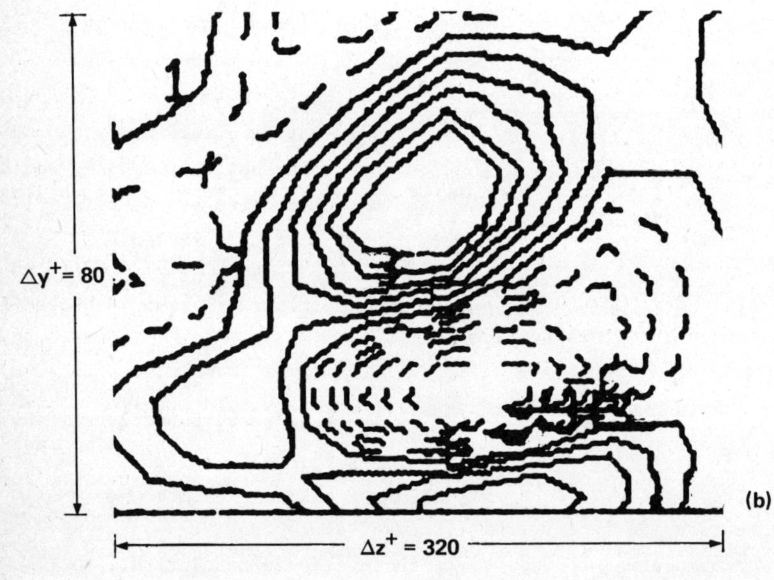

FIG. 22 Contours of streamwise vorticity in a y-z plane — channel-flow simulation of KM: (a) one seventh of computational plane; (b) blowup of region indicated.

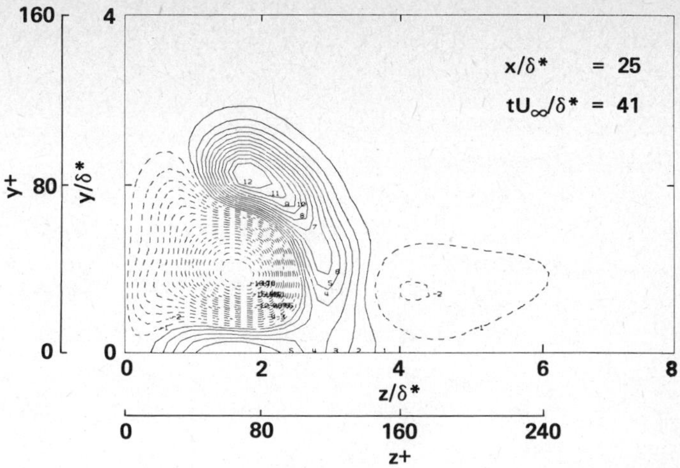

FIG. 23 Contours of streamwise vorticity in a y-z plane; spot simulation with scale in wall units.

III. EFFECTS OF SUBGRID-SUPERGRID SCALE MOTIONS

In the numerical simulation of fluid flows one can resolve only a limited range of length scales in the flow field. In most cases of interest, turbulent flows will contain important scales of motion outside this range. A major problem in the simulation of turbulent flows, therefore, is to model the effect of these subgrid or supergrid components on the resolvable scales. The objective of this modeling effort should be to minimize the error on any desired statistical information that is derived from the resolvable scale motions, including structural information. This problem appears particularly acute in the case of wall-bounded shear flows because of the importance of small-scale structures near the wall in the generation of turbulent energy and in turbulent mixing. In addition, the use of periodic boundary conditions will be a problem if scales larger than the computational domain are important.

In the first two studies discussed in the previous section, which were concerned with the transition of a laminar boundary layer, there was no explicit attempt to account for subgrid-scale effects. In the study by WH of the evolution of a wave-like disturbance, a Fourier representation was used for the dependence of the flow variables in the streamwise and spanwise directions. Thus, by the quadratic nonlinearity of the pressure and convective derivatives, the evolution of a given Fourier coefficient with two-dimensional wave number $\underline{k}_1$ within the resolved range ($k_{min} < |\underline{k}_1| \leq k_{max}$) will contain contributions from other wave numbers $\underline{k}_2$ and $\underline{k}_3$ satisfying

$$\underline{k}_2 + k_3 = \underline{k}_1 ,$$

if neither of the corresponding Fourier coefficients is negligible. If $\underline{k}_2$ and $\underline{k}_3$ are in the resolved range there is no problem. If either or both are out of the resolved range an error is made by neglecting this contribution. The assumption of streamwise periodicity leads to errors involving supergrid components. For the simulation of WH, this error (discussed in Section II) appears to be small. In flows where turbulent energy is simply cascading to higher wave numbers near the cutoff k_{max} (e.g., the decay of isotropic turbulence) the error involving subgrid components has the effect of turning off the cascade, resulting in a "pileup" of turbulent energy around $|\underline{k}| \lesssim k_{max}$ in the numerical simulation. Several schemes, based on a Fourier representation of the flow variables, have been proposed to model the effects of the unresolved flow field ($|\underline{k}| > k_{max}$) and, therefore, to allow the turbulent cascade process to pass energy on to the subgrid scales. See Leslie and Quarini [13] for a discussion of these possibilities. In wall-bounded shear flows, however, the production of turbulent energy may also occur near the cutoff wave number. This poses an additional challenge to the capabilities of a subgrid model. In this case the model must account for all the important unresolved instability mechanisms that break up the wall shear layer and produce turbulence.

Again, in the vortex simulation of a spot-like disturbance no explicit model for the subgrid scales has been used as yet. Fine-scales of turbulence are produced automatically in the simulation as the space curves representing the vortex filaments take on highly convoluted configurations under the continual action of local velocity gradients. There is a lower limit to scale of turbulence that can be resolved, however. From Eq. (5) we see that this scale is the core radius of the filaments, σ. (Note that the Fourier transform of ω contains the factor $\exp(-|\underline{k}|^2\sigma^2/4)$ so the maximum resolved wave number is $\sim 1/\sigma$.) In the vortex simulation discussed in this paper some of the computed space curves representing vortex filaments had become highly stretched and contorted by the end of the simulation. More computational node points would be required to accurately follow the subsequent motion of these filaments. A less costly, alternative way of proceeding with the simulation would be to remesh the vorticity field, perhaps by smoothing the filament curves and by changing the topology of some curves [5]. Of course this procedure would be required repeatedly at later times as more fine scales reappear. The method appears to have promise as a subgrid modeling technique. Fine scales are allowed to develop in a natural way, retaining some interplay between the large and small scales. The method seems analogous to computing with a fine mesh in Eulerian calculation and periodically filtering out the high wave-number components of the velocity or vorticity field [14].

In the channel-flow simulation, KM use an eddy viscosity model for the subgrid-scale stresses. This model has proved adequate in less demanding situations, for example, the decay of homogeneous turbulence [15]. The model is successful in these cases in spite of the fact that it does a relatively poor job of estimating the actual point-wise subgrid-scale stresses computed by filtering the results of a fine-mesh simulation of a homogeneous turbulent flow [16,17]. Such a capability is, of course, not a necessary requirement for a subgrid model, but it would be more encouraging if the correlation between the actual and modeled stresses was higher than the 30% level determined by McMillan and Ferziger [17]. A new model that does quite well in this respect has been proposed by Bardina et al. [18]. The idea is to model the subgrid stresses directly in terms of the small-scale component of the resolvable field.

In the simulation of KM, there is every indication the the eddy-viscosity model is quite satisfactory away from the wall ($y^+ \gtrsim 90$). The mean velocity and the three turbulence-intensity profiles agree well with experiment in that region. Near the wall, however, the mean velocity is low, as are the vertical velocity fluctuations. Apparently the eddy-viscosity model, being essentially dissipative, is not able to represent or account for important turbulent production processes near the wall that occur in the subgrid scales or in the smallest resolved scales where the interaction with the subgrid scales is of great importance. Evidence from the two studies concerning transition of a laminar boundary layer strongly suggests that that eventual breakup of the laminar layer involves structures that are $O(\delta^*)$ in the streamwise and spanwise directions. In KM's channel-flow simulation it is quite possible then that the lack of resolution in the x and z directions or other numerical effects or both inhibit the breakup of vorticity layers at the wall until they grow excessively thick. Indeed, in numerical experiments involving the filtering procedure, KM showed that substantial increases in turbulent shear stress near the wall are possible for short time intervals by modifying the treatment of the smallest resolved scales. It appears, therefore, that a subgrid model for the near-wall region should contain a destabilizing feature that would promote the breakup of vorticity layers rather than, or in addition to, the dissipative aspect of the eddy-viscosity model.

By using periodic boundary conditions in the horizontal directions, KM encountered errors due to unresolved supergrid components of turbulence. The magnitude of this error is not known, but is estimated to be small based on experimental two-point correlations.

IV. SUMMARY AND CONCLUSIONS

Numerical simulations of wall-bounded turbulent shear flows have confirmed the existence, and illuminated some of the characteristics, of cell-like substructures near the wall containing substantial vertical and streamwise vorticity. These structures appear to be responsible for or intimately related to many features of turbulent flows observed experimentally — intense vertical mixing, low-speed and high-speed streaks, and, hence, the streaky structures seen in photographs of dye-particle contaminants.

These structures were first revealed numerically in the simulation of turbulent channel flow. In this case, these structures seem to be related to strong $(-u'v)$ hotspots and, ultimately, the mean-velocity profile since planar averages of $u'v$

yield the turbulent shear-stress profile. In the case of a turbulent spot, span-averaging over one cell produces a streamwise profile qualitatively similar to ensemble-averaged measurements of experiment.

In the simulation of the vibrating ribbon transitional flow all the features of the corresponding experiment were reproduced including the development of secondary instabilities with length scales $O(\delta^*)$. At this writing, Wray is investigating the properties of a strong streamwise vortical structure that he has located off the peak plane in the simulation of WH.

The impetus to simulate more complex wall-bounded shear flows in reasonable computer times means a continued sparseness of grid points for the near-wall region. More sophisticated subgrid-scale models will be required to account for the fine-scale structures discussed above that appear to be so important to the turbulent shear stress and turbulent mixing, in general, near the wall. Many interesting possibilities exist. We speculate that successful models will enhance the breakup of wall-bounded vorticity layers while perhaps retaining some aspects of an eddy-viscosity model.

ACKNOWLEDGMENT

The author thanks J. Kim, P. Moin, and A. Wray for many helpful discussions.

REFERENCES

1. A. WRAY AND M. Y. HUSSAINI, "Numerical Experiments in Boundary-Layer Stability," AIAA Paper 80-0275, AIAA 18th Aerospace Sciences Meeting, Pasadena, Calif., 1980.
2. S. A. ORSZAG AND L. C. KELLS, J. Fluid Mech. 96 (1980), 159.
3. L. S. G. KOVASZNAY, H. KOMODA, AND B. R. VASUDEVA, "Detailed Flow Field in Transition," Proceedings of the 1962 Heat Transfer and Fluid Mechanics Institute, 1962, pp. 1-26.
4. A. LEONARD, "Vortex Simulation of Three-Dimensional, Spotlike Disturbances in a Laminar Boundary Layer," in Turbulent Shear Flows II, L. J. S. Bradbury et al., eds., Springer-Verlag, Berlin, 1980, pp. 67-77.
5. A. LEONARD, J. Comput. Phys. (1980).
6. I. WYGNANSKI, M. SOKOLOV, AND D. FRIEDMAN, J. Fluid Mech. 78 (1976), 785.
7. B. CANTWELL, D. COLES, AND P. DIMOTAKIS, J. Fluid Mech. 87 (1978), 641.
8. S. J. KLINE, W. C. REYNOLDS, F. A. SCHRAUB, AND P. W. RUNSTADLER, J. Fluid Mech. 30 (1967), 741.
9. M. GAD-EL-HAK, R. F. BLACKWELDER, AND J. J. RILEY, "A Visual Study of the Growth and Entrainment of Turbulent Spots," Proceedings of IUTAM Symposium on Transition, Stuttgart, Germany, 1979.
10. J. KIM AND P. MOIN, "Large Eddy Simulation of Turbulent Channel Flow — ILLIAC IV Calculation," Proceedings of AGARD Symposium on Turbulent Boundary Layer — Experiment, Theory, and Modelling, The Hague, Netherlands, Sept. 24-27, 1979.
11. P. MOIN, W. C. REYNOLDS, AND J. H. FERZIGER, "Large Eddy Simulation of Incompressible Channel Flow," Report No. TF-12, Mechanical Engineering Department, Stanford U., Stanford, Calif., 1978.
12. R. S. BRODKEY, J. M. WALLACE, AND H. ECKELMANN, J. Fluid Mech. 63 (1974), 209.
13. D. C. LESLIE AND G. L. QUARINI, J. Fluid Mech. 91 (1979), 65.
14. B. FORNBERG, J. Comput. Phys. 25 (1977), 1.
15. N. N. MANSOUR, P. MOIN, W. C. REYNOLDS, AND J. H. FERZIGER, "Improved Methods for Large Eddy Simulations of Turbulence," in Turbulent Shear Flows I, F. Durst et al., eds., Springer-Verlag, Berlin, 1979, pp. 386-401.
16. R. A. CLARK, J. H. FERZIGER AND W. C. REYNOLDS, J. Fluid Mech. 91 (1979), 92.
17. O. J. McMILLAN AND J. H. FERZIGER, "Direct Testing of Subgrid Scale Models," AIAA Paper 79-072, New Orleans, La., 1979.
18. J. BARDINA, J. H. FERZIGER AND W. C. REYNOLDS, "Improved Subgrid-Scale Models for Large-Eddy Simulation," AIAA Paper 80-1357, Snowmass, Colo., 1980.

EXPERIMENTAL METHODS IN TURBULENT STRUCTURE RESEARCH

J. Mathieu and G. Charnay

Laboratoire de Mécanique des Fluides (associé au CNRS)
de l'Ecole Centrale de Lyon
69130 Ecully, France

I. INTRODUCTION

For a long while statistical properties of turbulent flows were particularly emphasized. Such an approach was mainly supported by exhaustive measurements and theoretical developments related to isotropic turbulent fields. Indeed, due to the extreme complexity of the usual turbulent flows, such as jets, mixing zones, boundary layers, many specialists first investigated this more tractable case. Besides, they thought isotropic turbulence should exhibit the essential mechanisms acting in a turbulent field. Since the non-linearity of the basic equations give rise to interactions between turbulent structures of different sizes, the concept of an energy cascade was developed with the form of a "one way street" where big eddies generate smaller, eddies. It can also be viewed as a self interaction process associated with the stretching of the vortex lines. Hence, energy exchanges between turbulent structures principally held the attention of scientists. The statistical predominance of local interactions being supposed, a typical role is devoted to each range of the spectrum so that three zones should roughly be considered : large structures with long life-times correspond to small values of the wave number $|\vec{k}|$, within the inertial range scrambling effects widely prevail, whereas viscous effects are located within a limited range of the spectral space i.e. only for ($|\vec{k}| > |\vec{k}_o|$). Some limitations of this model are easy to see. In the absence of a mean velocity gradient, the turbulent motion cannot be fed, therefore the amount of energy of initially isotropic turbulence can only decrease ; the growing of the macroscale is not statistically a consequence of rearrangements of the turbulent eddies but the result of the action of the viscosity on the smallest structures, the energy of which is transformed into heat. To a large extent, these contributions play a decisive role in the understanding of the central mechanism. It could however be conjectured that bounded turbulent fields, subjected to mean velocity gradients or to driving forces, could display other typical behaviours.

Usual patterns of flows can be listed : boundary layers, wakes, mixing zones, jets, wall jets and so on. All these flows exhibit many significant differences when compared to isotropic turbulence. The interactions between the double velocity correlations and the mean velocity gradients, can feed the fluctuating motion. Moreover these fields are limited in extent and we can also expect the mean velocity gradient to play a dominant role in the orientation of the biggest turbulent structures. If so, some deterministic properties of the turbulence could be revealed whereas random characteristics should be confined in a more or less extended range of mean and small structures. From this point of view a gap could be bridged in investigating homogeneous fields associated with mean velocity gradients although they are not limited in extent. In such a configuration turbulent motion is fed from the mean flow and some classes of structures could be oriented by the mean velocity gradient. Besides, recent developments bring to light the complex nature of energy exchanges even in homogeneous flows ; the concept of a "one sided energy cascade" has to be fundamentally reexamined. In typical configurations, extended areas with "negative turbulent production" are revealed.

In usual patterns of flows which are limited in extent, the importance of the entrainement process connected with the contamination by turbulence of the engulfed packs of fluid is of course a typical and additional problem. So far, so complex a mechanism is not completely understood. The role of the big structures, the size of which can be compared to the overall thickness of the turbulent zones should be most important. On the whole, a lot of evidence can be collected showing the existence of coherent structures in patterns of flow subjected to particular experimental conditions. However it is questionable whether such ordered structures emerge from the surrounding flow in all cases. Conclusions of experts are different for many reasons. From measurements by probes located at fixed points in turbulent fields an Eulerian description of the flow can be extracted. Probably it is easier to capture the development of coherent structures through shadow pictures which give a Lagrangian description of the phenomenon. Accordingly the opinion of the specialists can be partially biased by both the method used to investigate turbulent fields and also by their initial point of view, i.e. if the existence of coherent structures is suspected, they will be more easily detected. Many configurations were studied ; on the whole, coherent structures are particularly apparent in mixing zones, and in jets. Investigations carried out in boundary layers display more complex patterns of flow ; turbulent structures in channel flows are even more intricate.

At this stage we have to make some general comments. In mixing zones, where ordered structures are particularly visible, the mean vorticity has a constant sign, hence turbulent motion is fed from a unique center, approximately located at the inflection point of the mean velocity profile. The growth of the mixing area is very fast, which supposes a very active mechanism for the entrainment of the surrounding fluid ; from this point of view a pairing process seems fairly convincing under any condition. For many reasons, mixing zones are only explored near the wind tunnel opening ; therefore upstream conditions should play an important role which is not always evoked. The shape of the mean velocity profile with an inflexion point should reinforce the role of two dimensional instabilities. The fluid behaviour is not so clear where round or plane jets are considered ; far from the exit in the quasi equilibrum region, we lack data on coherent structures. In these flow-patterns, the mean vorticity changes sign. The fluctuating motion is generated from two turbulent centers so that the axis of the jet should be a very complex region where large structures coming from two areas interact with each other ; on the average, the shear stress is zero but it is possible to identify the origin of the bulges through the sign of the transverse velocity and to recognize their special contributions to the mean shear stress. Even though overall properties of the turbulent field do not exhibit any peculiarity (the root mean square values of the three components of the velocity are almost equal), and a simple connection between the mean velocity gradient and the shear stress can be made in the zone next to the axis of the jet, in fact, such a situation results from a complex statistical balance.

Boundary layers are characterized by the presence of a wall which limits the transversal velocity fluctuations. The rate of entrainment is rather low. Disregarding the effect of the wall, the mean vorticity has a constant sign ; however vortex structures of opposite signs have to be introduced in order to account for the presence of the wall. To a large extent, the turbulence energy is produced very near the wall. The growth of large eddies can also be conjectured but the detection of organized structures is partially obliterated by secondary phenomena.

In channel flows the situation is most complex. At the entrance, boundary layers develop along each wall but downstream of the point of confluence the turbulent flow is limited in extent by the width of the channel so that all the turbulent scales should become independent of x. The development of large turbulent structures issued from a wall is hampered by those coming from the opposite side. This interference gives rise to a very intricate field in which some specific events can only be detected with difficulty.

From the available experimental evidence, it seems rather hard to imagine a unique mechanism acting in a similar way in any pattern of flow and generating coherent structures which would approximatively behave in the same manner. Any assesment of the current results is of course very difficult ; to begin with we will examine the available methods for investigation making comparisons in order to identify causes of disagreements. Then, starting with the simplest patterns of turbulent fields, such as homogeneous turbulence, it will be worthwhile to reexamine them with a view toward preparing the ground for an analysis of more complex and apparently more organized flows.

II. EXPERIMENTAL METHODS FOR DIAGNOSIS

In this section, the tools available for the detection and the description of coherent structures which appear in shear flows will be examined. First, the flow conditions which generate such structures are reviewed. Then the technical investigations and processing methods are discussed. Finally, two typical investigations including large structures are given as examples.

2.1. Coherent structures ; observations and experimental conditions

We do not intend to present a general statement of these problems (overall aspects were given by Laufer 1975, Davies and Yule 1975, Roshko 1976, Leibovich 1978, Comte-Bellot 1979). However, the existence of coherent structures in connection with the values of the Reynolds numbers of the flows under consideration will specially be reexamined herein. A general definition of coherent structures has been proposed by Yule : "Large eddies which may be repetitive in structure, remain coherent for significant downstream distances and which may be quasi deterministic with birth-life-death cycles" (Yule 1979). These structures, the existence of which is suspected by many authors in all practical shear flows, can be detected by means of the facilities described in section 2.2. In the last decade numerous patterns of flows have been investigated at rather low Reynolds numbers. Such investigations were encouraged by the relative capability of optical tools to detect organized structures within flows at low velocity, a fact with is illustrated in figure 1 for an impinging round jet. At Re_D = 1 300 the edges of eddies are well defined, but at Re_D = 50 000 the large structures are smeared by fine grained turbulence. Nevertheless other techniques, such as conditional sampling, provide information even at large values of the Reynolds number. Hence, investigations on coherent structures have been made under various experimental conditions. Investigated flows are listed in table I with their associated Reynolds numbers.

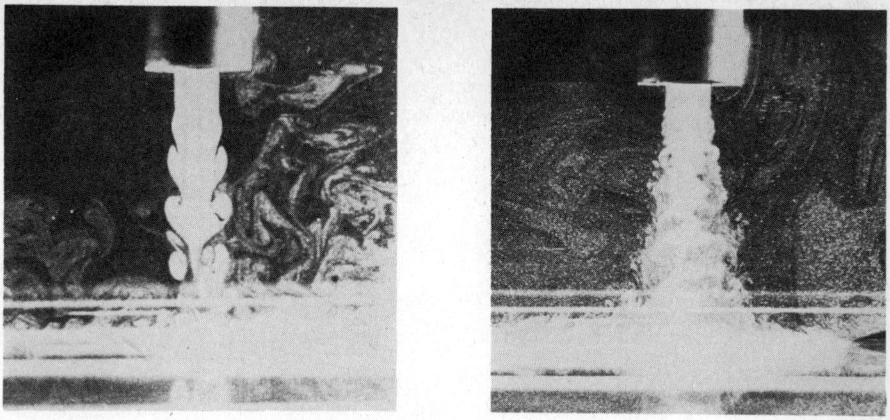

Fig. 1 : Impinging round jet. Visualization. a) Re_D = 1 300, b) Re_D = 50 000 (Zegadi R. 1980).

Kinds of flows	Reynolds numbers	Locations of the test sections with respect to the starting station
Jet (Round and Plane).	U_jD/ν = 300 to 300 000.	15 to 100 D
Wake (rod, sphere, disk, cone).	U_jD/ν = 500 to 100 000.	3 to 400 D
Mixing zone of jet with or without excitation.	U_jD/ν = 100 to 370 000.	0 to 16 D
Mixing zone behind a plate.	$U_je/\nu \sim$ 30 000 to 100 000.	0 to 10 e
Boundary layer (except for transition L.T).	$U_j\delta/\nu \sim$ 20 000 to 40 000.	20 to 100 δ

Table I - Investigated flows and associated Reynolds numbers.

(U_j = nozzle velocity or external velocity)

2.2. Techniques for detection and description of coherent structures

If the hot wire has first been used for the detection of large structures (Corrsin 1943, Townsend 1956), during the last decade the increasing interest in this research field is partially connected with the development of visualization methods. Observations at one or more spatial points of velocity and vorticity componants and of passive contaminants are not well adapted to emphasize the existence of coherent structures. On the other hand, with optical methods such organized structures are magnified, but no accurate information is collected except by using new techniques as digital image analysis. Even in this case however, the storage of numerical data is limited. It appears that the best approach to experimental investigation should be the simultaneous use of the two experimental methods. In the next paragraphs visualization methods and point measurements methods will be briefly presented.

a) Visualization methods.

Most of the papers devoted to coherent structures include photographs of the analyzed flow. From an overall observation many of these pictures seem to be very similar. In fact the visualizations habe been carried out with a wide variety of techniques and experimental facilities These visualization methods give an excellent description of the generation and development of coherent structures ; additionally they give quantitative data on eddy frequencies, time scales, length scales, trajectories and mean convection velocities.

Without tagging the flow, Schlieren spark and shadow spark can be used if the fluid density is not constant. But since the investigations of coherent structures have been especially concerned with subsonic flows, the density variations are mainly due to the mixing of two gases. For example it is possible to use pairs of gases such as Freon-Air (Bradshaw et al 1964), CO_2-Air (Crow & Champagne 1970), He-NO_2 (Brown & Roshko 1971 & 1974, Hernan & Jimenez 1979). Strong heating can also be considered (Hopfinger 1977. Schlieren and shadow methods introduce an integration throughout the flow and also emphasize fine details ; both techniques do not require a seeding process which is difficult to fit in.

In order to readily observe large patterns of flows, marking processes have successfully been introduced. For air flows, smoke or particle injection can be used ; whilst for water flows, dye, particle or Hydrogen Bubble injections (fig. 2) are preferred.

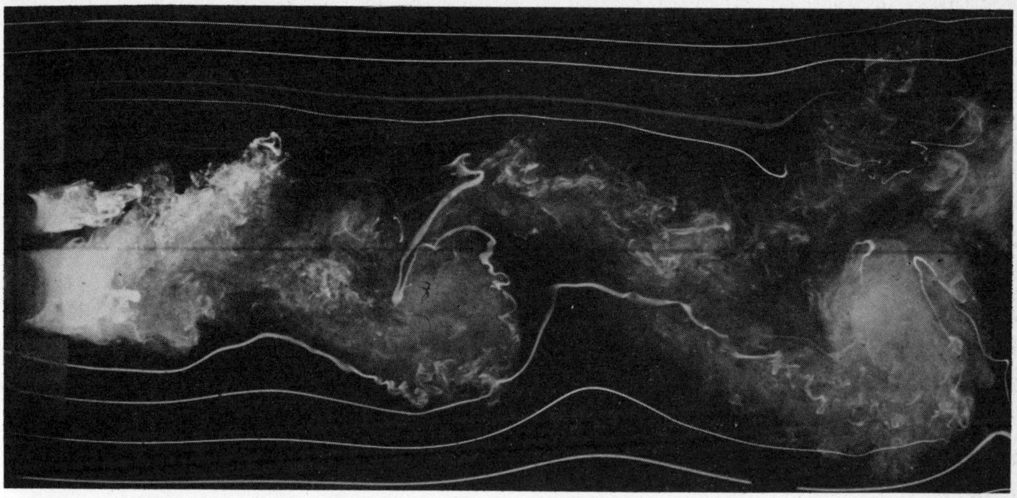

Fig. 2 : Visualization behind two rods of different diameters. Water. with Courtesy of R. Dumas, Institut de Mécanique Statistique de la Turbulence, Marseille, France.

Even though the basic technique is always the same, different methods are chosen where photography or localization of seeding are concerned. With a long time exposure no coherent structures appear ; however overall pictures of the flow are exhibited. For example through these photographies the spread rate of jets can be determined (Oster et al. 1977, Fiedler & Korschelt 1979). Ordered structures in shear flows may be revealed with a short time exposure. By means of a single photograph and by continuous tagging an overall view of all the structures within the flow comes to light {for instance the wake (Bevilaqua 1973), jets (Freman and Taularides 1979, Crow & Champagne 1971), mixing layer (Chandresea et al 1978), impinging jet (Yokobori et al 1979), buoyant shear flow (Hopfinger 1977)}. The results may be improved by stroboscopic light and for forced flows it is possible to observe the behaviour of vortices over a long time, the frequency of illumation being often equal to that of the forced excitation (Fiedler 1979, Perry & Lim 1978, Zaman & Hussain 1979). A more sophisticated tracking of structures with temporal information is given by high-speed cine film (Yule 1978, Rockwell et Knisley 1979, Praturi & Brodkey 1978, Cherdron et al 1978, Hernan & Jimenez 1979, Clark & Hussain 1979).

The tagging may be introduced within the entire flow or within a restricted area. Either a volume or a plane can be illuminated in the last case (plane convex lens or rotating mirror). In spite of some attractive aspects, the use of visualizations by marking is somewhat delicate. The time exposure must be in tight connection with both flow properties and time illumination, particularly where stroboscopic lights are used. More fundamental is the adequacy of tagging particles or bubbles which must have the flow velocity ; in order to limit the crossing trajectory such a condition is required. Through observations located in a plane it is difficult to reconstitute a motion which is typically three dimensional. Nevertheless more and more quantitative results are deduced from visualizations. For example the instantaneous velocity and vorticity is detected by a rotating mirror (Schon et al 1979) and the digital image analysis is capable of giving quantitative information (Hernan & Jimenez 1979).

b) Simultaneous records of signals.

Without processing, the records of signals delivered from a rake of sensors may be compared with a confined visualization. In general, the number of detection points is limited approximately to ten. The output signals can be made proportional to the velocity componants (Blackwelder & Eckelmann 1978), to the temperature (Sunyach & Mathieu 1971, Chen & Blackwelder 1978, Subramanian & Antonia 1979), and to both phenomena (Charnay & al 1976). The large coherent motion shows a similar signature, with back and front, on the signals of adjacent probes. From this type of observation one deduces that the size of organized structures is of the same order of magnitude as the width of the flow. Quantitative results may also be determined. For instance, the inclinations of the interface of thermal structures can be detected by staggering the four wires of a rake at an adequate angle to the wall so that the peak in the temperature cross-correlation curves should occur at zero time delay (Subramanian & Antonia 1979). Accounting for the rotation associated with the shear this angle should decrease with the age of the bulges (Blackwelder, 1970). Such an angle is of course a random variable.

c) Conditional sampling.

The conditional sampling technique is a sophisticated method for the detection and quantitative characterization of coherent structures. It can be performed with one or several probes and consists in the observation of a property only when some criterion is satisfied. The criteria can include properties relative to one point or more they can also be relative to a more or less extended volume. The difficulties of conditional sampling are not due to the detection of signals from an electronic processing (during the last decade improvements in digital treatment have been very efficient), but they are the consequences of the delicate choice of an adequate criterion.

Most of the studies using the conditional sampling methods have been developped by means of one probe located at a point M. This probe has two functions : at first it delivers a signal Q(t) which is proportional to a physical property ; afterwards a signal S(t) is extracted from Q(t) through an adequate processing. One of the simplest criterion is the inclusion or non inclusion of the point M in a turbulent peculiar region $\mathcal{D}$, hence the intermittency function at M(x) will be defined by :

$$I(\vec{x},t) = \begin{cases} 1 \text{ if } M(\vec{x}) \in \mathcal{D} \\ 0 \text{ if } M(\vec{x}) \notin D \end{cases}$$

From this definition the boundary of the area ($\mathcal{D}$) is supposed to be a geometric surface. More detailed criteria are obtained with another ponctual function P defined when M enters $\mathcal{D}$ (front) or leaves $\mathcal{D}$ (back). The distance of M to a frontier can also be considered as a criterion. The functions I, is obtained by means of S(t), (fig. 3). The zone averages in the region $\mathcal{D}$ are defined as (Kovasznay et al 1970)

$$\bar{Q}^i(\vec{x}) = \lim_{T\to\infty} \frac{1}{T\gamma} \int_{t_o}^{t_o + T} I(\vec{x},t)\; Q(\vec{x},t)\,dt$$

whereas the intermittency factor is given by

$$\gamma(\vec{x}) = \lim_{T\to\infty} \frac{1}{T} \int_{t_o}^{t_o + T} I(\vec{x},t)\,dt$$

The point average can be written

$$\hat{Q} = \lim_{T\to\infty} \frac{1}{T\gamma} \int_{t_o}^{t_o + T} P(\vec{x},t)\; Q(\vec{x},t)\,dt$$

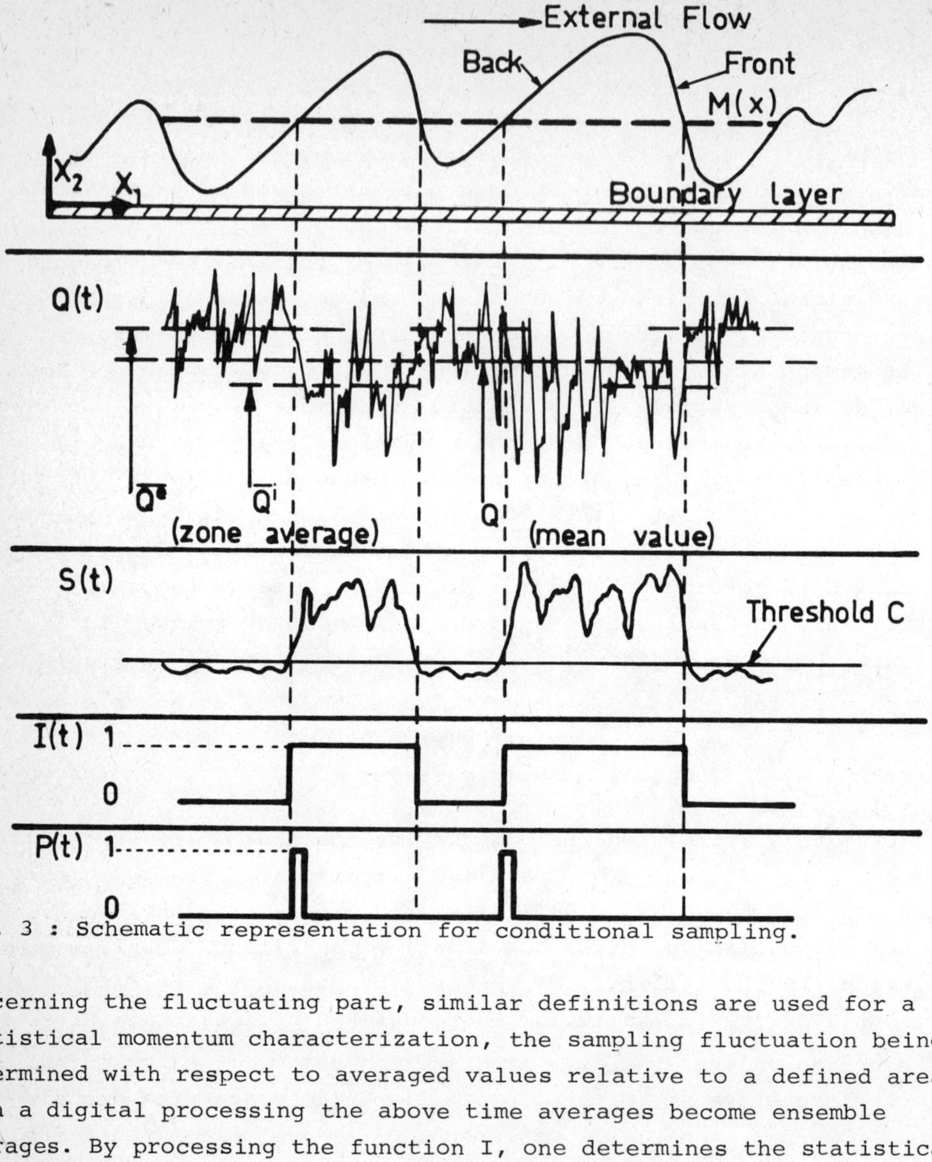

Fig. 3 : Schematic representation for conditional sampling.

Concerning the fluctuating part, similar definitions are used for a statistical momentum characterization, the sampling fluctuation being determined with respect to averaged values relative to a defined area. With a digital processing the above time averages become ensemble averages. By processing the function I, one determines the statistical properties of the frontier such as coherent structure rates, boundary slopes, average zone lengths and zone length probability functions (Kawall & Keffer 1978).

The main problem is to obtain from the detected signal S(t) the intermittency function. Usually S(t) is compared to an arbitrary threshold C. S(t) can be expressed by :

$$S(t) = I(t)S^i(t) + (1 - I(t))S^e(t)$$

where S^i and S^e are random signals whose amplitudes and frequency ranges are different. The detection seems possible if S^i and S^e are quite different. This question has extensively been considered in specific papers (Sunyach 1971, Bradshaw 1972, Kibens et al 1974, Hedley & Keffer 1974, Charnay & Schon 1976, Haverbeke et al 1979). The fitting of C is difficult. For the detection of the edge of turbulent flows, kinematic velocity signals are not very adequate because of the irrotational random motions induced by the frontiers outside the turbulent zones. The contrast between S^i and S^e may be made more significant by introducing a passive scalar in the region $\mathcal{D}$. As a rule, the characterization of two regions becomes possible with a thermal tagging (Sunyach & Mathieu 1969, Charnay 1974, Antonia et al 1975, Alcaraz & al 1976, Chen & Blackwelder 1978).

Conditional sampling with one probe has been used for the investigation of many typical flows. Examining undisturbed boundary layers Kovasznay et al (1970) propose for the external region a model of large three dimensional bulges which have both an upward and a rotating motion. This result is confirmed for boundary layers disturbed by an external random motion or by the existence of an external shear flow (G. Charnay et al 1976, Ahmad et al 1976). By using this technique a lot of flows have been considered such as jets (Chevray & Tutu 1978, Anderson & al 1979), wall jets (Alcaraz et al 1977), wakes (Bevilaqua & Lykoudis 1977, Fabris 1979, Kawall & Keffer 1979).

Using two probes, the first one for triggering and the second one for measuring, a spatial analysis can be made. In many cases the trigger point is fixed and the measuring point is variable. Where forced excitations are concerned the sampling process permits an observation over the time and provides information at the same frequency as that of the excitation (Oster & al 1977, Zaman & Hussain 1979). The sampling values are given at different stages of a period and are called phase averages. Without excitation, the triggering probe must detect physical events, for instance the passage of coherents structures. The criterion becomes more difficult and it is obtained through physical interpretation as, for example, the sign of the streamwise fluctuating velocity component (Yule 1978). With two probes, the problem of the kinematic interactions arises (Comte-Bellot et al 1971). In interpreting results, the random effect that is termed "phase scrambling appears and has extensively been considered by Yule (1979).

The quantitative behaviour of coherent structures, which are generally three dimensional, is improved by a sampling process defined at several points. In pratice, this basic procedure is rarely used owing to the great number of conditional criteria required. It is more convenient to compare the flow quantities and the characteristics of a physical model. Significant results have been obtained by Townsend (1979). This comparison is made through a fitting process. Such an analysis may be carried out through the correlations ; however the conditional aspect is introduced somewhat with the characteristics of physical models acting as an oriented sampling. One can consider the expression

$$A = \sum_{i}^{n} u(x,y)\, u_o\, (x,y) / \Sigma\, u_o^{\,2}$$

where the sum is taken over all the virtual positions of the triggering probes. u_o stands for a velocity induced by the conjectured model ; it plays a role similar to the intermittency function. The analytic expression of u_o (x,y) is imposed by a model which supposes the existence of a group of eddies, " the spacing of eddy centres in each group being uniform and varies considerably from one group to another". By means of a parametric study depending on wave-numbers it appears that the "sign of the energy transfer from the mean flow to the periodic eddy" changes at a fixed wave-number. A distinction is made between groups at late stages of their life history with negative growth rate and large structures which receive energy from the mean flow. The methods may be extended by a fitting of correlation functions weighted by the total intensity of the Reynolds stress within the effective volume which is swept by the array of sensors.

d) Correlation methods.

The following section presents only some examples of the correlation techniques applied to the analysis of coherent structures ; this type of processing has a very large range of applications. If we consider correlations of second order, the simplest information is obtained at the same point and at the same time. In the case of velocities they physically represent the normal and tangential shear stresses which are measured in all cases. However, for coherent structure detection and analysis it is more interesting to introduce spatial and temporal information.

With the correlations of two velocities detected at the same time and at two points, information is obtained on spatial coherence of the kinematic flow. In this way, Chandrusea et al 1978 (by means of spanwise correlations of longitudinal velocity components in a mixing layer), show that the spanwise scale decreases if the Reynolds number increases "which is consistent with the conventional development of threedimensionality". For turbulent near wakes Prasad and Gupta (1977), with streamwise, transverse and spanwise correlations, detect that the extent of eddies in the free shear layer is larger region than of those located in the central region.

Autocorrelation of the velocity measured at a single point with a time delay has successfully been used for the characterization of coherent structures. For a free mixing zone, Winant & Browand (1974), Dimotakis & Brown (1976), and Rockwell & Knisley (1979), observe a typical period of oscillation which is longer at the larger streamwise distance, implying occurence of vortex pairing ; this observation is not obliterated by the large variabilities of the autorcorrelations. When driving excitations contribute to the organization of the flow, autocorrelations do not detect variations in frequency. In free jets with periodic initial conditions the autocorrelation technique permits the detection of the Strouhal number corresponding to the maximum amplification (Fiedler & Korschelt 1979). Analysing velocity and pressure autocorrelations Yokobori andal 1979 drew similar conclusions for impinging jets.

Cross-correlations of two signals at different fixed points and for different time delays provide information on the overall properties of the flow. For instance, with two probes located on opposite sides of plan jet, it is possible to diagnose the puffing or flapping behaviour (Goldschmidt & Bradshaw 1973, Weir and Bradshaw 1975, Yule 1978). With a streamwise separtion of the probes we obtain the spatio-temporal correlations which give information on the life time of structures (Favre 1946, Sabot & Comte-Bellot 1976).

Correlations of higher orders obviously improve the description of coherent structures. Significant results for a turbulent boundary layer were obtained by Fulachierand al (1974) who measured spatio-temporal correlations of temperature signals detected at three points.

e) Spectral methods

Periodic or quasi periodic events are easily revealed by spectral methods. It is known that in jets, the power spectral distribution exhibits peaks.For instance Sunyach (1971) observes maxima in the spectral distribution which are more important for round jets than for plane jets, This property is confirmed by several authors, even for complex geometries such as annular jets (Chan & Ko 1978). As we move downstream the contribution of the low frequencies to kinematic power spectra becomes more and more important and whereas the spectral curve changes in shape. In the transitional region, one can detect two peaks which are separated by a factor of two in frequency and can be interpreted as the result of successive pairings (Yule 1978).

Many flows, such as wakes (Bevilagua & Lykoudis 1977, Budny et al 1979), mixing layer (Jimenez et al 1979), impinging flows (Rockwell & Knisley 1979) and duct flows with sudden expansions display similar characteristics. The discussions are essentially devoted to the survival distances of the large scales, the amalgamation process, the randomizing effect and the values of the Strouhal number. With functions linked to the wave number vector (Clark & Hussain 1979) additional information becomes also available ; they take the orientation of the large structures into account.

f) Probabilities.

The probability density functions are successfully used for convenient descriptions of coherent structures. This notion provides information which is more or less detailed according to the properties analyzed. Employing concentration probes for experiments conducted in gases and chemical-reactions, and in water, it is possible to infer the degree of mixedness. For example Bernal er al (1979) obtained a direct measure of the mixedness in a mixture of Helium and Nitrogen. This procedure gives a determination of the molecular state which depends on the Schmidt number. Nevertheless it is possible to locate the station from "which the mixedness inside the mixing layer is considerably increased by the action of small-scale and three-dimensional motions".

For velocity signals, the analysis is related to the amplitude. It is well known that in regions affected by coherent structures, which introduce internal intermittency, the probability density distribution is not gaussian. In boundary layers or channel flows this fact has been observed either in the external region (Gupta & and al 1972, Dumas et al 1972) or in the wall region (Comte-Bellot 1965, Kreplin et al 1979). The information is given in the form of histograms, flatness factors S, and skewness factors T. The large values of S or T may be estimated by means of a synthetic density distribution built with two independent gaussian distributions mixed through an intermittency factor (Charnay et al 1976). Such a method also provides information on the choice of the threshold used in conditional sampling (Sunyach 1971) or in imagine analysis (Hernan & Jimenez 1979).

To bring out further details on the coherent structures, two components of the velocity which produce the Reynolds stress are often considered. The method called the "u,v plane quadrant analysis" permits an accurate evaluation of the contribution of the two components u and v to correlation term : $\overline{uv}$, accounting for the sign of each component. This technique has extensively been applied for the observation of burst and sweep events in the wall region (Willmarth & Lu 1972, Wallace & al 1972, Sabot & Comte-Bellot 1976, Wallace et al 1977, Obermeier 1979, Blackwelder & Eckelmann 1979). The same approach may be extended to many pairs of properties such as instantaneous velocity and wall shear stress fluctuations and the results can be presented with joint probability density contours(Wallace and Brodkey 1977, Rajagopalan & Antonia 1979).

g) Other approachs

The above examples of experimental tools for coherent structures are not exhaustive. Many might be imagined either for the physical detection or for the processing of data.

The knowledge of the vorticity is of great importance for interpretation and for comparison with visual studies. However, vorticity measurements are difficult to perform. Nevertheless Kastrinakis et al (1978), have obtained significant results with hot wire probes.

With the recent development of digital processings, numerous computer algorithms are available to recognize coherent structures. In the field of conditional sampling we recall the patterns or signature recognition techniques (Wallace et al 1977, Comte-Bellot et al 1979). But simpler processings would be useful such as the scaling methods (Kim et al 1971, Rao 1971).

2.3 Two typical results

In spite of a great number of studies devoted to coherent structures, all experimental conditions have not been investigated. The flow conditions have often a great influence. For a boundary layer which develops along a flate plate, external disturbances play an important role in whatever form they may be superimposed i.e. random fluctuations (Charnay et al 1976) or shear stresses (Ahmad 1976). A more striking result is shown in figure 4 for an equilibrum boundary layer with a positive pressure gradient. The distribution of intermittency factors shows that the identations of the external interface extends up to the wall in the case of strong adverse pressure gradients.

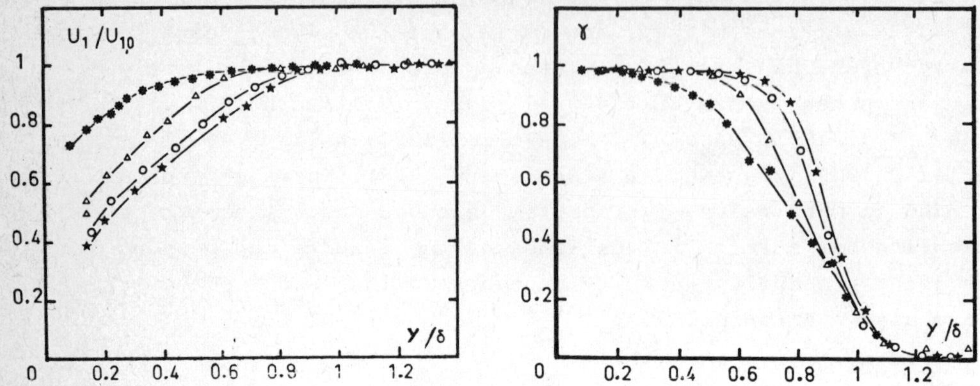

Fig. 4 : Boundary layer with negative pressure gradient (Charnay & Mélinand 1980). a) Mean velocity ; b) Intermittency factor γ.

Concerning the entrainment process convenient information is given by the transfer terms weighted by an intermittency factor. These equations have been derived by Dopazo (1977) and extended to two-phase flows by Lance and al (1979). Figure 5 presents some terms of mass entrainment and x momentum, determined in a boundary layer and a wall jet (Muszynski 1979). Estimation of these transfer terms by measurements carried out inside or outside the frontier are consistent with one another. These results are also in agreement with those of Dopazo & O'Brien (1977) in a wake.

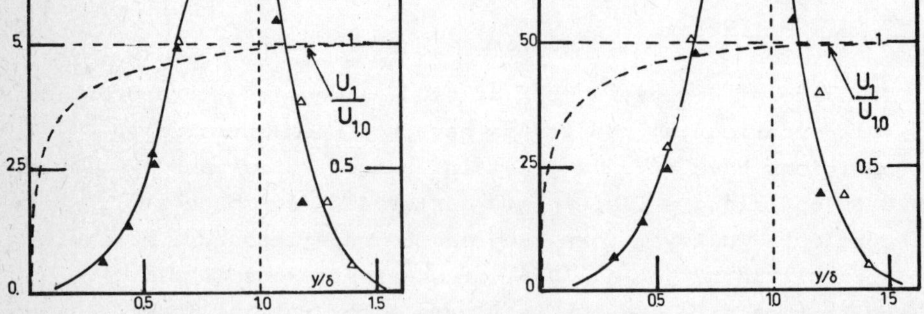

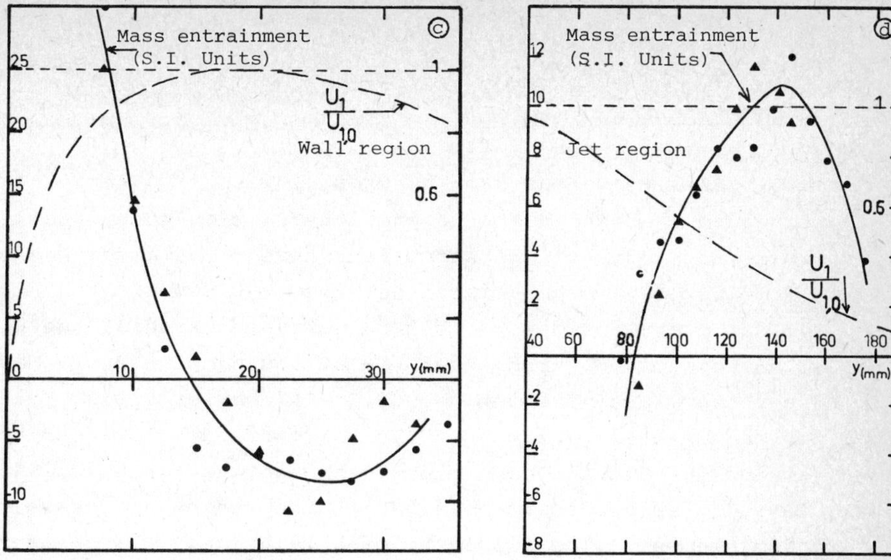

Fig. 5 : Terms of entrainment in conditioned equations (Muszynski 1979). a, c & d : mass entrainment (Δ : $\partial\gamma U^i/\partial x + \partial\gamma V^i/\partial y$; ▲ : $\partial(1-\gamma)U^e/\partial x + \partial(1-\gamma)V^e/\partial y$) ; b : x momentum entrainment ; a & b : boundary layer (Charnay 1974) ; c & d : wall jet (Alcaraz et al 1977).

III. ANALYSIS OF SHEAR FLOW PATTERNS

3.1 Historical insight

In recent years there has been a strong interest in large scale coherent structures. As shown in the previous section novel statistical techniques and significant observations have stimulated this trend so that in the last decade many meetings have been devoted to this problem. Typical contributions to this subjets presented at the Colloquim on Coherent structures in Turbulence held at Southampton 26-29 March 1974 have been excellently reviewed by P.O.A.L. Davies & A.J. Yule (1975).

However the existence of more or less organized structures had been conjectured in several forms many years ago. As early as 1948 Dryden wrote : "It is necessary to separate the random processes from the non-random element". The discovery of a relatively sharp interface between turbulent and non-turbulent fluid (Corrsin 1943) and the subsequent study (Corrsin & Kistler 1955) of its behaviour give the first information about both the existence and the role of large structures in a turbulent shear layer. In his book entitled "The structure of Turbulent Shear Flow 1956" Townsend conjectured that it was physically realistic to regard shear flow turbulence as an assembly of repetitive ordered structures which interact strongly with each other and coalesce as they move downstream ; this idea is supported by straightforward computations. Townsend also brought to light the role of the straining process imposed by the mean velocity gradient.

In the sixties, the attention of experts was drawn to the behaviour of the fine scale structures, the dynamical properties of which were suspected to be independent of the main flow except for the amount of energy which is fixed by the interaction of larger structures with the mean field. Irrespective of its complexity, a turbulent flow was expected to exhibit universal properties within the range of fine scale structures. Anyway, the scrambled effect of the inertial range should finally generate isotropic turbulence.

Several studies were carried out on the spatial distribution of these dissipative eddies, disregarding somewhat the role of large structures in these locations. Due to the improvement of the hot wire technique and of the associated electronic instrumention it was possible to study the spectral distribution of the turbulent kinetic energy with particular emphasis on the evaluation of the various rate terms occuring in the turbulent energy equation. This information is of crucial interest for modelling methods such as they have been introduced by Bradshaw et al (1967), and later extended by several authors (Hanjalic & Launder 1972, Jeandel 1976, Jeandel et al 1978, Cambon et al 1980).

In the past ten years at least three important observations were reported that are directly connected with the problem under consideration. It was observed, with rather simple optical techniques, that turbulent flows of simple geometry are not so chaotic as was previously presumed (Kline & Runstadler 1959, Corino & Brodkey 1969, Brown & Roshko 1971). An observable chain of events was detected which occur randomly but a statistical mean period can be defined. Sunyach (1971) looked for a more accurate definition of the moving interface previously detected by Corrsin, at first using classical hot-wire anemometry. A systematic study was carried out in order to reduce the role played by arbitrary threshold in the detection of the moving frontier ; comparisons were made between kinematic and thermal measurements (Sunyach & Mathieu 1969). Moreover an array of hot-wires located in the mixing zone of a plane jet revealed transversal motions of large amplitude. It was also found that energy spectra of two dimensional and axisymmetric jets exhibited peaks located at low wave numbers. The entrainment process was analyzed taking into account previous works (Corrsin & Kistler 1955, Fielder & HEAD 1966) and other reports published approximatively at the same time (Kibens 1969, Kovasznay et al 1970, Blackwelder 1970).

As pointed out by Laufer (1975), the interpretation of two point correlation measurements in terms of spatially coherent structures can be quite misleading. These structures have a history, thus observations at one or two spatially fixed points include realizations of a large number of structures at various stages of their"life history". More relevant information can be deduced from space-time correlations introduced by Favre (1949) many years ago. In a lecture series, given at Von Karman Institute, J. Mathieu (1971) pointed out the cumulative role of the mean velocity gradient to fix the correlation coefficient ; accounting for this time delay he physically explained that the shear stress may be different from zero at a point where the mean velocity gradient is equal to zero, a fact which invalidates a first gradient approximation, particularly in asymmetrical shear flows. Modelling methods developped by Hanjalic & Launder (1972) and Jeandel (1976) also predicted such a situation.

It certainly was the Stanford research group that strongly suspected the existence of coherent structures and worked out measurements at a few points fixed in space in order to obtain detailed information about the behaviour of coherent patterns (Kline et al 1967, Kim et al 1971). Visual observations were also performed. In particular situations, this work confirmed the existence of certain significant events in the flow occuring with a statistical mean period. The group proposed that the usual notion of a random velocity field superimposed on a mean flow should be reexamined (Offen & Kline 1973). These observations were corroborated by Winant & Browand (1974) who gave a detailed description of the generation and development of the quasi-ordered structures in a high-velocity water flow mixed with a lower velocity stream behind a thin splitter plate. Conditional sampling was performed and it was shown that significant Reynolds stress production was associated with the pairing interaction. Roshko gave an overall view of these problems in a paper entitled "Structure of Turbulent Shear flow, A New Look" (1976). Probably, one of the most striking remarks is that "such flows with organized large structures correspond in every way to what have been classicaly called turbulence mixing layers. Their mean velocity fields, the Reynolds shear stress calculated from them and the corresponding growth rates agree with what has already known about these flows". It clearly appears that ordered structures pointed out by several authors by using more or less adequate experimental methods exist under circumstances which are not clearly stated. In an important paper on the effect of free stream turbulence on large structures in turbulent mixing layers Chandrsuda et al (1978) reexamined the experimental features emphasized by Bradshaw et al (1967), Wygnanski & Fielder (1970), Crow and Champagne (1971), Brown and Roshko (1974), Winant & Browand (1974) Dimotakis & Brown (1976), ... and "seeked to reconcile the received view of classical fully three dimensional turbulence with the more recent observations and to show that the two dimensional Brown-Roshko structure is rare". It had already been suggested from Wille's experiments that perhaps the flow reached a self-preserving fully turbulent state independent of the Reynolds number after a considerable time (Wille 1963).

The received view of mixing layer transition was that the mixing layer became unstable at a very low value of the Reynolds number based on local thikness, to two-dimensional or axisymmetric disturbances. However, the two dimensional "vortex roll" disturbances which are the eigenmodes of instability of a laminar mixing layer could become persistent under typical circumstances. These disturbances could grow and pair once, twice or even more often before appreciable three dimensionality appeared. Initial and boundary conditions could play a dominant role, for example the upstream turbulence level of the two streams could be predominant. It was conjectured that a fine grained turbulence acting as an additional viscosity was capable of inhibiting the breakup of two dimensional instabilities. Bradshaw showed that the vortex pairing process is sensitive to small three dimensional disturbances in the early stages, i.e. where the vortex roll amplitude is small. The characteristic parameters of the turbulent field such as turbulent kinetic energy and Reynolds shear stress are not significantly altered by the existence of vortex rolls. Bradshaw found that the maximum shear stress could reach a value as high as 1.5 times that in the fully turbulent flow. To a large extent the three-dimensional character of the flow was supported by the evolution of the spanwise correlations of longitudinal component velocity fluctuations in mixing layers in "still air" (fig. 6)[(x)]. Bradshaw's observations led him to conclude that vortex pairing in the presence of naturally occuring three dimensional disturbances led to a double helix pattern before turning into turbulence.

Bradshaw concluded that : "Experiments in plane mixing layers and the foregoing review of recent works suggest that the two dimensional vortex-roll disturbances arise in the transition region and would appear only in plane mixing layers with laminar boundary layers at the nozzle exit and low turbulence in the external flow". The problem was open to debate so that in the Congress which was held in London (1979) on Turbulent shear flows, several papers were devoted to this topic (Bernal & al, Jimenez & al, Clark & Hussain ...). Roshko and co-authors remarked that during the transition "the mixedness inside the mixing layer is considerably increased by the action of small scales which are three dimensional motions which themselves result from an earlier instability that produces organized secondary motions with components of vorticity in the streamwise direction". References were made to Bradshaw's experiments and it was commented that. "This is an important point in the discussion as to whether the final equilibrium state of a mixing layer is still dominated by the large, spanwise organized structures".

(x) It could also be argued that in previously mentioned papers (Sunyach 1971) the mixing zone also develops in still air, nevertheless the existence of two dimensional structures seems persistent.

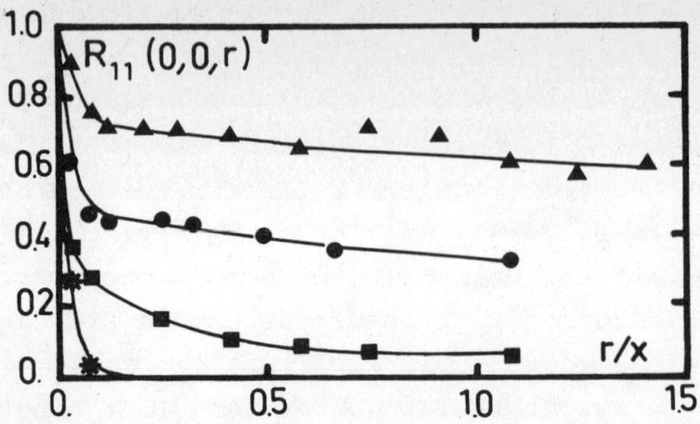

Fig. 6 : Spanwise correlation of longitudinal component velocity fluctuations in mixing layer in still air (Chandrsuda et al 1978).

The above discussion emphasizes the importance of large eddies in free turbulent flows. In what follows it will also prove illuminating to briefly examine the state of this problem where turbulent boundary layers are concerned. At the present time it is admitted that typical events significantly contribute to the development of boundary layers. The generaly accepted view of a boundary layer divided into two parts somewhat encouraged workers to focus their attention on one of these two regions. In the inner region turbulence production and dissipation are both large and of similar magnitude so that this part of the flow could be considered to be approximately in a state of equilibrium ; from the viewpoint taken in this colloquim this inner region is principally a domain where successive events termed "bursts" and "sweeps" occur. The turbulent structures which are observed in the wall region $u^{+} < 100$ were extensively analyzed by Kline et al (1967), Corino & Brodkey (1969), Praturi & Brodkey (1977), Eckelmann et all (1977). The phenomenon of bursting is associated with most of the turbulent energy production and may be due to instantaneously counter rotation axial vortex pairs. The detection of this sequence of events is obtained by quadrant probability analysis, conditional sampling with pattern recognition and visualisation. The large structures in the outer region of boundary layers are experimentally easier to characterize than those in the wall region. They generate the intermittent phenomenon which particularly affects the free edge of the boundary layer. A good deal of study has been devoted to the description of the external bulges (Fielder & Head 1966, Sunyach 1971, Antonia 1972, Hedley & Keffer 1974), their motions appear to be typically three dimensional. The introduction of external disturbances notably modify both the shapes and the sizes of the bulges in connection with the entrainment. As a consequence of the engulfment of external isotropic turbulence through the porous region of the bulges, the values of the correlation coefficients decrease, the turbulence which is generated by the interaction of the shear stress with the velocity field is immediately marked by the gradients whereas the injected turbulence has to be subjected to a velocity gradient during some time before reaching a more or less ordered situation. An extensive study of the mechanisms occuring in such patterns of flow has been carried out by Charnay (1974).

We have thus arrived at the situation where both events in the wall region and motions in the outer area have been reasonably well documented. What is required to close the picture, is to establish the overall mechanisms which connect these two phenomena on the one hand it may be surmised that the kinematic behaviour of the whole boundary layer is dominated by the bursting effects located next to the wall ; on the other it may be conjectured that fundamental instabilities in connection with the general shape of the velocity profile are predominant. This alternative will be discussed after we have examined the role played by some driving effects in the development of turbulent fields.

3.2 The role played by driving effects on the development of turbulence structures

Except for the isotropic turbulence which is only subjected to its own influence, most of the flow patterns are influenced by external forces the role of which can be more or less predominant. Disregarding initial conditions which under typical circumstances could generate turbulent structures directly marked by transitional modes, we may wonder whether the final state of a turbulent flow is still dominated by large organized structres. With magnetic forces, dissipative structures can be so oriented as to transform the overall turbulent field into a quasi two dimensional field. Coriolis or buoyancy forces can also influence the development of turbulent fields. Bertoglio (1980) has recently considered a homogeneous turbulent field associated with a mean velocity gradient ; this turbulence is subjected to a rotating motion about the axis OX_1. The double velocity correlations at a given point can increase or decrease according to the sign of the ratio Ω/ω (the rotation of the frame is termed Ω whereas the rotation induced by the shear is called ω, (fig. 7))

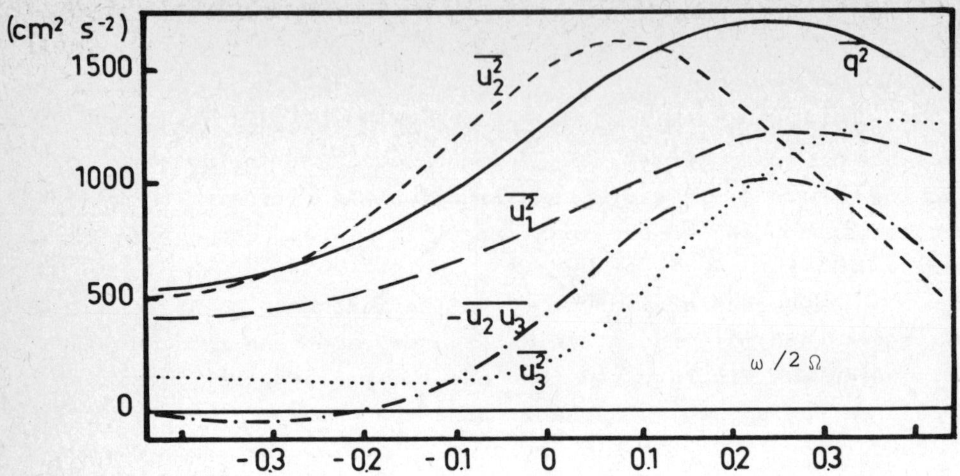

Fig. 7 : Influence of the rotation on the Reynolds stress tensor (Bertoglio 1980).

The driving forces generated by the rotation of the frame can be strong enough to cancel randomizing effects, so that a waving motion may prevail. Inspection of the spectrum shows that the Coriolis effect decreases as the wave number increases. Accordingly the importance of random motions increases with wave number, a fact which can also be extended to the case of turbulent fields subjected to the influence of buoyancy forces (Hopfinger 1977, Gence et al 1978). The action of a mean velocity gradient on a turbulent field is perhaps not so effective as the previous effects but it is the only mechanism capable of orienting turbulent structures during all the final stage of the flow development through the typical role of the straining process (Mathieu 1971). Accordingly let us now discuss in detail some typical situations which support this point of view.

Disregarding the simple case where a turbulent field is subjected to pure straining we consider at first the more usual situation where the turbulence is subjected to a shear flow. For the sake of simplification we shall examine homogeneous fields such as those treated by Burgers and Mitchner (1953) and Craya (1958). Spectral analysis is well adapted to carry out computations even though such an approach gives rise to much criticisms. Where homogeneous flows are concerned the role played by the mean fied can be considered separately since, among other reasons, the turbulent motion does not significantly affect the mean field.

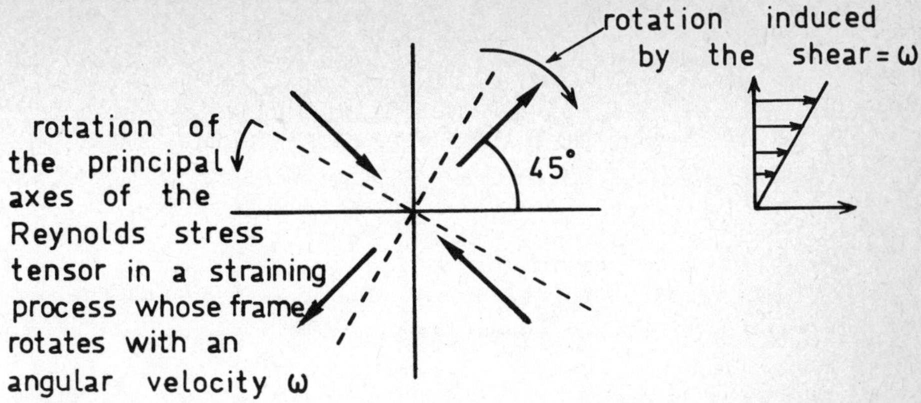

Fig. 8 : The rotation of principal axes of the Reynolds stress tensor as opposite to the mean rotation associated with the shear.

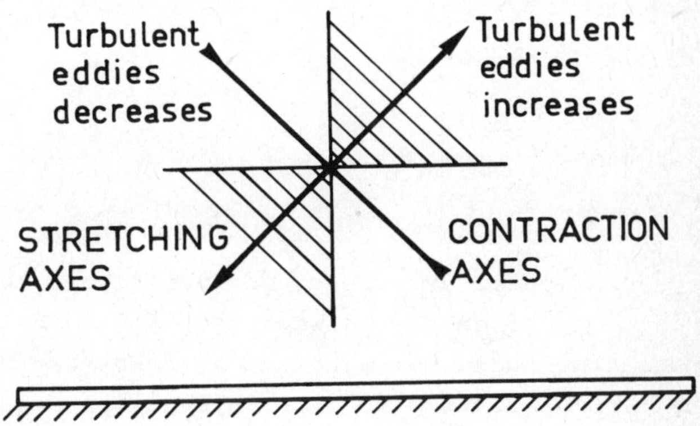

Fig. 9 : Direction of amplification and of contraction of the turbulent eddies in a wall flow.

In all turbulent flows where shear effects are dominant straining and rotating processes are mingled, so that the behaviour of turbulent structures is more difficult to identify than in the case of a pure strain where the principal axis of the mean strain and those of the Reynolds shear stress coincide. However we can conjecture that the development of larges eddies is subjected to a selection effect. Eddies, the axis of which coincide with that of the stretching for the main flow, will be tracked and strongly amplified. In a paper entitled : "On the application of successive plane strains to grid generated turbulence", Gence & Mathieu (1979) experimentally examined how much the turbulence production is dependent on the relative position of the principal axis of the strain with respect to those of the Reynolds stress tensor. In the case under consideration the angle α which characterizes the relative position of these axes can be adjusted experimentally. After some algebraic manipulation the term describing the exchange of energy can be written immediately after the sudden change in position of the principal axes of the strain :

$$\underline{-} \underline{D \overline{q^2} (b_{22} - b_{33})} (2 \cos^2 \alpha - 1)$$

where $\overline{q^2}$ and b_{ij} stand for turbulent kinetic energy and the non-dimensional deviatoric tensor respectively.

$$b_{ij} = \frac{u_i u_j}{\overline{q^2}} - \frac{\delta ij}{3}$$

The gradient of the velocity field is represented in cartesian coordinates by the matrix

$$\begin{vmatrix} O & O & O \\ O & D & O \\ O & O & -D \end{vmatrix}$$

In this fundamental experiment the underlined terms are positive so that the exchange of energy between mean and turbulent fields can be adjusted through the angle α. Pathological cases can be singled out : for $\alpha > \frac{\Pi}{4}$ the term accounting for the interaction between the mean and the fluctuating field, which is generally called "production term", becomes negative ; in this case the fluctuating motion gives energy to the mean motion. The ability to apply a second strain independent of the direction of the first one is essential to this sort of experiment. However the work gives information about the complex situation encountered in usual shear flows such as boundary layers. In that case a double mechanism acts, e.g. a plane strain combined with a rotation. When both a straining process, and a rotating motion $\vec{\omega}$ are simultaneously applied to a turbulent field the principal axes of the Reynolds tensor turn in the opposite direction ($- \vec{\omega}$) so that their final position is determined by a delicate balance between two effects : a rotation induced by the shear flow which tends to rotate these axes with the sign of $\vec{\omega}$, and a counter rotating effect. An equilibrium situation takes place which does not correspond to a maximum rate of turbulence production. Positive and negative areas are brought to light, as shown in figure 8. Rolling structures are not subjected to stretching effects but transversal disturbances should generate three dimensional motions and permit the development of "hairpins" in a selected way. The shape of these hairpins was thoroughly described. Their growth is linked to the stretching mechanism but also to the "velocity" field induced by very close hairpins. If we accept that hairpin growth is essentially due to induced velocities, it is reasonable to ask what mecahanisms can limit their growth. At first we can remark that the hairpin can grow only as long as it remains within a pre-existing region of shear. Besides, the zone of stretching is limited in extent so that each hairpin should stay during a limited time in a geometrical position favourable to its growth. Head & Bandyopadhyay (1978) illustrated the behaviour of hairpins in zero shear and also recognized the significance of a 45 % angle to the wall ; during the first stages of their development the role of the viscosity on turbulent structures is very weak provided the Reynolds number is high enough. It can be conjectured that hairpins whose limbs initially make an angle greater than 45 % stay for a longer time in the zone of stretching ; accordingly they are the most amplified (fig. 9). The role played by a shear was also emphasized by Cambon (1979) for a homogeneous field (x), (fig. 10). Initial conditions are introduced in a spectral form which exhibits a peak located in the inertial range. The amount of energy located within the peak is conveyed to the other regions of the spectrum but a large amount goes to the range of the smaller wave numbers. At this stage it is reasonable to admit that turbulent structures are selected by shear effects ; the rate of strain in a shear flow being similar everywhere, in the overall behaviour of the turbulent structures should be dominated by such a mechanism. Therefore a selected amplification takes place scrambled by random motions.

The spreading of turbulent flows is a typical property of non-homogeneous fields. It is clear that the scale of the contortions which give rise to the intermittency and to the engulfment is an order of magnitude larger than the scale of the turbulence and is comparable with the width of the whole flow. Where quasi two-dimensional fields are concerned, such as in Browand's experiments, the interaction processes are given by pairing modes which can be easily reproduced by

---

(x) Cambon starts from Craya's equation. The closure of the equation is obtained by an assumption similar to that called E.D.Q.N.M. The role of the mean velocity gradient is introduced at all the stages of this computation except where the departure from the quasi-normal law through a damping term is taken into account.

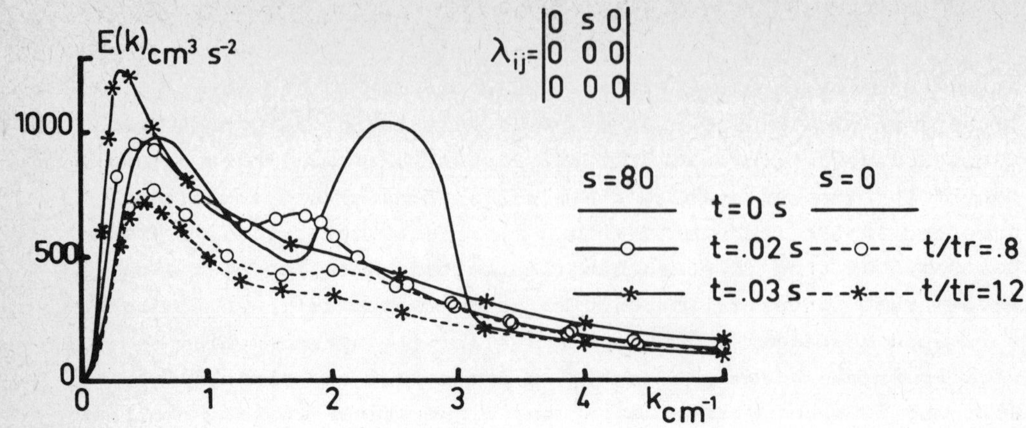

Fig. 10 : Evolution of a power spectrum: comparison without shear ($s = 0$) and with shear ($s = 80\ s^{-1}$), (Cambon 1979).

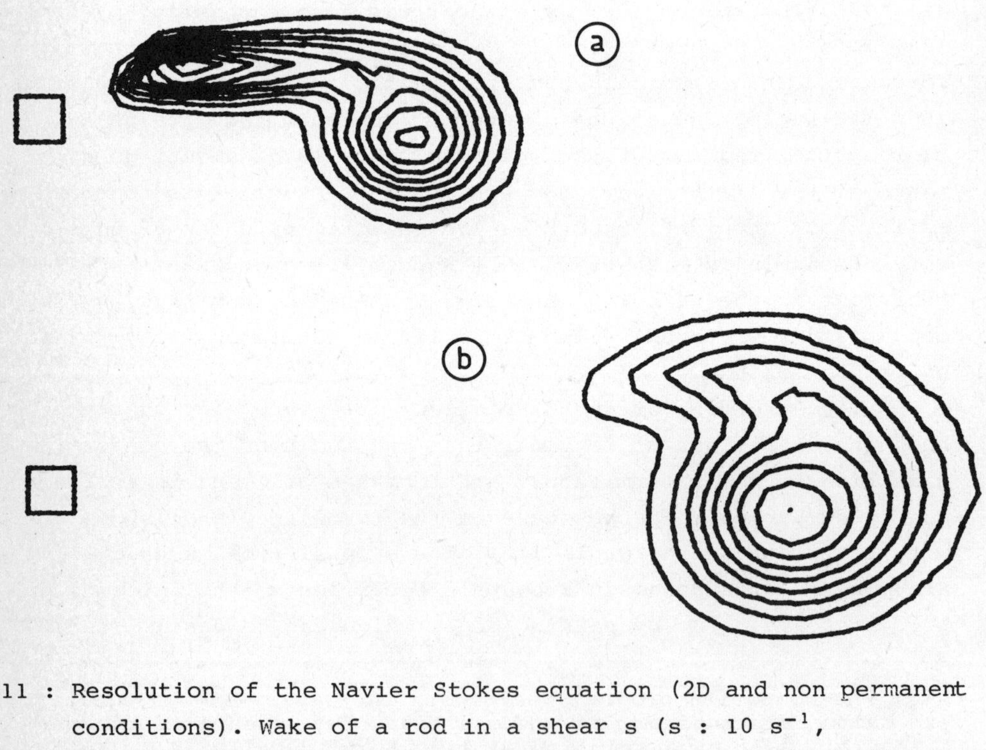

Fig. 11 : Resolution of the Navier Stokes equation (2D and non permanent conditions). Wake of a rod in a shear s (s : $10\ s^{-1}$, $Re_D = 25$, $U_o = 1$ m/s). a) $t = 0{,}02$ s ; b) $t = 0{,}185$ (Cellard 1980).

computations (xx). The pairing of two vorticies is shown in figure 11 ; it is rigourously similar to that given by Browand and Weidman (1973). Even though ordered structures can be detected in boundary layers, similar mechanisms should only play a very small role in the spreading of this flow. However the growth of the turbulent zone has to be explained. Except for the contamination of irrotational zones, temporarily imbedded in the turbulent fluid, a nibbling process does not seem adequate. All the experimental data display the role of the moving frontiers, however in the case of boundary layers these motions of large amplitude do not seem very effective for the entrainment. A very limited pairing process cannot be excluded.In any case no simple relation has been established between the degree of identation of the frontier and engulfment.

(xx) The computing program dealt with the Navier-stokes equation. In this finite element program the two dependent variable were ψ and ω. Computation was carried out by Cellard at the E.C.L. Laboratory.

3.3. Towards a splitting of the turbulent motion

Direct simulations of the Navier-Stokes equation have been extensively developped during the last decade. The mathematical properties of these equations are not precisely known, and one can wonder whether the results, which are necessarily obtained through numerical schemes which introduce their own instabilities, are conclusive. Moreover, to resolve the smallest scales of turbulence in a grid based calculation, the mesh size has to be smaller than the dissipation scale which is of the order of the Kolmogorov microscale $\eta = (\nu^3 / \varepsilon)^{1/4}$. Accordingly, with available computers it has been shown that a full simulation is only possible for low turbulent Reynolds numbers. If we admit that ordered structures exist in turbulent flows subjected to a mean velocity gradient it seems more convenient to simulate the more or less deterministic structures directly leaving the smaller structures being treated in a statistical manner.

Large scale structures have conventionally been defined in several ways that disregard the properties of the basic equations (x). Deadorff (1971) applied approximately the method involving averaging over the surface of grid boxes the sizes of which determine the scale of the large structures. The "filter function" approach of Leonard (1974) can be linked more directly to the spectral approach. It has been thoroughly applied to isotropic fields by Reynolds and his co-workers. This approach should be more applicable where driving forces come into play. However this splitting has been used in recent work to study isotropic turbulence, hence some information has come to light.

Leonard separates the velocity field $u_i(x,t)$ into a grid scale component ($\overline{u_i}$ (x,t) and a subgrid component $u_i'(x,t)$ by a filtering operation

$$\overline{u}_i(\vec{x},t) = \int G(\vec{x},\vec{x}')\, u_i(\vec{x}',t)\, dx'$$

$$u_i = \overline{u}_i + u'_i$$

The overbar precisely denotes the filtering operation which is not to confuse with a statistical averaging indicated by angular brackets. A filtered quantity has to be considered as a typical event which occurs

---

(x) For instance the uniqueness of the solution of Navier-Stokes equation is not educed in this kind of approach.

under any circumtances. When a set of realizations are considered, it is possible to introduce statistical quantities by averaging. This event, when it happens, develops in an environment whose properties have to be considered in connection with that principal event which should introduce conditional sampling supplemented by an averaging operation.
When a turbulent field is supported by a mean field, such as in Craya's equation, this basic field has to be considered as an event, the probability of which is equal to unity (it is represented by a Dirac located at the origin of the axes in spectral space). whereas the power spectrum of the fluctuating motion has to start from zero. From this viewpoint the usual properties of turbulent fields, conventionnaly split into mean and turbulent parts.

$$u = \bar{u} + u' \quad ; \text{ where } \langle u' \rangle = 0,$$

can be found by starting from a more general approach.
The filter G is independent of time and $\int G(x)\,dx = 1$.
If the physical properties of the fluid are constant the result of filtering the Navier Stokes equation is :

$$\frac{\partial \overline{u_i}}{\partial t} + \frac{\partial}{\partial x_j} \overline{u_i u_j} = - \frac{1}{\rho} \frac{\partial \overline{p}}{\partial x_i} + \nu \Delta^2 \overline{u_i}$$

This equation is supplemented by the continuity equation

$$\frac{\partial \overline{u_i}}{\partial x_i} = 0$$

For the advection term it is found that :

$$\overline{u_i u_j}(x_o, t) = \overline{\bar{u}_i \bar{u}_j} + \overline{u_i' u_j} + \overline{u_i u_j'} + \overline{u_i' u_j'}$$

It would be convenient to have the evolution of the filtered term controlled by an equation similar to that usually encountered when $\overline{u_i}$ is equal to the mean field. For a regular function f(x), Leonard (1974) has shown that it is possible to develop f(x) in Taylor series expansion from a point x_o located at the center of the grid model, so that

$$f(\vec{x}) = f(\vec{x}_o) + \frac{\Delta^2}{\gamma} \frac{\partial}{\partial x_k} \frac{\partial}{\partial x_k} f(\vec{x}_o) + O(\Delta^4)$$

For a Gaussian filter $\gamma = 24$. The subgriscale is termed Δ

and $$\frac{\partial}{\partial x_j}\overline{\bar{u}_u\bar{u}_j} = \frac{\partial}{\partial x_j}(\bar{u}_i\bar{u}_j) + \underbrace{\frac{\partial}{\partial x_j}\frac{\Delta}{\gamma}\frac{\partial}{\partial x_k}\frac{\partial}{\partial x_k}\bar{u}_i\bar{u}_j)}_{L_{ij}} + O(\Delta^2)$$

The Leonard term L_{ij} can be written

$$L_{ij} = \overline{\bar{u}_i\bar{u}_j} - \bar{u}_i\bar{u}_j$$

which takes into account the difference between two terms $\overline{\bar{u}_i\bar{u}_j}$ and $\bar{u}_i\bar{u}_j$. The product of filtered quantities is considered with and without additional filtering. The "cross term"

$$C_{ij} = \overline{\bar{u}_i u'_j} + \overline{u'_i\bar{u}_j}$$

represents interactions between the residual field and the filtered field. "A priori" this sort of interaction should not be located in spectral space. On the contrary the subgrid Reynolds tensor

$$\overline{u'_i u'_j} = R_{ij}$$

emphasizes the role of small structures. However it can be proved that such interactions can occasionally generate turbulent structures far outside the subgrid scale. Finally we get

$$\frac{\partial\bar{u}_i}{\partial t} + \frac{\partial}{\partial x_j}\bar{u}_i\bar{u}_j = -\frac{1}{\rho}\frac{\partial\bar{p}}{\partial x_i} - \frac{\partial}{\partial x_j}(L_{ij} + T_{ij}) + \gamma\Delta^2\bar{u}_i$$

with

$$T_{ij} = \overline{\bar{u}_i u'_j + u'_i\bar{u}_j + u'_i u'_j}$$

As is done with the usual splitting, it would be possible to write for the residual terms, u' and p', an equation which should be similar to that generally introduced for the fluctuating quantities and which could provide additional information in a modelling process. However additional difficulties would be encountered at this stage ; in particular, residual quantities would become sensitive to the rotation of the frame and representation theorems cannot be applied.

Notwithstanding real successes in predicting complex flows, closures introduced in the filtered equations are, in principle, very questionable. At first, all the developments are carried out on filtered quantities without introducing an averaging process. For example all subsequent workers have followed Smagorinsky (1963) and have represented the subgrid drain T_{ij} by an eddy viscosity model.

$$T_{ij} = -\gamma_n \left(\frac{\partial \bar{u}_i}{\partial x_j} + \frac{\partial \bar{u}_j}{\partial x_i}\right) \qquad \text{(x)}$$

so that the complete grid scale equation should be written as

$$\frac{\partial \bar{u}_i}{\partial t} + \frac{\partial}{\partial x_j} \bar{\bar{u}}_i \bar{u}_j = -\frac{1}{\rho}\frac{\partial \bar{p}}{\partial x_i} - \frac{\partial}{\partial x_j} L_{ij} + \frac{\partial}{\partial x_j}\{(\gamma + \gamma_n)\left(\frac{\partial \bar{u}_i}{\partial x_j} + \frac{\partial \bar{u}_j}{\partial x_i}\right)\}.$$

Besides information has not been extracted from the equations governing the residual quantities.

Leslie & Quarini (1979) have examined this problem for an isotropic field. The correspondance between physical and spectral space is found for each term starting from the equation

$$\left(\frac{d}{dt} + \gamma k^2\right) u_i(\vec{k},t) = M_{ijm}(\vec{k}) \frac{\Sigma}{\Delta} u_j(\vec{p},t) u_m(\vec{r},t) + F_i(\vec{k},t)$$

where

$$M_{ijm}(\vec{k}) = -\frac{i}{2}\{k_m P_{ij}(\vec{k}) + k_j P_{im}(\vec{k})\}$$

$$P_{ij}(\vec{k}) = \delta_{ij} - \frac{k_i k_j}{k^2}$$

and $\frac{\Delta}{\Sigma}$ implies

$$\iint \delta(\vec{k} - \vec{p} - \vec{r})\, d^3p\, d^3r$$

F_{ij} represents the driving force which sustains the turbulence ; it is conjectured that this term should not play an important role in

---

(x) This connection between simultaneous events could roughly be considered as introduced through conditional sampling the structure responsible for the cross terms are supposed immediately adaptable to the principal event represented by the filtered field. However if conditional averages must be introduced, the complexity of the problem would be considerably enlarged and the turbulent field should be treated simultaneously.

transferring energy from the large eddies to the small ones (x).

The equation of motion for the grid scale velocity $\overline{u_i}(\vec{k},t)$ is formed by filtering from the previous equation :

$$\left(\frac{\partial}{\partial t} + \nu k^2\right)\overline{u}_i(\vec{k},t) = M_{ijm}(\vec{k}) \overset{\Delta}{\Sigma} C(\vec{k})\, u_j(\vec{p},t)u_m(\vec{r},t) + C(\vec{k})F_i(\vec{k},t)$$

The inertial term on right hand side can be brought in agreement with the analysis of Leonard as proposed by Leslie by the expression

$$M_{ijm}(\vec{k}) \underset{\Delta}{\Sigma} \{\underbrace{G(\vec{p})G(\vec{r})}_{} - \underbrace{\{1 - G(\vec{k})\} G(\vec{p})G(\vec{r})}_{} + \underbrace{G(\vec{k})\{1 - G(\vec{p})G(\vec{r})\, M_{ijm}(\vec{k})\, u_j(\vec{p})\, um(\vec{r})\}}_{}\ u_j(\vec{p},t)\ u_m(\vec{r},t)$$

where term (1) corresponds to $\overline{u_j}\ \overline{u_m}$

term (2) corresponds to Leonard stress L_{ij}

term (3) is the subgrid stress.

The final purpose of subgrid modelling should be to represent the unknown terms as a functional of the filtered velocity $\overline{u}$. The equation is treated in order to obtain the evolution of the filtered power spectrum.

$\overline{E}(k,t) = G^2(k)\ E\ (k,t)$

At least at this stage an averaging process is involved. By using classical closures (E.D.Q.N.M.) it should be possible to identify what sort of triadic interactions are involved for each previous terms. With this assumption the transfer term T can be divided into two parts and can be written :

$$T = \iint \theta_{kpq}\ a_{kpq}\{\ k^2\ pE(\vec{k})E(\vec{q}) - p^3\ E(\vec{p})E(\vec{q})\}\ \frac{dpdp}{pq}$$

which roughly corresponds to a normal, and to a reversal transfer in wave number space.

(x) In spectral space the filter acts in a multiplicative way so that we get

$u_i(x,t) \quad v_i(k,t)$

$\overline{v}_i = C\ v_i\ ;\ v'_i = (1 - C)v_i$

From this computation the Smagorinsky hypothesis would be verified approximatively.

In spite of the crude hypotheses, the splitting method seems fairly well supported by these results, at least for isotropic fields. However the necessity to match a deterministic approach for the large structures with a statistical one for the small eddies introduces a challenging problem. In the case of very complex field where a mean velocity gradient is associated with rotation of the frame computations have just been carried out with both a spectral and a splitting method respectively (Weil[(1)] and Courty[(2)]). The results are in good agreement which tends to prove that the basic criticisms previously evoked do not prohibit the method.

(1) Ecole Centrale de LYON

(2) Avions Marcel Dassault.
The two methods developed by the two teams have been worked out simultaneously

---

a_{kpq} depends on the geometry of the triad under consideration and on the filter function G.

CONCLUSION

In a survey such as this one, all the facts are so compressed as to risk distortion ; moreover it is impossible to review all the relevent papers devoted to the problem under consideration. In fact, interesting papers dealing with coherent structures which are forced or excited are just quoted. Also, the discussion is certainly biased by the opinion of the authors whose experience is necessary focussed within more limited fields.

Recent developments about he nature of coherent structures in turbulent fields have been discussed. The lack of an accurate definition of turbulence invites varying views. If what is called turbulence is the final stage of the flow evolution at high Reynolds numbers, only three dimensional flows should be considered. In support of this opinion it can be argued that two-dimensional fluctuating fields are the consequence of organized mechanisms which are due to the action of driving forces or to a sequel of initial instabilities. If we set aside this formal discussion, we can assume that in most of the industrial flows large coherent structures exist, the consequence of which can be fundamental for entrainment, mixedness and so forth, although mean velocity gradients are not very effective in ordering turbulent structures. Even though the treatment of large structures in a deterministic way is actually a questionable approach in the case of isotropic turbulence, it seems reasonable to take a great interest in the method developed by Leslie and also in any fundamental work adapted to this treatment (Basdevant et al 1978) and in particular to flow fields subjected to driving forces.

ACKNOWLEDGEMENTS

We are very indebted to Professor Bataille, Professor Jeandel and Doctor Cambon for very helpful discussions.

REFERENCES

AHMAD Q.A., LUXTON R.E. & ANTONIA R.A., 1976, J. Fluid Mech., 77, 369.

ALCARAZ E., CHARNAY G. & MATHIEU J., 1977, Phys. Fluids, 20, 203.

ANDERSON P., LARUE J.C. & LIBBY P.A., 1979, Phys. Fluids, 22, 1857.

ANTONIA R.A., 1972, J. Fluid Mech., 56, 1.

ANTONIA R.A., PRABHU A. & STEPHENSON S.E., 1975, J. Fluid Mech., 72, 455.

BASDEVANT C., LESIEUR M. & SADOURNY R., 1978, J. Atm. Sci., 35, 1028.

BERNAL L.P., BREIDENTHAL R.E., BROWN G.L., KONRAD J.H. & ROSHKO A., 1979, Symposium on Turbulent Shear Flows, Imperial College, London, p. 8 - 1.

BERTOGLIO J.P., 1980, AIAA, 13th Fluid Plasma Dynamics, Snowmass, Colorado.

BEVILAQUA P., 1973, Thesis, Purdue University.

BEVILAQUA P.H. & LYKOUDIS P.S., 1977, AIAA Journal, 15, 1194.

BLACKWELDER R., 1970, Doctoral Dissertation, John Hopkins University.

BLACKWELDER R.F. & ECKELMANN H., 1979, J. Fluid Mech., 94, 577.

BRADSHAW P., 1972, I.C. Aero Report 72-18.

BRADSHAW P., FERISS D.H. & ATWELL N.P., 1967, J. Fluid Mech., 28, 593.

BRADSHAW P., FERISS D.H. & JOHNSON R.F., 1964, J. Fluid Mech., 19, 591.

BROWAND F.K. & LAUFER J., 1975, Turbulence in liquids, Missouri Rolla, p. 333.

BROWAND F.K. & WEIDMAN P., 1973, Bull. Am. Phys. Soc., 18, 1469.

BROWN G. & ROSHKO A., 1971, Symposium on Turbulent Shear Flows, Imperial College, London, p. 23.1.

BROWN G. & ROSHKO A., 1974, J. Fluid Mech., 64, 775.

BROWN G. & ROSHKO A., 1979, J. Fluid Mech., 64, 775.

BUDNY R.S., KAWALL J.G. & KEFFER J.F., 1979, Symposium on Turbulent Shear Flows, Imperial College, London, p. 7.20.

BURGERS J.M. & MITCHNER M., 1953, Proc. Kon. Ned. Akad., 56, 228.

CAMBON C., 1979, Docteur Ingenieur Thesis, Université Lyon I.

CAMBON C., JEANDEL D. & MATHIEU J., 1980, J. Fluid Mech., to be published.

CELLARD P., 1980, D.E.A. Université Lyon I.

CHAN W.T. & KO N.W.M., 1978, J. Fluid Mech., 89, 515.

CHANDRSUDA C., MEHTA R.D., WEIR A.D., BRADSHAW P., 1978, J. Fluid Mech., 85, 693

CHARNAY G., 1974, Thesis, Université Lyon I, 264

CHARNAY G., MATHIEU J & COMTE-BELLOT G., 1976, Phys. Fluids, 9, 1261

CHARNAY G. & MELINAND J.P., 1980, to be published

CHEN C.H.P. & BLACKWELDER R.F., 1978, J. Fluid Mech., 89, 1

CHERDRON W., DURST F. & WHITELAW J.H., 1978, J.Fluid Mech, 84,13

CHEVRAY R. & TUTU N.K, 1978, J. Fluid Mech, 88, 133

CLARK A.R. & HUSSAIN A.K.M.F., 1979, Symposium on Turbulent Shear Flows, Imperial College, London, p. 2.30.

COMTE-BELLOT G., 1965, Pub. Sci. Tech. Min. Air, N° 419

COMTE-BELLOT G., AGARD Conference N° 271

COMTE-BELLOT G., SABOT J. & SALEH I., 1979, Dynamic Flow Conf., Marseille.

COMTE-BELLOT G., STROHL A. & ALCARAZ E., 1971, J. Applied Mech, 767.

CORINO E.R. & BRODKEY R.S., 1969, J. Fluid Mech., 37, 1

CORRSIN S., 1943, NACA, ARC 3L23.

CORRSIN S. & KISTLER A.L. 1955, NACA Rep.1244.

CRAYA A., 1958, Pub. Sci. M. Air, 345.

CROW S.C. & CHAMPAGNE F.H., 1971, J. Fluid Mech. 48, 547

DAVIES P.O.A.L. & YULE A.J., 1975, J. Fluid Mech. 69, 513.

DEADORFF J.W., 1971, J. Comp. Phys. , 7, 120

DIMOTAKIS P.C. & BROWN G.L., 1976, J. Fluid Mech., 78, 535.

DOPAZO C., 1977, J. Fluid. Mech. 81, 433

DOPAZO C. & O'BRIEN E.E., 1977. 1st Symposium Turbulent Shear Flow, University park, Pennsylvania

DUMAS R., FULACHIER L. & ARZOUMANIAN E., 1972, C.R. Acad. Sc. Paris., A. 275, 555.

ECKELMANN H., 1974, J. Fluid Mech., 65, 439

ECKELMANN H. & NYCHAS S.G., Brodkey R.S. and Wallace J.M, 1977 Phys. Fluids, 20, S225

EVERITT K.W. & ROBINS A.G., 1978, J. Fluid Mech., 88, 563

FABRIS G., 1979, J. Fluid Mech., 94, 673

FAVRE A.J., 1946, VI Congres Int. Mec. Appl., Paris

FIEDLER H. & HEAD M.R., 1966, J. Fluid Mech., 25, 719

FIEDLER H. & KORSCHELT D., 1979, Symposium on Turbulent Shear Flows, Imperial College, London, p. 8.18.

FREEMAN R.W. & TAVLARIDES L.L., 1979, Phys. Fluids, 22, 782

FULACHIER L., GIOVANANGGELI J.P., DUMAS R., KOVASZNAY L.S.G. & FAVRE A., 1974, C.R. Acad. Sc. Paris, A 278, 683

GENCE J.N & MATHIEU J., 1979, J. Fluid Mech., 93, 501

GENCE J.N, SCHON J.P & MATHIEU J., 1978, J. de Mécanique, 17, 299.

GOLDSCHMITT V.W. & BRADSHAW P., 1973, Phys. Fluids, 16, 354

GUPTA A.K. & KAPLAN R.E., 1972, Phys. Fluids, 15, 981

HANJALIC K. & LAUNDER B.E., 1972, J. Fluid Mech., 52, 609

HAVERBEKE A., WOOD D.H & SMITS A.J., 1979, Symposium on Turbulent Shear Flows, Imperial College, London, p. 11.1

HEAD M.R. & BANDYOPADHYAY P., 1978, AGARD 271

HEDLEY T.B. & KEFFER J.F., 1974, J. Fluid Mech., 64, 625

HERNAN M.A. & JIMENEZ J., 1979, Symposium on Turbulent Shear Flows, Imperial College, London, p. 7.7

HOPFINGER E.J., 1977, Lecture Notes in Phys., 75, 65

JEANDEL D., 1976, Thesis, Université Lyon I, 37

JEANDEL D., BRISON J.F. & MATHIEU J., 1977, Phys. Fluids, 21, 169

JIMENEZ J., MARTINEZ VAL R. & REBOLLO M., 1979, Symposium on Turbulent Shear Flows, Imperial College, London, p. 8.7

KASTRINAKIS E.G., WALLACE J.M, WILLMARTH W.W., GHORASHI B and BRODKEY R.S. , 1978, Lecture Notes in Phys. 75, 175

KAWALL J.G. & KEFFER J.F., 1979, Phys. Fluids, 22, 31

KIBENS V., 1968, Doctoral Dissertation, John Hopkins University.

KIBENS V., KOVASZNAY L.S.G. & OSWALD J., 1974, Rev. Sci. Instrum. 45 1138

KIM H.T., KLINE S.J & REYNOLDS W.C., 1971, J. Fluid Mech., 50, 133

KLINE S.J., REYNOLDS W.C., SCHRAUB F.A. & FUNSTADLER P.W., 1967, J. Fluid Mech., 30, 741

KLINE S.J & RUNSTADLER P.W., 1959, J. Appl. Mech., 26, 166

KOVASZNAY L.S.G., KIBENS V. & BLACKWELDER R.F., 1970, J. Fluid Mech., 41, 283

KREPLIN H.P & ECKELMANN H., 1979, Phys. Fluid, 22, 1233

LANCE M., MARIE J.L., CHARNAY G. & BATAILLE J., 1979, C.R. Acad Sc. Paris, A 288, 957.

LEONARD A., 1974, Adv. Geophys., A 18, 237.

LESLIE D.C. & QUARINI G.L., 1979, J. Fluid Mech., 91, 65.

MATHIEU J., 1971, IVL Lecture series 36.

MUSZYNSKI M., 1979, D.E.A., Université Lyon I.

OBERMEIER F., 1979, Symposium on Turbulent Shear Flows, Imperial College, London, P. 18.22.

OFFEN G.R. & KLINE S.J., 1973, Rep MD 31, Stanford University.

OSTER D., WYGNANSKI I., DZIOMBA B. & FIEDLER H., 1977, Lecture notes in Phys., 75, 48.

PERRY A.E. & LIM T.T., 1978, J. Fluid Mech., 88, 451.

PERRY A.E., LIM T.T., WATMUFF J.H. & CHONG M.S., 1979, Symposium on Turbulent Shear Flows, Imperial College, London, p. 8.30.

PRASAD J.K. & GUPTA A.K., 1977, AIAA Journal, 15, 1569.

PRATURI A.K. & BRODKEY R.S., 1977, J. Fluid Mech., 80, 215.

RAJAGOPALAN S. & ANTONIA R.A., 1979, Phys. Fluids, 22, 614.

RAO K.N., NARASIMHA R. & BADRI NARAYANAN M.A., 1971, J. Fluid Mech., 48 339.

ROCKWELL D. & KNISELY C.G. 1979, J. Fluid Mech., 93, 413.

ROSHKO A., 1976, AIAA Journal, 14, 1349.

SABOT J. & COMTE-BELLOT G., 1976, J. Fluid Mech., 74 767.

SCHON J.P. & CHARNAY G., 1977, Measurement of unsteady Fluid Dynamic Phenomena, Hem. Pub. Corp.

SCHON J.P., DANEL F., MELINAND J.P., REY C. & CHARNAY G., 1979, Sixth biennanl symposium on turbulence in liquids, Missouri-Rolla.

SMAGORINSKY J.S., 1963, M. Weath. Rev., 91,99.

SUBRAMANIAN C.S. & ANTONIA R.A., 1979, Symposium on Turbulent Shear Flows, Imperial College, London, p. 4.18.

SUNYACH M., 1971, Thesis, Université Lyon I, 37.

SUNYACH M. & MATHIEU J., 1969, Int. J. Heat Mass Trans., 12, 1679.

TOWNSEND A.A., 1956, The structure of Turbulent Shear Flow, Cambridge, Univ. Press.

TOWNSED A.A., 1979, J. Fluid Mech., 95, 515.

WALLACE J.M., BRODKEY R.S. & ECKELMANN H., 1977, J. Fluid Mech., 83. 673.

WALLACE J.M., ECKELMANN H. & BRODKEY R.S., 1972, J. Fluid Mech., 54,39.

WEIR A.D. & BRADSHAW P., 1975, I.C. Aero Report 75.07.

WILLE R., 1963, Flugwiss, 11, 222.

WILLMARTH V.W. & LU S.S., 1972, J. Fluid Mech., 55, 65.

WINANT C.D. & BROWAND F.K., 1974, J. Fluid Mech., 63, 237.

WIGNANSKI I. & FIEDLER H.E., 1970, J. Fluid Mech., 41, 327.

YOKOBORI S., KASAGI N., HIRATA M., NAKAMARU M. & HARAMURA Y., 1979, Symposium on Turbulent Shear Flows, Imperial College, London, p. 4.12.

YULE A.J., 1978, J. Fluid Mech., 89, 413.

YULE A.J., 1979, Symposium on Turbulent Shear Flows, Imperial College, London, p. 7.1.

ZAMAN K.B.M.Q. & HUSSAIN A.K.M.F., 1979 Symposium on Turbulent Shear Flows, Imperial College, London, p. 8.24.

ZEGADI R., 1980, D.E.A. Université Lyon I.

INVESTIGATIONS OF EDDY COHERENCE IN JET FLOWS

by

A.J. YULE

Department of Chemical Engineering and Fuel Technology
University of Sheffield, Mappin Street, Sheffield S1 3JD, England

---

ABSTRACT

In turbulent shear flow the term "Coherent Structures" refers to eddies which are both spatially coherent, i.e. large eddies, and also temporally coherent, i.e. they retain their identities for times which are long compared with their time scales in fixed point measurements. In certain cases, for example transitional flows, the existence of such structures is evident from flow visualisations. However in many other flows, and in the complex flows usually found in practical situations, such structures are not so evident although some indications of their possible existence have been found. In this paper an evaluation of the reasons for the existence of these two classes of flows is first given. Attention is then focused upon the more difficult flows, particularly the round turbulent jet, where coherent structures are not so evident, and upon techniques by which the existence (or non-existence) of such structures in these flows, can be established from point measurements, backed up by flow visualisations. A major difficulty is the need to discriminate between real losses in eddy coherence and apparent losses in coherence introduced by 'phase scrambling' effects which 'smear' multipoint correlations. The analysis of multiprobe time dependent data in cold and reacting round turbulent jets is described and it is shown how evidence of strong eddy coherence can be extracted from data in spite of small values of the classical statistical cross-correlations.

1. INTRODUCTION

It is useful to define *Coherent Structures* as large eddies which, (i) are repetitive in structure, (ii) remain coherent for distances downstream very much greater than their length scales, (iii) contribute greatly to properties of the turbulence; in particular, turbulent energy and shear stress, entrainment and mixing. The recent emphasis on the possibly dominating importance of the large eddies has resulted from the dramatic results of flow visualisations in certain specific types of flow, notably the two-dimensional mixing layer experiments of Roshko[1] and his co-workers.

Following these experiments it has been queried whether such large eddies were a universal feature of all turbulent shear flows. This question remains unanswered. Yule[2] described how a failure to recognise and measure coherent structures in a flow by using point measurements, might be attributable to so called "phase scrambling effects". This phenomenon is also outlined below in the light of recent multi-probe data in jets and flames. If coherent structures are indeed important features for a given flow, there arises the question of how essential it is to include descriptions of these structures in models for such flows. This problem is addressed in the final section of this paper which discusses the roles of coherent structures in jets and jet-type flows connected with turbulent combustion.

2. THE RECOGNITION OF COHERENT STRUCTURES

One can conveniently divide experimental turbulent flows into those in which coherent large eddies have been unambiguously identified, both in flow visualisations and point measurements (Class 1), and those in which such eddies are not so immediately obvious (Class II). The questions arise:

(i) Are the Class I flows, or the experiments which investigated them, *pathological* in the sense that special peculiarities introduce or enhance such structures?

(ii) Are coherent structures in fact present in all, or some of the Class II flows but these are not immediately identifiable due to obscuring phenomena?

These questions should be answered before the phenomenon of coherent structures is included in physical models for turbulence. The present experimental evidence is so incomplete, and often contradictory, that it is not possible to give firm answers to either question; even when one considers a narrow class of flow such as the plane mixing layer which has recently been subjected to many investigations. However, below a checklist is given of phenomena which may contribute either to the patholog-ical nature of an investigation or to obscuration of coherent structures. Reference is made only to jet and mixing layer flows although the general comments apply also to other turbulent shear flows.

2.1 Contributions to the possible 'Pathological' Nature of Experiments Turbulent or Transitional Flow?

Unforced, fully developed turbulent flow is recognised from point measurements by:-

(i) Three-dimensionality, so that u, v, and w fluctuations are of the same order. This requires that the larger eddies (which contain most of the turbulence energy), cannot be completely two-dimensional. This does not however prohibit these eddies from being coherent in the azimuthal, or spanwise, sense while for example, having 'wrinkles' along their lengths.

(ii) A wide and continuous frequency or wavenumber, spectrum without sharp peaks and harmonics. (For this situation the 'energy cascade' process is generally accepted to exist).

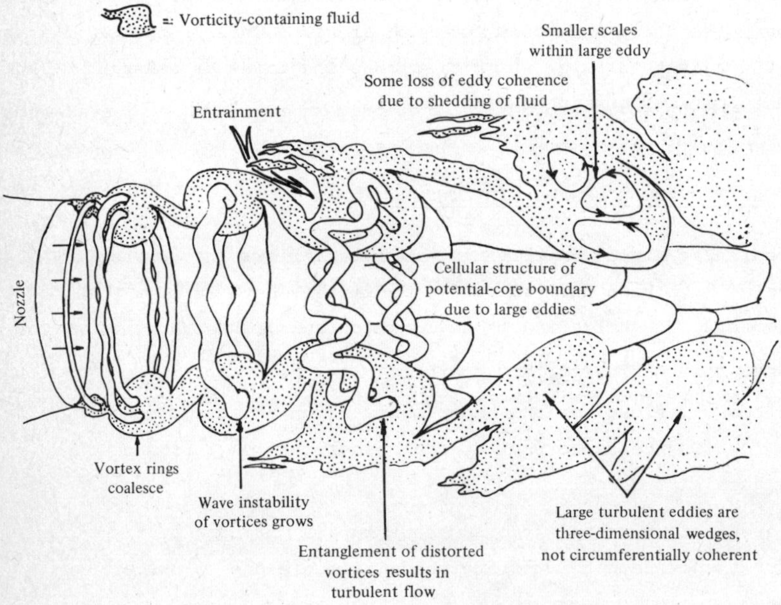

FIG. 1 Physical structure of transitional jet

For jet and mixing layer flows the problem of defining the distance downstream required for fully developed turbulence to be established remains unsolved. As sketched in Fig. 1, the transition process is complex and involves an inviscid initial instability mechanism. However viscosity influences the rolling-up of the mixing layer into vortices and also influences the core sizes of these rings and thus the mode of the subsequent three-dimensional instability.[3] The coalescing of the transitional vortices is basically an inviscid process but viscosity can again have a stabilising influence.

It is certain that the large eddies in *transitional flow* are both highly coherent and also 'two-dimensional'. It is thus necessary to ensure that a flow is locally fully turbulent before studying the local eddy structure. The only method of ensuring this is by making sufficient local point fluctuating velocity measurements to check that the flow satisfies the above criteria. Measurements of mean velocity distributions and spreading rate are not sufficient.

Another aspect of turbulent flow is the existence of a random small scale turbulence structure which ensures rapid mixing once fluid has been entrained into the turbulent flow (by the large scale structure). Some plane mixing layer experiments[4] show that this small scale structure can be established when the large eddies are still two-dimensional. It thus appears that the various attributes of fully turbulent flow need not be established at the same position downstream.

Pressure Field and Rig Effects

There are many examples of experiments in which the rig dimensions strongly influence the turbulence structure in an unexpected and unwanted manner. For example acoustic feedback is well known in compressible and combustion flows. Bradshaw[5] has proposed that even for incompressible flow, the plane mixing layer, and particularly the irrotational velocity field outside it, can be strongly influenced by the test section floor significantly before the attachment position. It has also been suggested that wall and floor effects may enhance the spanwise coherence of the irrotational flow, while the turbulent flow may not be particularly coherent. Going even further, others have suggested that the spanwise coherence of the turbulent eddies themselves, is enhanced or prolonged by the pressure fields introduced by rig effects. Unfortunately these proposals remain unresolved either way. The typical plane mixing layer experiment can have up to four rig length scales: Heights of the initial primary and secondary flows. Width of test section. Length of test section. When the characteristics of the splitter plate boundary layers, the test section boundary layers and the initial turbulence characteristics are also considered, it is not surprising (in hindsight) that observations made using different rigs do not always agree. However the various experimenters have proposed their flows to be fully developed and the question has been raised whether the mixing layer can have two self-preserving forms; with and without two-dimensional large eddies.

It might be considered that the plane mixing layer is a relatively 'simple' turbulent flow which is an ideal candidate for experiments designed to gain insight into the physical structure of the turbulence. Unfortunately this has not proved be be the case due to the unresolved problems described above. There is thus a strong case for the study of the round free jet: With properly designed nozzle and settling chamber, the round free jet does not suffer from rig effect, a large body of 'classical' data is in existence and the jet is closer in character to the flows which exist in most practical mixing systems.

The Effects of 'Forcing' the Flow

In many jet-type flows experiments have been carried out by using a periodic 'forcing' of the jet; using sound fields, splitter plate vibration, pressure fields or sparks. One can divide these experiments into those which have the objective of measuring the effects on the turbulence structure of the forcing and those which have the intention of making coherent structures more amenable to measurement. These latter experiments have problems in their methodology and interpretation. In particular it is assumed, at least implicitly, that the turbulence structure is basically similar with and without forcing and that the eddies are merely made more easily measured. However in practice it is found that the jet can be drastically changed by forcing, with differences found in the velocity profiles, rates of spread, turbulence intensity, shear stress etc. It therefore does not seem to be reasonable to deduce information regarding 'natural' turbulence by studying such forced jets. One can propose, on the basis of the available evidence, that a result of forcing the jet is to prolong the existence of the relatively orderly transitional flow, perhaps by delaying the onset of three-dimensionality in the large eddies. One can also surmise that an essential feature of the 'real' turbulent flow is the natural dispersion of the strengths, wave numbers, separations and trajectories of eddies as they move downstream. Forcing can delay the dispersion and thus delay 'real' turbulence.

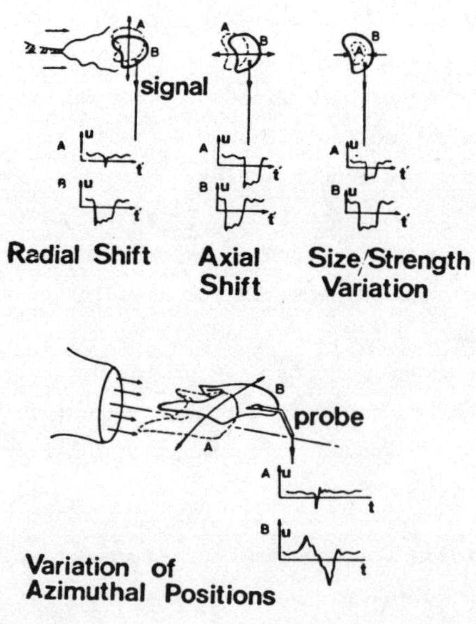

FIG. 2 Different modes of jitter and their effects on signals

2.2 Possible Contributions to Difficulties in Observing Coherent Structures

Yule[2] described how the so-called phase scrambling effects can make the recognition of coherent structures from point measurements very difficult. In a typical conditional sampling experiment structures are sampled and ensemble averaged at times measured relative to trigger events. As indicated in Fig. 2, quite small variations in the structures and their positions at the time of sampling, can cause large changes in the signals which they produce. This can cause the signals received by the conditioning sampling technique to be quite small in amplitude, although the actual signal amplitudes corresponding to individual eddies can be maintained. Thus conditional sampling experiments should indicate the approximate average structures of eddies but they are not necessarily reliable for indicating the coherence of eddies nor for showing details of the eddies. On the other hand if coherent structures are present some indication at least of their presence should be found in signal time histories, conditionally sampled signals and cross-correlations.

Flow visualisation of coherent structures is often difficult in fully turbulent jet flow because of: (i) the three-dimensional structure of the turbulence, (ii) the rapid diffusion of marker species by the small scale turbulence and (iii) an outer 'sheath' of weak small scale turbulence which can obscure the main motion inside the jet. In spite of this, indications of these eddies can be seen in schlieren photographs. In particular the potential flow entrainment regions between eddies can be seen, even though the eddies themselves are not as clear as those seen in visualisation of the transitional flow. It remains true that visualisations of the eddies in turbulent regions of round jets have never approached the clarity of the plane mixing layer visualisations of Roshko et al.

3. STRUCTURE OF COLD JETS

Figure 1 shows the structure of a jet developing from a laminar nozzle boundary layer, drawn on the basis of numerous flow visualisations. The general flow structure is qualitatively the same for a wide range of Reynolds numbers. The laminar mixing layer rolls up periodically into vortex rings. These coalesce and sometimes coalescence can occur several times. Azimuthal core waves develop until the distorted vortex rings coalesce and turbulent large eddies are formed.

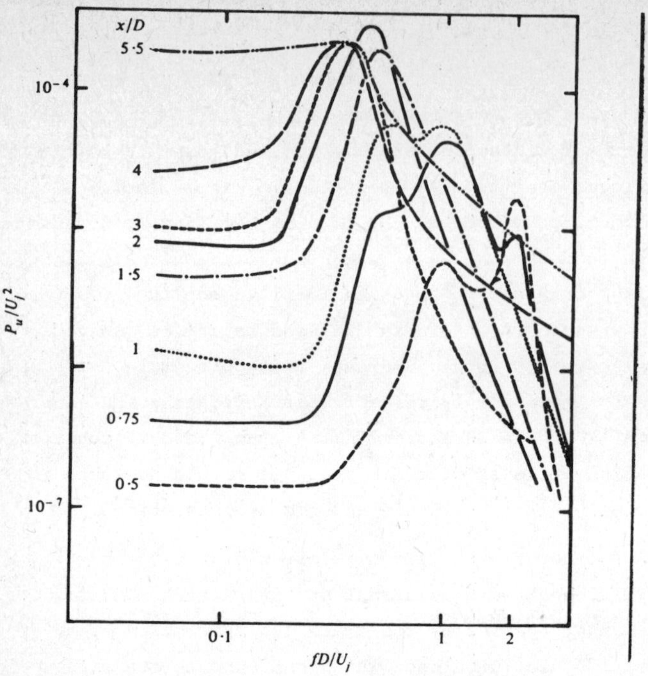

FIG. 3 Spectra of u in 50.8 mm diameter cold jet at η = 0.15 for Re = 2.1x 10^4

It is important to discriminate between the transitional and turbulent flow regions. Fig. 3 shows spectra along the inside edge of the mixing layer inside the potential core for a 50.8 mm diameter jet with Re = 2.1 x 10^4. The appearance of subharmonics is a characteristic of the transition region and this indicates that the vortices coalesce fairly consistently at given positions. The jet represented by Fig. 3 has three distinct coalescing regions, and other jets have been found to have two or only one clear coalescing regions in the transitional flow. Spectra in the centre of a transitional mixing layer show fundamental and subharmonic frequencies which agree with the potential core values; however there are also often higher harmonics present. These are caused by modifications to the basic cosine-type signal, which result from the cores of the vortices passing the hot wires. Numerous experiments [3, 6, 7, 8] have shown that the peak frequency f_p corresponding to the natural shear layer instability agrees well with the most amplified frequency predicted by Michalke[9] so that the Strouhal number is,

$$St_\delta \equiv f_p \delta_j / U_j \approx 0.07$$

for the typical experimental range of δ_j/D : $0.001 \leq \delta_j/D \leq 0.01$.

where: U_j is the jet velocity, D is the nozzle diameter and δ_j is the laminar nozzle boundary layer shear thickness,

$$\delta_j \equiv - U_j \, (\partial U/\partial r)^{-1}_{max}$$

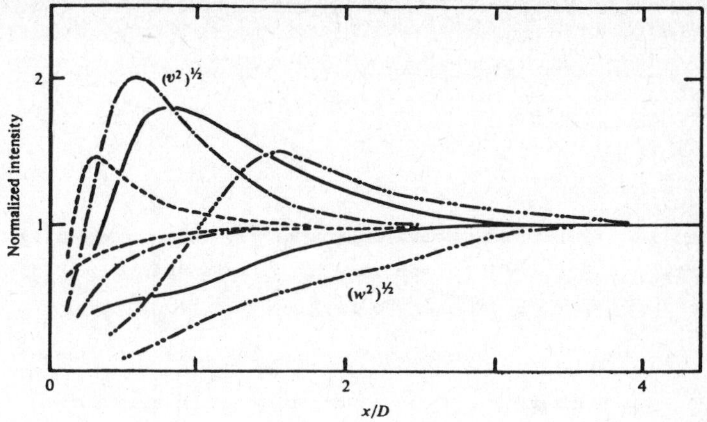

FIG. 4 Longitudinal distributions of turbulence intensity at r = 0.5D, ··———·· Re = 9 x 10^3, ______ Re = 3.5 x 10^4, ·———·———·Re = 5 x 10^4, ------- Re = 10^5.

Yule proposed that the transitional round jet mixing layer can be assumed to end where all three fluctuating velocity components have reached similar, nearly asymptotic values and this also coincides approximately with the attainment of turbulent forms for correlations and spectra. Fig. 4 shows measurements of

$$(\bar{v}^2)^{\frac{1}{2}} \quad \text{and} \quad (\bar{w}^2)^{\frac{1}{2}}$$

along the centre of the mixing layer for different jets. It is interesting to note the gradual increase in w fluctuations in the transition region and this corresponds to the gradual growth of waves in the vortex cores.

FIG. 5 Jet transition distances as a function of nozzle boundary layer shear thickness. ▲ Yule 25.4 mm dia. jet, ● Yule 50.8 mm dia. jet, O from data of Bradshaw.

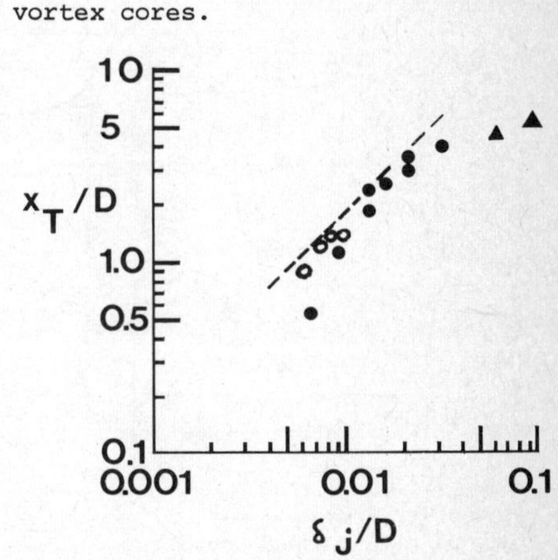

By using this type of data the transition distance x_T can be estimated for various jets and the results are given in Fig. 5 Bradshaw suggested the proportionality

$$x_T \approx 1000 \; \Theta_j$$ where Θ_j is the nozzle boundary layer momentum thickness. This gives $x_T \approx 250 \; \delta_j$ which gives a fair approximation to the data for most of the range.

The initial instability frequency given by Michalke's theory is a weak function of δ_j/D so that $f_p \; \delta_j/U_j \approx 0.07$

The subharmonic and quarter-harmonic peaks approximately (but not exactly) halve each time during coalescing. One would expect some correlation of the various data on peak frequencies in jets when plotted in the form

$f_p \; \delta_j/U_j$ against x/δ_j and there is a reasonable collapse of the various vortex passing frequency data when plotted in this form.

Thus, although more detailed information is still required and models cannot yet cope with all of the secondary instabilities, the transitional jet is relatively orderly and is amenable to conditional sampling studies. A typical arrangement for conditional sampling of the transitional vortex-coherent structures is shown in Fig. 6. Signals from recovery probes are ensemble averaged at times measured relative to events located in the signal from the trigger probe. Bruun[10] has used this technique in a jet transition region to educe the averaged velocity fields of coalescing vortex pairs. As shown in Fig. 7 the proportion of the signal recovered by this technique was very high in the transitional jet and this indicates the repetitiveness of the vortices and their movements.

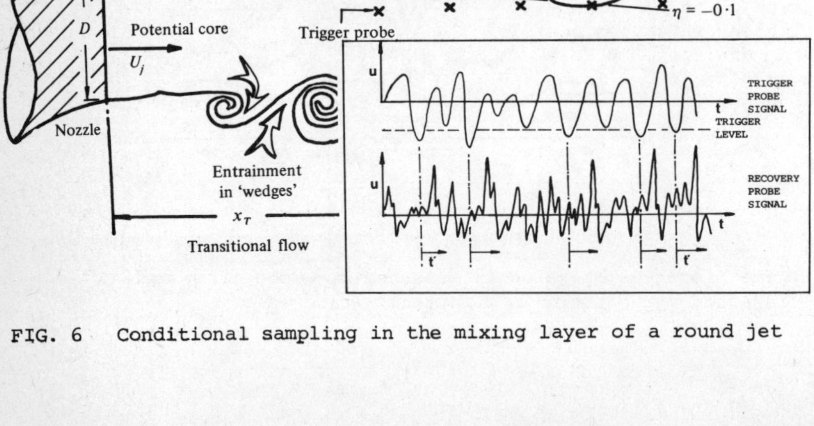

FIG. 6 Conditional sampling in the mixing layer of a round jet

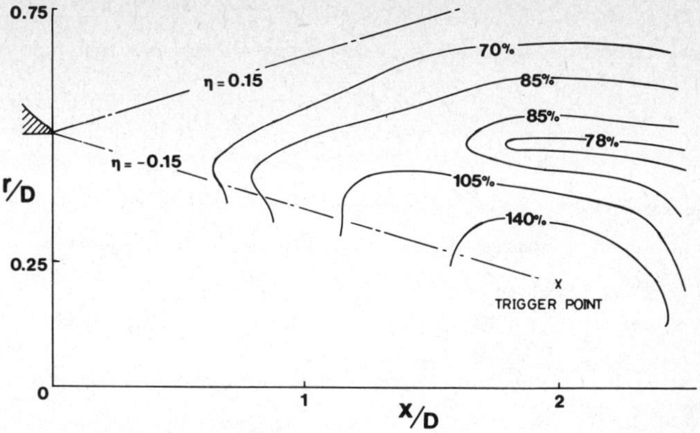

FIG. 7 Percentage of local signal recovered by ensemble averaging eddies in round jet, Re = 10^4

When one moves downstream into the turbulent region of the round jet (i.e. either the turbulent mixing layer or the turbulent fully developed downstream regions), the situation becomes much more difficult with regard to the application and interpretation of conditional sampling experiments. A useful technique for relating observed structures to point signals, is to film the flow at the same time as probe signals are recorded. Figure 8 shows the signals accompanying the passing of a typical large eddy, as delineated by dye patterns in a submerged water jet. It is found that - ve potential core 'u' peaks occur after the eddy has passed, strong - ve peaks (or 'spikes') are distinctive of the region inside the eddy.

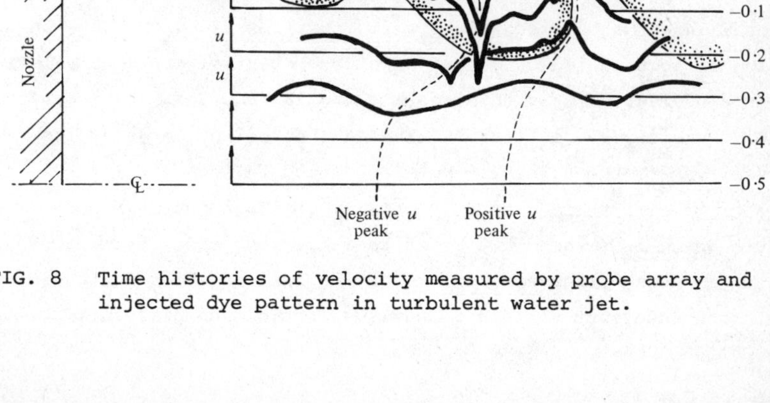

FIG. 8 Time histories of velocity measured by probe array and injected dye pattern in turbulent water jet.

By using, for example, the - ve spikes in the inner side of the mixing layer as triggers one might hope to educe the average eddy structures. In fact the ensemble averaged signals [11] which results at the recovery probes are then of low amplitude and thus are barely comparable, in their information content, with the corresponding cross-correlations. For example Fig. 9 shows the peak amplitudes of recovered signals in the turbulent jet as a proportion of the local r.m.s. signal. It is seen that the amplitude of the recovered signal falls very quickly with increasing distance from the trigger position, in comparison with the conditional sampling results in the transitional flow. As has been pointed out[2] an objective of future conditional sampling experiments must be to discover how much of this loss in amplitude can be attributed to the various phase scrambling effects of basically repetitive coherent eddies, and how much represents a real loss in amplitude caused by a loss in coherence (break down) of the sampled eddies. This question is unresolved quantitatively, however, the indications of several flow visualisations[3,7] are that the sampled eddies do remain coherent for very considerable distances downstream.

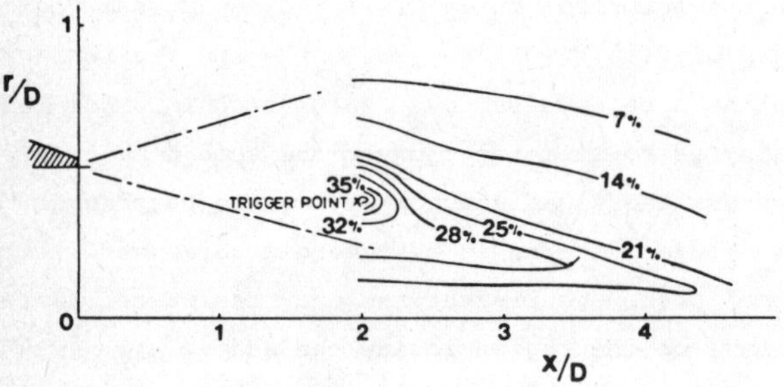

FIG. 9 Percentage of local r.m.s. signal recovered by ensemble averaging eddies in turbulent jet, Re = 2×10^5

A major contribution to the smearing or scrambling of signals comes from the growth of three-dimensionality in the large eddies. This can be seen in Fig. 4 and also in azimuthal cross correlations[3] with probes separated by π, in the centre of the mixing layer which show that the growth of w fluctuations along the jet in the transition region is accompanied by a loss of circumferential cross-correlation. However it remains unclear whether this results from a true loss or circumferential coherence of the transitional vortices (so that the turbulent large eddies are perhaps rather similar to the horseshoe vortices found in boundary layers), or alternatively the turbulent eddies are circumferentially coherent but have an azimuthal 'lobed' structure which is smeared out by phase scrambling effects. Front views of the turbulent jet using transverse slit lighting give no clear indication of circumferential coherence, but certain flow visualisations do give an impression of this and helical turbulent eddies have even been proposed for certain flows.

By using very large large - ve potential core u fluctuations as trigger signals Yule[3] found that a certain class of turbulent large eddy could be educed. There is considerable evidence that the inner side of the turbulent round jet mixing layer contains eddies which either retain their coherence longer than in the central and outer part of the mixing layer, or are more repetitive in their structures and movements. This can be seen in the contours of recovered signal amplitude in Fig. 9. By sampling particular examples of these inner eddies which are unusually close to the centre of the jet, the phase scrambling effects are reduced and substantial values of educed signals can be recovered downstream of the trigger probe.

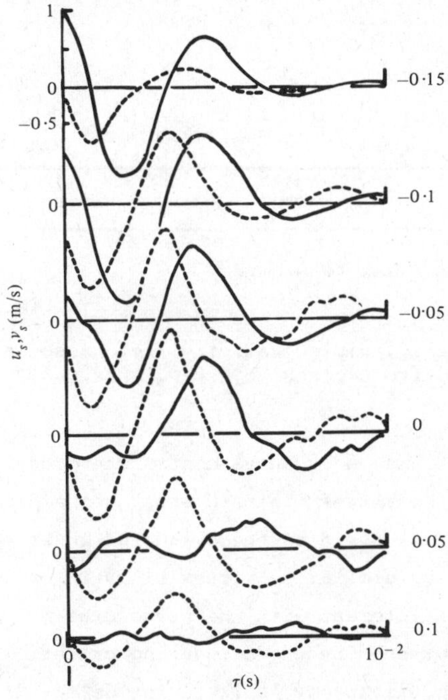

FIG. 10 Conditionally sampled time histories of u and v at x/D = 2.5, Re = 4.3 x 10^4. Triggered by large negative u peaks at x/D = 2, η = - 0.2

———————— u_s - - - - - v_s

For example Fig. 10 shows recovered u and v time histories at x/D = 2.5 for triggering in the potential core at x/D = 2, r/D = 0.1. Recovered u signals are small at the outer part of the mixing layer (η > 0) but the recovered v signal is significant for the complete width of the mixing layer. This is common to all conditional sampling experiments in round jets which show a much greater spatial coherence for the transverse velocity component than for the other velocity components. To some extent, at least, this might be attributed to the smearing effects of the differing convection velocities from one eddy to the next. The educed signals for these inner eddies show that although the magnitudes of u and v are large, there is nowhere within the turbulent mixing layer where the peaks of these sampled signals are in phase. When this sampling procedure is repeated, but with the trigger probe at the centre of the mixing layer, one finds a rapid decrease of educed signal amplitude with increasing probe separation, but u and v are, on average, now in phase in the central part of the mixing layer.

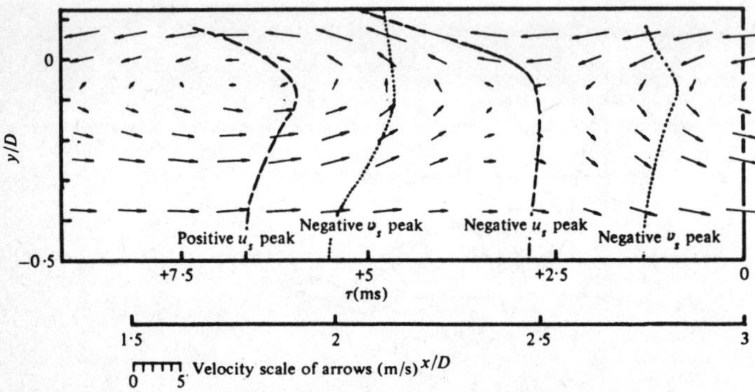

FIG. 11 Conditionally sampled velocity field associated with large negative u peaks at $\eta = -0.2$, $x/D = 2$, $Re = 4.3 \times 10^4$

Thus although these sampled inner structures contribute greatly to the turbulence intensity, they are not by themselves the main shear containing feature of the turbulence. The averaged velocity field of these sampled inner eddies is shown in Fig. 11. They are seen to be similar to vortex rings in cross section. It is considered that these eddies interact with the structures at the outer part of the flow, which are not being detected by the triggering criterion used. This interaction produces the shear containing uv 'spikes' inside the mixing layer.

The potential core fluctuations, adjacent to the turbulent mixing layer, do not exhibit the subharmonics found near the transitional region. However, a single clear frequency peak is found near the turbulent region, which indicates a strong periodicity of the inner turbulent eddies. The frequency of this peak gradually reduced in value[3] with increasing downstream distance. This may indicate that the inner turbulent eddies have a high degree of randomness in their movements.

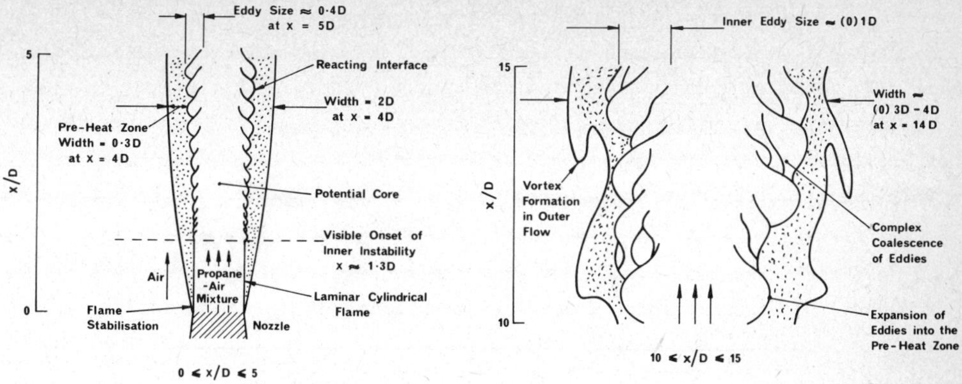

FIG. 12 Sketches of structures of jet flame taken from schlieren films $0 \leq x/D \leq 5$ and $10 \leq x/D \leq 15$ respectively. $Re = 1.05 \times 10^4$ equivalence ratio = 10.4

4. COMBUSTION FLOWS WITH COHERENT STRUCTURES

Experiments with burning round jets of pure fuel (propane), and various fuel/air mixtures, have been carried out in Sheffield[12, 13, 14]. As in the cold jet experiments described above, the gas jet is carefully designed to have a uniform velocity profile at the nozzle, with a turbulence level less than 0.5% and a laminar nozzle boundary layer. Many of the features found in cold jets are found in these 'jet flames' although there are, quantitatively, significant modifications. Fig. 12 shows the typical appearance of the jet flame near the nozzle and further downstream. Two distinct vortex-like instabilities are seen. An inner high velocity, high frequency instability appears to be an acoustic resonance of the nozzle which is driven by combustion processes, whilst an outer, slower moving instability corresponds to the vortex rings resulting from a Kelvin-Helmholtz instability, similar to the cold jet. Reaction occurs in interface layers, which are distorted by the vortices and thus provide clear visualisations of these vortices in schlieren films.

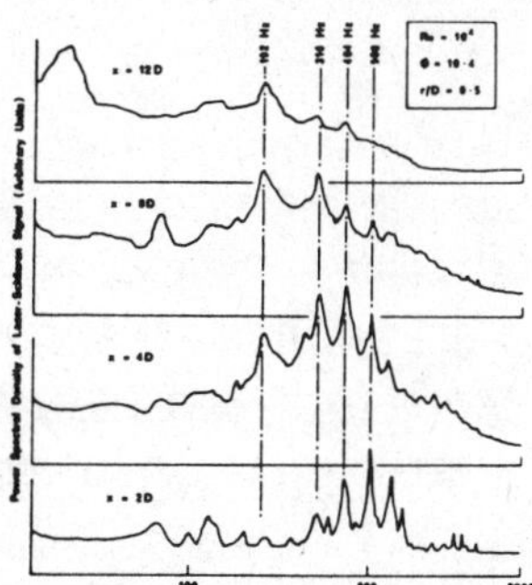

FIG. 13 Spectra of density gradient fluctuations measued by laser beam deflections (laser schlieren) in jet flame, $Re = 1.05 \times 10^4$, equivalence ratio = 10.4

A laser schlieren technique measures the deflections of a 1 mm laser beam passing diametrically through the flame. The technique is particularly sensitive to high temperature zones which are circumferentially coherent. Figure 13 shows spectra obtained by using this technique at different distances along the centre of the mixing layer of the flame. Discrete peaks are found in the spectra and there is a transfer of energy to lower frequencies. The laser schlieren peaks correspond to the high frequency eddies found at the inside of the jet flame mixing layer. These peak frequencies have been compared[12] with the average passing frequencies of the inner eddies measured from high speed schlieren films with good agreement.

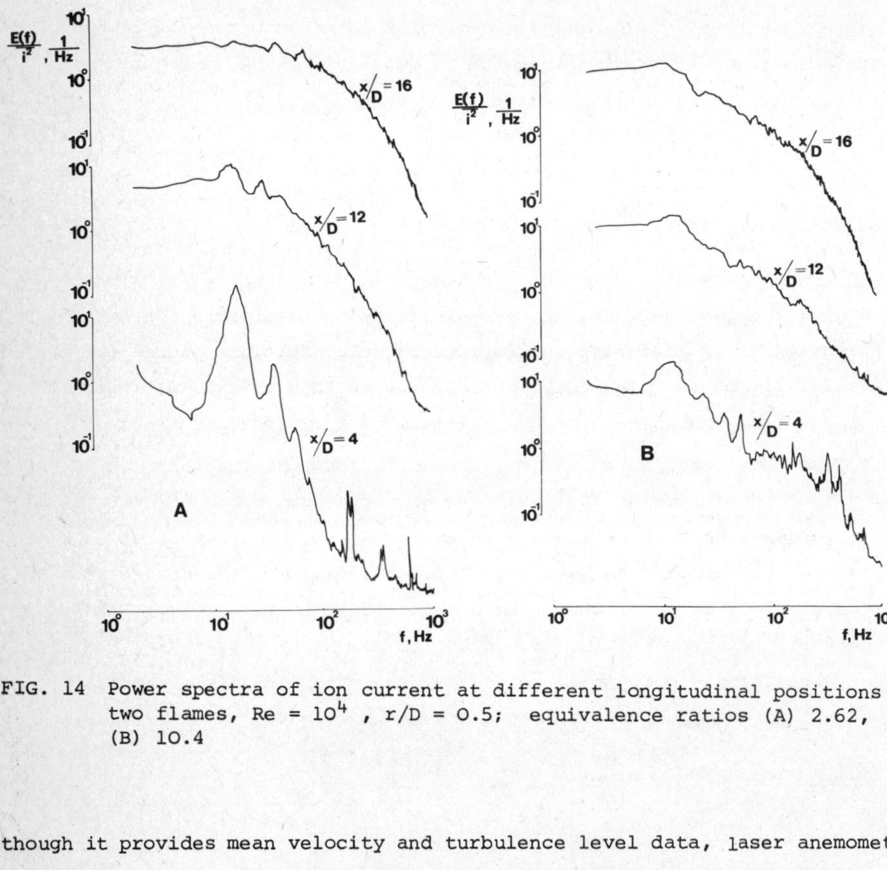

FIG. 14 Power spectra of ion current at different longitudinal positions in two flames, Re = 10^4 , r/D = 0.5; equivalence ratios (A) 2.62, (B) 10.4

Although it provides mean velocity and turbulence level data, laser anemometry is not a useful technique for point measurements of the frequency spectra of velocity in flames, because of the restrictions introduced by the need to seed the flow. Ionisation probes[14] have a good frequency response up to several kH_z and these are sensitive to the local occurrence of reaction. These probes are found to be sensitive to both the inner and the outer eddies and spectra are produced (Fig. 14) which are rather similar to those which have been measured by hot wires in transitional and turbulent cold jets. Examination of spectra such as these shows that the transition distance

x_T is typically 3 to 5 times longer in a jet flame than in a cold jet with the same exit velocity. L.D.A. measurements[12] show that the jet potential core can lengthen by a factor of more than 5 for the flame case. Similar to the use of hot wires in cold jets, two ionisation probe signals can be cross-correlated with variation in probe separation and time delay[14]. Both the slow convection of the outer eddies, and the fast convection of the inner eddies have been measured by this technique. The results show that the inner, acoustically driven eddies, move with a velocity close to U_j. The outer eddies are found to accelerate as they move downstream, due to expansion accompanying heat release.

The transitional jet flame, like the transitional cold jet, is relatively orderly compared with the turbulent flame region and is thus amenable to investigation. Processes such as the growth of the azimuthal core waves and coalescence can also be seen[13] in the transitional flame. Large eddies can also be seen in the turbulent region of the jet flames, and it has been observed that certain individual eddies can exist for almost the full length of visible flame (100 D). Sections of reacting interface have been identified[13] as coherent structures from their formation near the nozzle to at least 20 D downstream. The large eddies, their structures and their coherence, have important repercussions with respect to mixing, reaction and pollutant formation. In line with suggestions by Yule[2] multiple probes are now being used in order to attempt to track individual eddies as they move downstream. In this way the existence times and retention of coherence of individual eddies can be examined and the smearing of cross-correlations by jitter can be quantified.

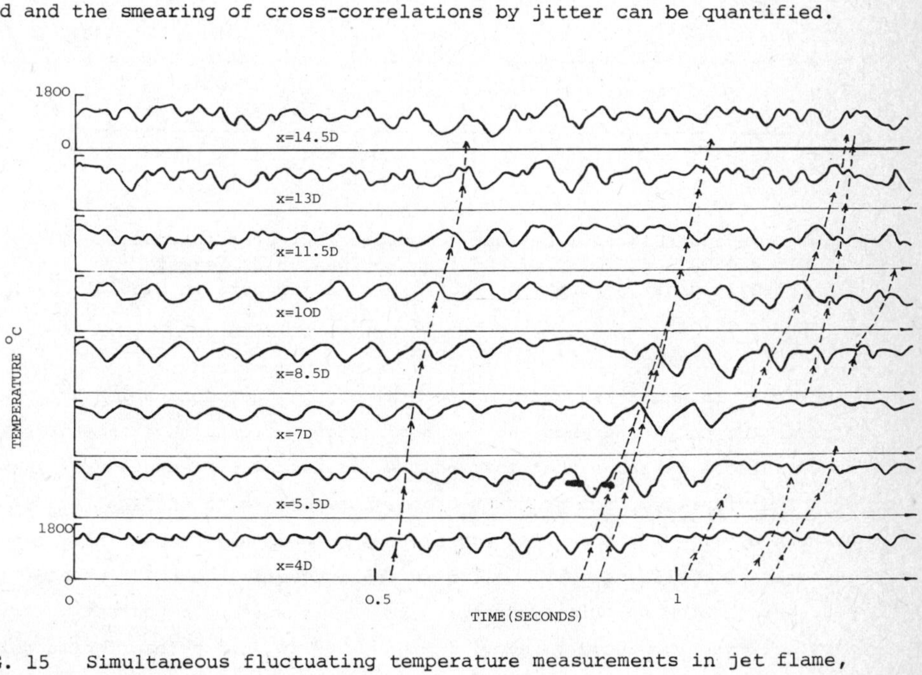

FIG. 15 Simultaneous fluctuating temperature measurements in jet flame, r = 0.71D

As an example Fig. 15 shows simultaneous time histories from 8 miniature thermocouples separated in the x (streamwise) direction by intervals of 1.5 D, with the probe nearest the nozzle at x = 4D. The probes are near the centre of the mixing layer, at r = 0.71D. It should be noted that the jet mixing layer region (i.e. where a potential core is found), can stretch beyond x = 15D for the flame case. The reacting round jet mixing layer is not 'self preserving' and the mixing layer tends to be curved so that the 'η' co-ordinate system is not useful.

Fig. 16 shows the space-time cross correlations between the probe at x = 4D and each of the other probes.

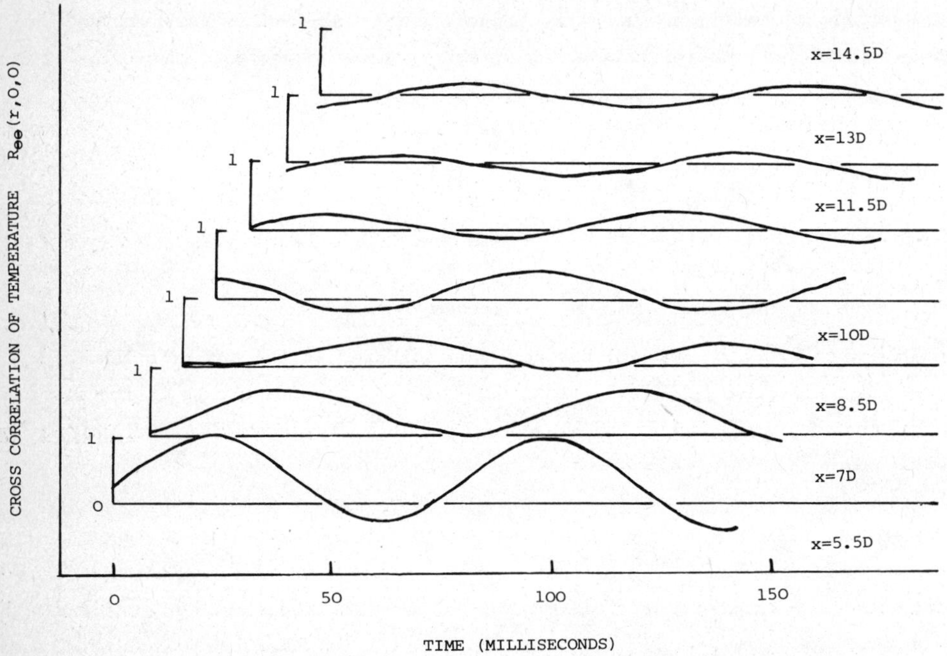

FIG. 16 Cross-correlations between temperature fluctuations at x = 4 D, r = 0.71 D, and fluctuations at different longitudinal separations (corresponding to data in Fig.15)

Examination of Figs. 15 and 16 illustrates several of the phenomena which have been discussed, concerning eddy coherence and the obscuration of coherence in statistically averaged data. Figure 15 shows that most eddies (as marked by temperature peaks) can be tracked individually, from probe to probe, between x = 4D and x = 14.5D. However, as shown by Fig. 16, the cross-correlation between signals at these two positions is small in amplitude. Variation in the convection times, from eddy to eddy, gives a contribution to this smearing. However a further important contribution arises from variations, from eddy to eddy, in their radial (and perhaps azimuthal) movements so that a probe 'sees' a different part of each eddy. This is particularly clear nearer the nozzle in Fig. 15. It is considered that the double-peaked structure of eddies found near the nozzle (e.g. x = 4D in Fig.15)may be indicative of a hot reacting region wrapped around a relatively cool eddy core. The signals at x = 4D and x = 5D in Fig. 15 clearly demonstrate how the relative magnitudes of these double peaks vary from eddy to eddy, presumably because of radial variations in eddy positions. An interesting phenomenon is the almost complete disappearance of indications of some eddies at certain times and positions, but the subsequent reappearance of these eddies at probes further downstream. This can be seen in Fig. 15 at the end of the time history for x = 7D. Multiprobe experiments of this kind show that a one-dimensional probe array gives indications of the existence of eddy coherence and also indicates the smearing, or phase scrambling effects. However, two-dimensional arrays are required to more clearly elucidate and quantify these effects.

5. CONCLUDING REMARKS

Examples can be given of the necessity for understanding coherent structures if satisfactory models are to be developed for turbulent combustion flows. Considering firstly the case of two phase flow, such as a turbulent liquid fuel/air spray, considerable photographic evidence has now been gathered in the author's laboratory showing that coherent large eddy structures play an important role in the dispersion, mixing, vaporisation and burning of sprays. Small droplets closely follow the gas flow and give a good visualisation of the large eddies. However, larger droplets with their smaller drag/inertia ratios, are seen to leave these eddies and penetrate the outer potential flow. Realistic modelling of droplet environments and thus vaporisation and burning, requires modelling of the large eddies and their interactions with the droplets. The second case concerns the field of pollution modelling and control. The existence of eddies, or of reacting interface layers, for distances comparable with the total flame length has important repercussions on the production of pollutants such as soot and NO_x. One can envisage flows in which individual eddies may be considered as reactors which are continually entraining, mixing and burning air, fuel and products. Current modelling approaches, based on time average flux equations, should preferably be adapted to include information derived from knowledge of the spatial fields within these eddies and also their time histories. A major objective of future research must be the development of techniques for including this 'Lagrangian' coherent structure information within computer models in a computable, yet physically meaningful, manner.

The round gas jet flame has been used as an example of a practically interesting flow in which coherent structures have been shown to be present and to have important influences on various flame parameters. Basic similarities have been found between the physical structures of reacting and non-reacting round jets, although quantitatively the two flows are very different. Experiments to measure the detailed structures of flames continue. However instrumentation problems are immense for the reacting flow, compared with the cold jet. Further work is required in the cold turbulent jet to discover the structures of the large coherent eddies which have been observed. It is likely that satisfactory progress will be made, only when very large multi-probe arrays can be used to acquire simultaneous information on large volumes of the flow.

ACKNOWLEDGEMENTS:

Research in the area of coherent structures in turbulent flames is supported at Sheffield by the Air Force Office of Scientific Research /AFSC, United States Air Force under Grant AFOSR-77-3414, and the U.S. Army European Research Office.

REFERENCES

1 Roshko A., 'Turbulent Mixing in Nonreactive and Reactive Flows' (ed. S.N.B. Murthy), pp 295 - 311, Plenum, 1975.

2 Yule, A.J., 'Phase scrambling effects and turbulence data analysis' Proceedings of Second Int. Symp. on Turbulent Shear Flows (ed. Bradbury et al), pp 263 - 281, Springer-Verlag, Heidelberg, 1980.

3. Yule, A.J. J. Fluid Mech., 89, pp 413 - 432, 1978.

4. Bernal, L.P., Breidenthal, R.E., Brown G.L., Konrad, J.G. and Roshko A. 'On the development of three dimensional small scales in turbulent mixing layers.' Proceedings of Second Int. Symp. on Turbulent Shear Flows, Springer-Verlag, 1980

5. Bradshaw P., (Personal Communication)

6. Bradshaw, P., J. Fluid Mech. 26, pp 225 - 236, 1966

7. Browand, F.K., and Laufer, J., Proc. 4th Biennial Symp. on Turbulence in Liquids Univ. Missouri - Rolla, pp 333 - 344, Princeton, New Jersey : Science Press, 1975

8. Davies, P.O.A.L. and Baxter, D.R.J., 'Transition in free shear layers' Internal Memo .I.S.V.R.,University of Southampton, 1978.

9. Michalke, A., J. Fluid Mech., 22, 2, pp 371 - 383, 1965

10. Bruun, H.H., J. Fluid Mech., 83, 4, pp 640 - 671, 1977

11. Yule, A.J., Bruun, H.H., Baxter, D.R.J., & Davies, P.O.A.L., University Southampton, ISVR Memo. No. 506, 1974.

12. Yule, A.J., Chigier, N.A., Ralph, S., Boulderstone, R., and Ventura, J.M.P. Paper AIAA-80-0077, AIAA, 18th Aerospace Sciences Meeting, Pasadena, Ca., 1980

13. Chigier, N.A., and Yule, A.J., Project Squid Techn. Rept. US-1-PU. Purdue University, March 1979

14 Ventura, J.M.P., Suzuki, T., Yule, A.J., Ralph, S., and Chigier, N.A., "The investigation of time dependent flame structure by ionisation probes." Proceedings 18th Int. Combustion Symposium, Waterloo, Canada, August, 1980

THE PLANE MIXING LAYER

FLOW VISUALIZATION RESULTS AND THREE DIMENSIONAL EFFECTS

A. Roshko
California Institute of Technology

Pasadena, Calif. 91125

The turbulent mixing layer between two streams of different velocities continues to play a central role in research aimed at improved understanding of turbulent shear flows in general. At present, not all researchers are in agreement as to what various experiments imply about the structure of mixing layers at high Reynolds number. The views which are held differ on the question as to how and to what extent three dimensionality develops in these flows and whether the characteristic spanwise organized large vortex structures (rollers) continue to be a dominant feature. The traditional view, as extended to the contemporary scene, is that ultimately (i.e., sufficiently far downstream or at sufficiently high Reynolds number) the flow will be completely disorganized. The view put forward by "eddy chasers" is that such vortex structures are primary elements, characteristic of the underlying mean vorticity field, which is particularly simple for the mixing layer, and that, as long as the velocity difference is maintained, there is a mechanism to regenerate these primary structures by what, for convenience, may be called a Kelvin-Helmholtz instability. The heart of the controversy then is whether, or to what extent, secondary and higher instabilities will ultimately break down, completely disorganize or prevent formation of organized primary structures. In a plane mixing layer, the primary structures would, ideally, be two dimensional, containing the basic single component of vorticity while secondary and higher modes of instability would introduce three dimensionality and the other two components of vorticity into the flow. An interesting question follows: to what extent do such secondary instabilities change the properties (e.g., the growth rate; the Reynolds stress) that the mixing layer would have in ideal two dimensional development? In this paper we examine several aspects of this question and discuss some recent relevant experiments.

The Ideal of a Two Dimensional Flow

To lend perspective to the problem and provide a gauge against which to evaluate experiments it would be nice to know how an ideal two dimensional mixing layer would develop. A picture of this could in principle be obtained from a computation of the two dimensional Navier-Stokes equations for suitable initial and boundary conditions.

Then growth rates and Reynolds stresses as well as details of the structure could be compared with measured ones in real flows. An exact calculation of such a flow, for boundary conditions appropriate to the experiments, is not available. Ashurst's (1979) approximate calculation by the method of discrete vortices is so far the most ambitious attempt to calculate a spatially developing mixing layer. The qualitative similarities between the computed two dimensional flow and the experimental ones are striking but inconclusive. It is impossible to say whether the differences, especially at the higher Reynolds numbers which were simulated, are due to limitations of the computation (for example, definitions of initial conditions and of Reynolds number are problematic) or to differences between an ideal two dimensional flow and the real one. Nevertheless, such calculations are useful and provide instructive insights. Of significance for the present discussion is that the large clumps of vortices which evolve and interact with each other and which show similarities to the experimentally observed vortex structures exhibit some statistical features qualitatively similar to those in real flows. In particular, scales and lifetimes of the vortex structures passing a given spatial location are found to be broadly distributed (dispersed). To this extent the flow develops a disorganized or turbulent character without intervention of any three dimensional effects.

The Initial Region and the Mixing Transition

To further explore the question of the role of three dimensionality in mixing layers it is instructive to compare the well known measurements by Bradshaw (1966) of Reynolds stress in the initial development region of a mixing layer and recent studies by Konrad (1976), Breidenthal (1978) and Jimenez, Martinez-Val and Rebollo (1979) of the development of mixedness in that region. The two results are schematically compared in Fig. 1. The important result found by Bradshaw is that when the free shear layer originates from an initially laminar boundary layer the development of Reynolds stress $(\overline{u'v'})$ is very different from that which occurs when the boundary layer is initially turbulent. In the laminar case the Reynolds stress was found to overshoot the final, asymptotic value while in the turbulent case it increased monotonically to approximately the same value. In a more recent investigation of the final state of development of a mixing layer, Foss (1977) concluded that "the flow fields from both the laminar and the turbulent initial conditions are essentially identical". Thus it seems fair to assume that the state downstream of the point marked B in Fig. 1 is the same in both cases and to describe it as "fully developed turbulent". (Previously to Bradshaw's work the point labelled A had been called "the end of transition").

Breidenthal (1978) studied the development of "mixedness" by measuring the amount of chemical product formed in the mixing layer between two streams of water

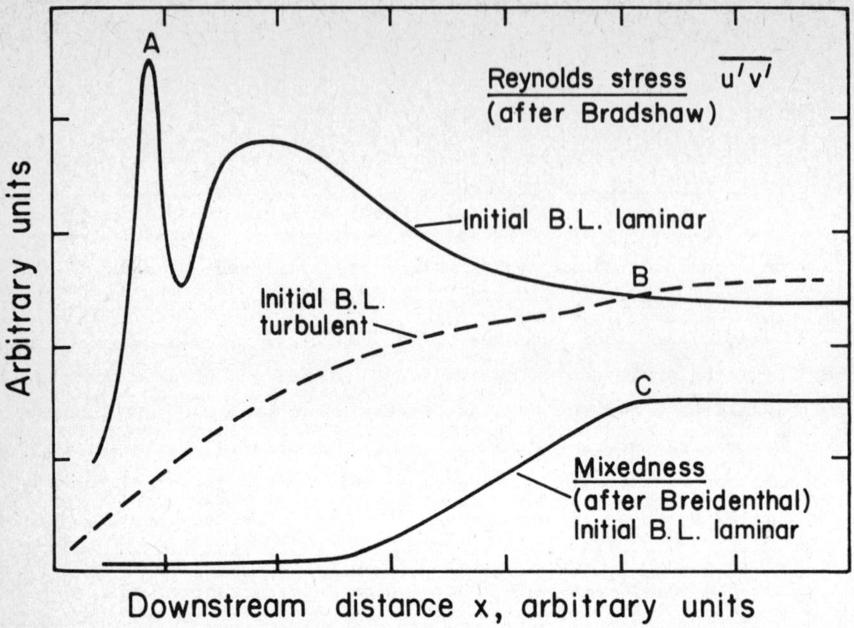

Fig. 1. Development of Reynolds stress and mixedness

carrying appropriate reactants. His result, for initially laminar boundary layers, is also shown in Fig. 1. From low values the mixedness increases, by a factor of more than 10, through a region which he called the "mixing transition" and reaches an asymptotic value at some downstream distance which, in Fig. 1, is indicated by C.

The juxtaposition of these curves in Fig. 1 is schematic and subjective and presupposes that the downstream distances to B and C are the same, in terms of a single parameter such as x/θ_o, where θ_o is the initial momentum thickness, or xU_1/ν, where ν is the kinematic viscosity. Actually, as argued by Bradshaw and by Breidenthal, both parameters or a corresponding pair are needed to uniquely define the development. The development is further dependent on other parameters such as the ratio of free stream velocities, U_2/U_1. In the examples from which Fig. 1 was constructed, point B for Bradshaw's initially laminar condition (with $U_2 = 0$) corresponds to x/θ_o from 500 to 1000 and to U_1x/ν about 7×10^5 while for Breidenthal's experiment (with $U_2/U_1 = 0.38$) point C corresponds to $x/\theta_o = 650$ and $U_1x/\nu = 1.3 \times 10^5$. Clearly there is need for a definitive experiment in which Bradshaw's and Breidenthal's measurements are made on the same flow. Nevertheless there is considerable indirect evidence that points B and C are the same. For example, Jimenez et al. (1979) found a range of downstream locations (approximately

$3 < 10^{-4} \frac{U_1 x}{\nu} < 6$, with $U_1/U_2 = 0.5$) in which the dependence on frequency of the power spectrum of velocity fluctuations in the inertial subrange changed from -3 to $-5/3$ power, suggestive of a change from two to three dimensional turbulent structure of the small scales. In the same transition range the amplitude of velocity fluctuations decreased, as at the end of Bradshaw's transition range. Those results are linked in turn to the measurements of Konrad (1976) and Breidenthal (1978) who associated the increase of mixedness with the development of small scale, three dimensional motions.

One more piece of this picture is important. Several methods of flow visualization in the experiments of Konrad and Breidenthal show that the small scale structure downstream of the mixing transition is superimposed on primary vortices which still have good spanwise coherence in the large. An example from Konrad (1976) is given in Fig. 2, which shows simultaneous edge and plan views of a mixing layer in flow with uniform density and $U_2/U_1 = 0.38$. Other examples may be found in papers by

Fig. 2 Edge and plan views of a mixing layer
$U_1 = 1000$ cm/sec, $U_2 = 380$ cm/sec, ρ_1 = density of N_2 at $p = 4$ atm., $\rho_2 = \rho_1$ = density of He/Ar mixture.
Scale: streamwise dimension of picture = 15 cm.

Bernal et al. (1979) and Breidenthal (1979). This figure is quite significant because it shows the existence of spanwise organized primary vortices well downstream of the mixing transition, which is located at the left hand side of the picture in the region where, in the edge view, there is an obvious qualitative change in the small scale pattern. The pattern of streamwise streaks which, in plan view, is superimposed on the spanwise pattern of primary vortices is important for the discussion in the following section.

On the basis of the above, to some extent circumstantial, evidence the following picture of developments in a mixing layer can be inferred.

(1) In an initially laminar shear layer, the Kelvin-Helmholtz instability produces a pair of two dimensional vortices which merge to form new pairs with twice the initial scale (Freymuth, 1966). These processes, as noted by Bradshaw, are occurring in the region where $\overline{u'v'}$ has its maximum values. It may be expected that in this region the distribution of scales (vortex spacings) is centered around the initial Kelvin-Helmholtz value, is possibly bimodal with respect to that value, and in any case must be quite different from the broad distribution, with shifting center, which develops further downstream (Brown and Roshko, 1976; Winant and Browand, 1976; Bernal, 1980) after several more pairings have occurred. As mentioned earlier in connection with Ashurst's computation, this redistribution could occur in two dimensional flow and the decrease in shear stress from peak values to the asymptotic value, if connected to this redistribution, would be a correspondingly two dimensional affair. Thus it is not clear whether the small scale, three dimensional motions, which in real flows develop in the same region of redistribution, are incidental or necessary for the decrease in stresses to the asymptotic value.

(2) What is quite clear is that the small scale, three dimensional motions are necessary for the mixing transition. This fact is especially well brought out in Breidenthal's experiments in water, which show that mixedness remains very low (at the initial laminar value) in the region of the first one or two vortex pairings, where shear stresses reach peak values.

(3) It follows that the mixing processes for momentum and for scalars are quite different in the developing region.

(4) Since the final, fully developed turbulent state is the same, whatever the initial conditions, it follows that for initially turbulent boundary layers the primary, spanwise organized vortices must emerge from the initially three dimensional structure. More specifically, the primary structures must develop from a Kelvin-Helmholtz instability of the initial vorticity layer, which in this case consists of "turbular" fluid (the term proposed by Liepmann, 1962). Evidence for the emergence of primary structure from initially turbulent or highly three dimensional conditions may be seen in pictures obtained by Hussein (1979) and by Breidenthal (1980).

Three Dimensional Structure

The turbulent mixing layer might be viewed as a synthesis of basic structures connected with a hierarchy of instabilities (Corcos, 1979, 1980). As discussed in the preceding, the primary structure would be the spanwise vortex resulting from the Kelvin-Helmholtz instability. The next mode might be either a spanwise instability or a secondary, internal instability or possibly a combination of the two.

By spanwise instability we mean waviness or other deviation of the vortex from a straight cylindrical structure. Instabilities of this kind have been studied by Hama (1963) and have been observed by Chandrsuda et al. (1978). Such disturbances would contribute to loss of spanwise phase coherence in individual vortices and in interaction processes (such as pairings) between them. Examples of the latter (spiral or bifurcated pairings) may be seen in pictures obtained by Chandrsuda et al. of a mixing layer with $U_2 = 0$. On the other hand, there is not much evidence of such spanwise instability at finite values of U_2/U_1, as in the case shown in Fig. 2. It is reasonable to suppose that any spanwise instability will be competing with the primary instability which continually regenerates the primary structure and changes the scale. Thus a spanwise instability may not develop if its rate is slow compared to the primary one. The relative development rates may depend on parameters such as U_2/U_1, as the cited experiments suggest.

Spanwise imperfections resulting from such instabilities will tend to degrade conventional spanwise correlations. Thus Browand and Troutt (1980) found that in a well developed mixing layer the conventional average correlation coefficient for velocity was down to 0.2 at a spanwise separation of three vorticity thicknesses ($\Delta z = 3\delta_\omega$). On the other hand, by deploying twelve hot wires spanwise they obtained instantaneous "pictures" of velocity correlation which exhibited spanwise well oriented contours, not perfectly two dimensional of course but suggestive of the spanwise organized structures seen in flow pictures.

Whether spanwise instabilities would greatly alter the Reynolds stresses, as compared to those in a two dimensional development, is part of the question we posed earlier. It seems fairly certain that they would not greatly enhance mixedness which, we believe, is increased mainly by the action of what we will call secondary, internal instability that produces strong secondary motions <u>inside</u> the primary structures. In the mixing layer this secondary instability creates pairs of streamwise vortices which are embedded in the primary vortices and in the connecting vortex layers or braids (the term used by Patnaik et al., 1976) which connect the latter.

The first evidence for these streamwise vortices was seen in pictures such as that in Fig. 2 obtained by Konrad (1976) in gas mixing layers, subsequently by Breidenthal (1978) in experiments in water. It was conjectured that the streaks mark the edges of streamwise vortex pairs but direct evidence for this has only recently been obtained by Bernal (1980) who, using a visualization technique developed by Dimotakis, obtained pictures of the flow through planes normal to the stream direction. Two examples are shown in Fig. 3. In Fig. 3a the visualized plane

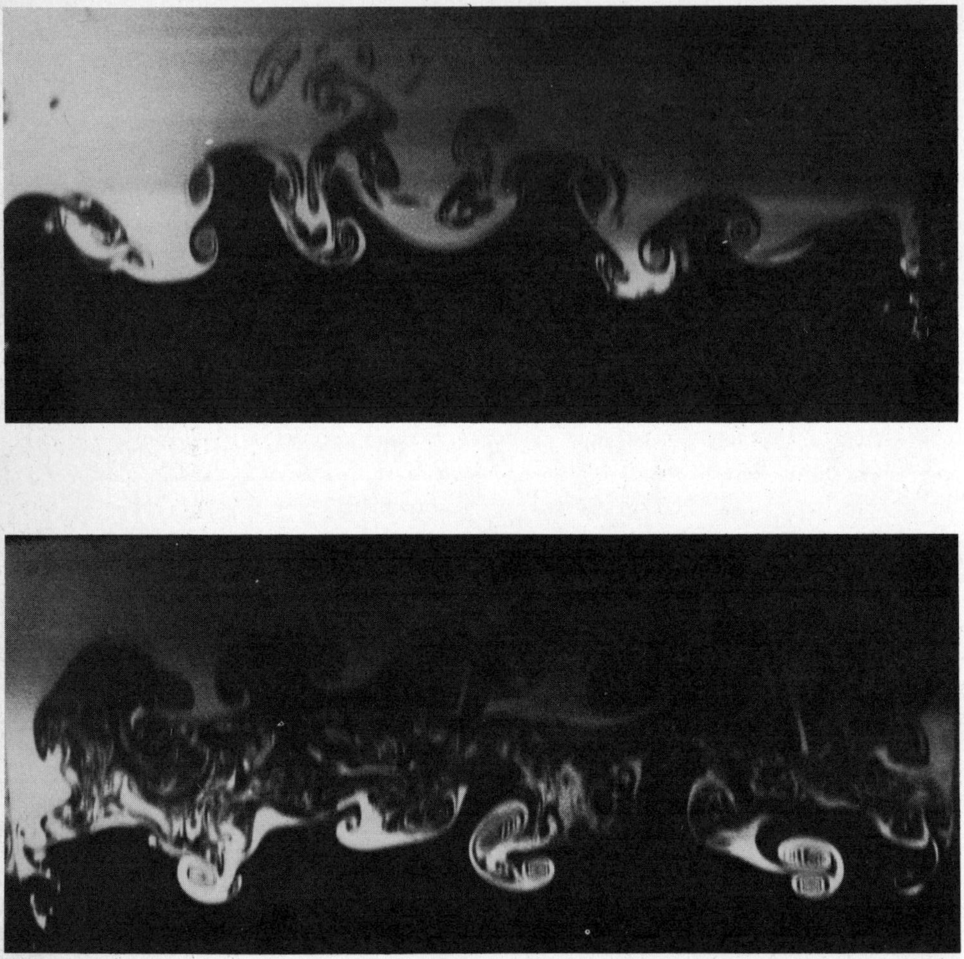

Fig. 3 Cross sectional views normal to flow direction of mixing layer in water.
$U_1 \doteq 32$ cm/sec, $U_2 \doteq 11$ cm/sec, x = 15 cm.
a) Section between primary vortices
b) Section through a primary vortex

cuts through the vortex sheet (braid) between primary vortices. It may be seen that the vortex sheet is highly distorted by the streamwise vortices, whose cross sections are imaged in the plane of the picture. In Fig. 3b the plane of view is through a primary vortex, whose spanwise cylindrical structure is highly sculptured by streamwise vortex pairs on both sides of the mixing layer. It is not difficult to suppose that the secondary motions induced by those streamwise vortices will be much more effective in promoting internal mixedness than would the spanwise instabilities discussed above. But again, it is not clear how much they modify the Reynolds stresses connected with the primary structures.

The existence of the secondary vortex pairs poses further, interesting questions. Upstream of the mixing transition region their spanwise spacing is fairly regular and approximately equal to the Kelvin-Helmholtz spacing in the initially laminar shear layer (Breidenthal, 1978; Bernal 1980). If they are an essential part of the structure of a turbulent mixing layer then, it might be argued, their scales should increase as the mixing layer grows downstream. Indeed, there is some evidence for this from spanwise correlation measurements (Jimenez et al., 1979a; Bernal, 1980) and from flow pictures such as the one in Fig. 2. In the plan view of Fig. 2 the streaky pattern in the initial part of the mixing layer upstream of the mixing transition repeats itself, at a larger scale, near the right side of the picture well downstream of the mixing transition. This reorganization to a larger scale is reminiscent of a similar phenomenon observed by Taneda (1959) in vortex streets in the wakes of cylinders.

The intriguing question is how such a change of scale is accomplished. If it is by amalgamation processes, these would be rather more complex than those between the primary vortices because the streamwise, secondary vortices occur in pairs of opposite sign and are deployed on either side of the mixing layer. Some hint of an interaction process is seen in Fig. 3b, where two vortex pairs at the bottom right appear to be rotating in opposite directions. Possibly the streamwise vortices are regenerated locally, forming streamwise elongated loops, which line up by mutual interaction to produce the pattern of extended streamwise streaks, and changing scale only when the primary scale has developed to a sufficiently large value.

Summary and Conclusion

We have posed the possiblity that development of a mixing layer is largely determined by two sets of organized structures: the primary, spanwise vortices and the streamwise, counter rotating vortex pairs. The Reynolds stress and the growth of the layer are controlled mainly by the primary vortices while the secondary set provides internal mixing, and possibly modifies the stress. With increasing Reynolds number, higher order, smaller scale structure will be necessary, and available, for accomplishing internal mixing. It does not seem likely that the higher order structures will be perceived to be organized and it may be sufficient to view them in terms of the classical cascade and to model them appropriately.

Due to loss of phase coherence, the turbulent or random character of the flow appears already in the primary structure; this produces a broad spectrum of scales (wave lengths) about the mean value appropriate to any downstream position x. It apparently results from the effects of noise in the initial conditions and in the external flow (Delcourt and Brown, 1979). In contrast, a small, periodic initial or free stream disturbance tends to dominate and make coherent that portion of the shear layer whose scale is commensurate with the imposed wave (Wygnanski, Oster and Fiedler, 1979).

The secondary set of vortices, which is superimposed on the primary set, has a curious downstream development. The spanwise spacing tends to remain constant while the primary scale is increasing, through at least one amalgamation, but it ultimately readjusts to a larger scale. The persistence of spacing is probably connected with the streamwise orientation. Whether the readjustment to larger spacing occurs by amalgamation or by regeneration is not yet clear. Possibly those two processes are simply aspects of one and the same instability process. In fact such a complementarity is suggested in the results obtained by Patnaik et al. (1976) for the primary instability.

The motion generated by the system of primary and secondary vortices would evidently be complex even if phases were coherent. Adding to this the loss of phase in the primary system it is clear that these two sets of organized structures would generate a "turbulent" flow. For Reynolds number tending to infinity, a cascade of smaller structures would be needed to accomplish internal mixing of scalars, but it seems likely that the main features of momentum exchange and corresponding growth rate of the layer may be determined by the system consisting of the primary and secondary structures.

Acknowledgments

I am indebted to L. Bernal, R.E. Breidenthal, J.E. Broadwell, G.L. Brown and P.E. Dimotakis for discussions of the problems explored here and for use of material in the figures. The research work on which this paper is based was made possible by the financial support of the Office of Naval Research, through Project SQUID and through its Fluid Dynamics Program, and of the Air Force Office of Scientific Research through the Air Force Weapons Laboratory.

References

Ashurst, W.T. 1979 "Numerical simulation of turbulent mixing layers via vortex dynamics", Turbulent Shear Flows, Durst et al. (eds.), Springer Verlag, Berlin, 402.

Bradshaw, P. 1966 "The effect of initial conditions on the development of a free shear layer", J. Fluid Mech. 26, 225.

Breidenthal, Robert E., Jr. 1978 "A chemically reacting, turbulent shear layer", Ph.D. Thesis, California Institute of Technology, also AIAA Journal 17, 310-311.

Browand, F.K. and Troutt, T.R. 1980 "A note on spanwise structure in the two-dimensional mixing layer", J. Fluid Mech. 97, 772.

Chandrsuda, C., Mehta, R.D., Wier, A.D. and Bradshaw, P. 1978 "Effect of free-stream turbulence on large structure in turbulent mixing layers", J. Fluid Mech. 85, 693.

Corcos, G.M. 1979 "The mixing layer: deterministic models of a turbulent flow", Univ. of Cal., Berkeley, College of Eng. Report No. F. M-79-2.

Corcos, G.M. 1980 "The deterministic description of the coherent structure of free shear layers", Published in present proceedings,

Delcourt, B.A.G. and Brown, G.L. 1979 "The evolution and emerging structure of a vortex sheet in an inviscid and viscous fluid modeled by a point vortex method" Proc. Second Symp. on Turbulent Shear Flows, July 2-4, Imperial College, London.

Freymuth, Peter 1966 "On transition in a separated boundary layer", J. Fluid Mech. 25, 683.

Hama, Francis R. 1963 "Progressive deformation of a vortex filament" Physics of Fluids 6, 526.

Jimenez, J., Martinez-Val, R. and Rebollo, M. 1979 "The spectrum of large scale structures in a mixing layer", Proc. Second Symp. on Turbulent Shear Flows, July 2-4, Imperial College, London.

Jimenez, J., Martinez-Val, R. and Rebollo , M. 1979a "On the origin and evolution of three dimensional effects in the mixing layer" Universidad Politécnica de Madrid Report.

Konrad, John H. 1976 "An experimental investigation of mixing in two-dimensional turbulent shear flows with applications to diffusion-limited chemical reactions" Ph.D. Thesis, California Institute of Technology. Also Project SQUID Tech. Rep. CIT-8-PU.

Liepmann, H.W. 1962 "Free turbulent flows" Mecanique de la Turbulence, C.N.R.S., Paris, 211-226.

Patnaik, P.C., Sherman, F.S. and Corcos, G.M. 1976 "A numerical simulation of Kelvin-Helmholtz waves of finite amplitude" J. Fluid Mech. 73, 215.

Taneda, S. 1959 "Downstream development of the wakes behind cylinders" J. Physics Soc. Japan, 14, 843.

Wygnanski, I., Oster, D. and Fiedler, H. 1979 "The forced, plane, turbulent mixing-layer: a challenge for the predictor" Proc. Second Symp. on Turbulent Shear Flows, July 2-4, Imperial College, London.

USEFUL STRATEGIES IN TURBULENCE

EXPERIMENTS AND MODELLING

D. Coles

California Institute of Technology
Pasadena
U.S.A.

Note: Not available at the time of publication.

Initiation, Evolution and Global Consequences of Coherent Structures in Turbulent Shear Flows.

H.E. Fiedler, B. Dziomba, P. Mensing, T. Rösgen+
Hermann-Föttinger-Institut für Thermo- und Fluiddynamik,
Technische Universität Berlin

1. Introduction

The observation of a general tendency to forming organized and to some degree deterministic structures in turbulent shear flows - as first reported by Brown & Roshko [1] - started a whole series of investigations into this phenomenon which added considerably to our understanding of turbulence and in some cases to our confusion.
Some of the results and concepts that evolved were summarized by Roshko [2] by Liepmann [3], and by Laufer [4].
The structures are understood to be of large scale - typically comparable to the overall dimensions (width) of the flow - and of time scales at least of the order of time necessary to travel its own length.

Existence of large scale structures and their influence on the flow behaviour has already been postulated by Townsend [5]. His "big eddies" were, however, conceptualized as being fully random constituents of the turbulent motion.
Modern concepts of coherent structures include as an essential character its determinism. They seem to some extent to control the flow, in particular its mixing properties and its acoustic radiation.
Their detailed analysis may therefore provide a key for understanding as well as manipulating turbulent flows and provide a basis for their modelling.

Some of the specific questions to which we have to address ourselves are:

1) What typical flow-specific coherent "modules" are formed and how are they related to the mean flow?

2) What is the detailed internal structure of the coherent module?

3) What causes its repeated and spontaneous creation?

4) Is it a transient or an equilibrium phenomenon and what is its "natural" strength?

+Presently at California Institute of Technology/USA

5) What is its consequence regarding e.g. noise radiation, skin friction, mixing, combustion?

2. Experimental Techniques

For the analyisis of coherent structures in a turbulent flow special experimental techniques have to be applied and - in part - had to be developed according to the objectives of the investigation. While it is still possible to detect the presence and to some extent the strength, the geometry and the occurrency of a coherent structural constituent directly or by inference by simple statistical methods, those are insufficient for a detailed analysis.

For this aim the following tools may be successfully employed:

I short exposure visualization
II analysis of multi-probe signals
III pattern recognition
IV complex two-point cross power spectral density analysis (Fuchs, Mercker, Michel [6].)

For the investigation described herein methods I and III were primarily used.

Visualization of the flow will primarily serve as a qualitative tool; it may, however, in conjunction with a digital computer also yield to quantitative data (e.g. image processing etc.). It requires a very narrow sheet of light to prevent obliteration by depth-integration and on the other hand very short exposure times to ensure sharp contours of the tagged flow (e.g. smoke). This was obtained by a sweeping laser beam: while its total light intensity is in the test section for, say, 1/100 sec; each flow particle is illuminated only during the extremely short time while the beam passes. By this method, which is described by Mensing & Fiedler [7] , very sharp images are obtained.

A pattern recognition scheme was applied for analysis of coherent vortices in the excited shear layer; it was then, however, for the time being abandoned in favour of a complete flow-field mapping by phase averaging (phase of excitation sinewave). This is a considerably less pretentious way of extracting the desired characteristics of the stabilized structure. Its shortcomings are, however, to some extent counterbalanced by certain weaknesses of the pattern recognition method and it is well capable of supplying "mean" data of the vortex structure.
In short: While by phase averaging no other alignment procedure is employed than its linkage to the excitation frequency, pattern recognition methods try to align events by the local measurement itself. The former method thus yields only to a structure

"smoothed" by the jitter that is caused by frequencies essentially lower than the excitation frequency and has therefore no sharp contours (e.g. no "umbilical cord"). The randomness of the low frequencies, intrinsically no constituent of the structures to be educed show up as a pseudo-turbulence and cause a distorted distribution. This problem is in principle eliminated by the pattern recognition methods which, on the other hand, need a number of criteria which to some extent either drastically limit the selection or artificially modify the extracted pattern to fit a preset frame. Thus a certain artificiality is introduced and the structure finally obtained may not be an average - but an average "conditioned" structure. (See also Yule [8]).
True pattern recogntion was used to analyse the temperature field of a heated shear layer by Rösgen [9]: The high velocity side (u_1 = 10 m/s) was heated to 15° Celsius above room temperature whereas the zero-velocity side remained unheated. The resulting fluctuations were measured by temperature probes (.6 μm hot wire, low overheat ratio) with a frequency response of about 5 kHz (see [10]).
The signals were digitized at 2.56 kHz sampling rate (1 kHz cutoff) and stored directly on the disc of a minicomputer. The processing capacity of the machine restricted the data aquisition to two simultaneously sampled data channels. Multichannel results were then obtained by performing a coincidence test between the structure traces in the simultaneously sampled channels, determining their mean relative time shift and placing the single channel ensemble averages in this mean shift position.

The principal goal of the date processing was to produce a "noise-free" picture of a typical mean structure by ensemble-averaging over a large number of observed single realisations.
The detection algorithm for the structure signatures worked on each data channel separately. It was based essentially on the checking of various amplitude-, slope- and intermittency criteria. Since all results were obtained at a fixed measuring station in x (600 mm) the typical features of a structure could easily be defined and the algorithm was "calibrated" for their recognition. Various tests showed that the resulting ensemble averages were independent on the actual detection levels chosen. This independence is of special importance for self-triggering algorithms, since it proves that the detection is at least locally (in pattern feature space) non selective.

The temporal length of each recognized pattern was used to classify the structures. A statistical evaluation yielded the mean pattern length which was chosen to select the signal traces entering the final ensemble averaging process. In general, a deviation of ± 10% from the mean was allowed for a "good" signal trace, bringing the overall efficiency of the detection/ averaging algorithm to about 5% (1000 structures detected of about 2000 possible, 100 averaged). To suppress distortion of leading and trailing edges in the ensemble mean, each pattern was normalized to the computed mean length before the averaging.

This way of proceeding is questionable but it was assumed that in restricting the normalization to essentially similar structures (same length), the possibility of averaging physically not-comparable single realisations was effectively eliminated. Results of this investigation will be described later.

3. Excited shear flows - general

It has been observed frequently that by periodic stimulation those coherent structures may be strengthened and stabilized, which points to an instability-mechanism for their creation and on the other hand provides a tool for their detailed analysis and handling. These considerations have led to a number of investigations which have shed some light on the questions stated above.

Excited structures in wall bounded flows are treated elsewhere (e.g. Wygnanski [11], Eckelmann & Wallace [12], Coles [13]).
Of the free shear flows the jet has been the first configuration at which excitation was studied in great detail (e.g. Wille et al [14], Hussain [15]).

Recently D. Korschelt [16] investigated the reaction of a two-dimensional turbulent jet to low-frequency periodic excitation at the nozzle. The investigation was concentrated on the region of approximate self-preservation. Some of his observations were:

1) Only flapping (counterphase) excitation at the nozzle results in a strong non-linear widening of the flow at a distance x_E(>30d)

   where $$\frac{(x_E - x_o)}{\bar{u}_{m_E}} \approx 1.0$$

   for all frequencies investigated,

   where $\bar{u}_{m_E}$ = mean velocity at x_E in non-excited flow

   f_E = excitation frequency and d = nozzle width.

2) At $x < x_E$ the jet is widening by almost 100%; all turbulence intensities are increased and the coherent energy constituent of the basic frequency reaches approximately 80% of the total turbulent energy. The intermittency distribution is considerably wider than in the neutral case and remains below 1 in the jet center.

3) Excitation beyond a certain level is of little effect.

Apart from its analytic value this investigation again demonstrates the powerful possibility to influence a flow via coherent structures, - an observation, which for a flame is classical.

In the following we concentrate on the plane shear layer, being a flow module of greatest simplicity yet of considerable practical interest. (For a summary see Murthy [17]). Its understanding is also of importance for other, more complex shear flows, since in most cases those flows develop from simple shear layers.

Two fundamentally equivalent configurations of practical relevance are usually investigated:

1) The two stream layer: $u_2/u_1 > 0$,
it is the more general configuration, its properties are easier measured yet it requires a longer test section for full development.

2) The one stream layer: $u_2/u_1 = 0$,
being a special case of 1) it is of greatest simplicity and requires shortest evolution-lengths. Difficulties and thus uncertainties arise on its zero-velocity side where measurements are no longer reliable because of very low mean velocity and strong variations of flow angle.

Certain scales of one of these configurations can be transformed into the other by the relation:

$$\frac{\delta \cdot (u_1 + u_2)}{x \cdot (u_1 - u_2)} = \text{const, where } \delta \sim \text{transverse scale and } x \sim \text{longitudinal scale,}$$

Thus the spread may be expressed as

$$\sigma_{(u_1, u_2)} = \sigma_{(u_1, u_2 = 0)} \frac{(u_1 + u_2)}{(u_1 - u_2)}$$

Global effects of excitation were investigated in both configurations. Thereby it is found that certain properties, like the turbulent normal stresses, develop differently in the two cases and can therefore not be transformed in this simple way. Thus e.g. the strong double peak in the u' -distribution observed in the excited two-stream flow becomes less prominent for decreasing u_2/u_1 and is no longer observed for $u_2/u_1 = 0$ (Oster [25]).

A number of investigations into the flow- and scalar- structure of these flows were performed during the past years (e.g. Fiedler, Korschelt, Mensing [18]; Wygnanski, Oster, Fiedler [19]; Münch [20]; Mensing [21]; Wygnanski, Oster, Fiedler, Dziomba [22]; Oster, Wygnanski, Fiedler [23]; Hussain [24]).

A question of prime importance is the spacial evolution of a periodic disturbance leading eventually to a coherent vortical structure. This was, however, studied only recently: Oster [25] and Wygnanski investigated the development of a periodic excitation in the two-stream layer developing from a laminar boundary layer at the splitter plate. Excitation was introduced by a vibrating ribbon.

A similar analysis was performed in the one-stream layer yet with turbulent boundary layer at separation by Fiedler and Rösgen [26].
Both investigations yielded to essentially similar results. While, however, Oster observed vortex amalgamations already in the first - i.e. excitation region - no irregularities of this kind were found in the one-stream case. This difference may be attributed to the different separation conditions, where in the laminar case (Oster) the natural "transition"-frequency is superimposed on the excitation. ("Collective interaction" - Ho [27]). In the following the investigation of the one stream layer is described.

4. Vortex evolution in the excited shear layer

4.1 Experimental arrangements and general considerations

The test arrangement used is described in detail in [18]. The nozzle has a square outlet of 300 x 300 mm. The test section is 1160 mm long being extendable to 1660 mm. It has rectangular side walls fitted with rounded upper edges for smooth entrainment inlet. For flow excitation at the trailing edge a loudspeaker system is mounted by which the separating boundary layer can be oscillated. The bottom of the test section is closed. The velocity in the nozzle exit is u_1 = 11 m/s with a turbulence level of < 0.15%. Maximum Reynolds number is thus:

$$Re_L = u_1 \cdot L/\nu = 0.85\ (1.3) \cdot 10^6$$

The boundary layer is turbulent (sand paper) with $\Theta \approx 1.0$ mm. For reasons of desired disturbance-free "neutral" flow this test section is mounted between two large rooms between which a constant pressure difference was maintained. Correlation measurements along the axis of the flow $(y = 0)$ yielded to the following results:

1) The points x_E of maximum amplification (saturation) of the excitation frequency f_E scale according to a Strouhal number

$$S_E = f_E\,(x_E - x_o)/u_1 \approx 1.0$$

where x_o is a virtual origin.

2) The periodic constituents of the auto-correlations reach a saturation level for increased driving amplitude A_o.

A frequency of 20 Hz was chosen for the further investigations, as it was hoped that the full range of excitation was confined within the length of the test section, with minimum influence of the boundary layer thickness at the trailing edge.

The excitation amplitude A_o was chosen such that

$$\sqrt{\overline{v_o'^2}} \;/\; u_1 \approx 0.0025,$$

where $\sqrt{\overline{v_o'^2}}$ is the maximum rms-value of the lateral periodic fluctuation at the trailing edge. This amplitude represents a saturation value whereby 99% of the maximum possible deterministic energy content at x_E are measured. It is extremely weak and fully embedded in the turbulence of the separating boundary layer and barely discernible in the spectrum at $x \to o$.

Before discussing the results we shall consider some general points concerning the excitation:

1) The flow exhibits the same excitation behaviour for all frequencies applied. On the lower end the frequency is limited only by the length of the test section and the proximity of the walls; an upper bound exists as a consequence of finite thickness of the initial boundary layer. Since the length of the excitation region is proportional to the excitation wavelength the degree of randomization by the random turbulence constituent is the same for all frequencies.
   Defining a measure for the relative randomization

   $$l^*/\Lambda \sim l^* f_E / u_1$$

   where Λ = wavelength and
   l^* = deformation $\sim u_1 \cdot \Delta t$

   then $l^* \sim x$, since $\Delta t \sim \Delta x / u_1$

   and $l^*/\Lambda \sim \dfrac{x_E \, f_E}{u_1} = S_E =$ const.

   This is only valid for self-preserving flow in the range of randomization, i.e. for large x. Thus the maximum correlation of the excited wave will be the same for any frequency at x_E. The maximum v'-correlation of the periodic constituent found in the shear layer is

   $$R_{f_E(v')}\Big|_{max} \approx 0.8$$

   therefore $\overline{v_R'^2} / \overline{v'^2} \approx 0.2$ (random) and $\overline{v_p'^2} / \overline{v'^2} \approx 0.8$ (periodic)

The same was observed for the excited plane jet [16].

2) Since in a diverging flow there is only a certain length from the locus of excitation (origin at trailing edge) to the width of highest amplification and since there is a defined amplification rate $A \sim x^{k \cdot x_E}$, a certain finite excitation is necessary to achieve full amplification. Larger excitation amplitudes will result only in a somewhat widened, slightly upstream-shifted region of a maximum amplification yet non-linearities in the amplification range. For different frequencies a relationship $A_0 \sim e^{-k\, x_E}$ is approximately satisfied.

3) Concerning the mode of excitation we may distinguish between a local disturbance and a field disturbance and on the other hand between a longitudinal and a transverse disturbance.
Since the only integral scale of the shear layer is its width b, a displacement of width Δy will have its highest effect at the position of smallest b. Field excitation will therefore be most effective near the trailing edge, where $\Delta y/b = \max$.
Furthermore it is easily shown, that longitudinal and vertical excitation are effectively equivalent. The mode of excitation applied is therefore considered irrelevant regarding the flow development.

4.2 Results of investigation

Some characteristic global effects of excitation are shown in the following diagrams:

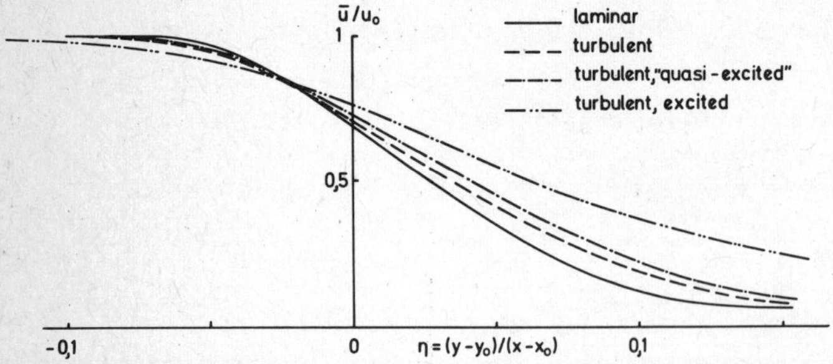

Fig. 1: Mean velocity profiles at large x for different initial conditions

Fig. 1 shows mean velocity profiles at large values of x for different boundary conditions.
In the case "turbulent, quasi-excited" an acoustical resonance frequency of the test chamber with otherwise neutral flow was present, which influenced the spreading of the flow. On the other hand a strong widening of the flow due to periodic excitation is obvious.

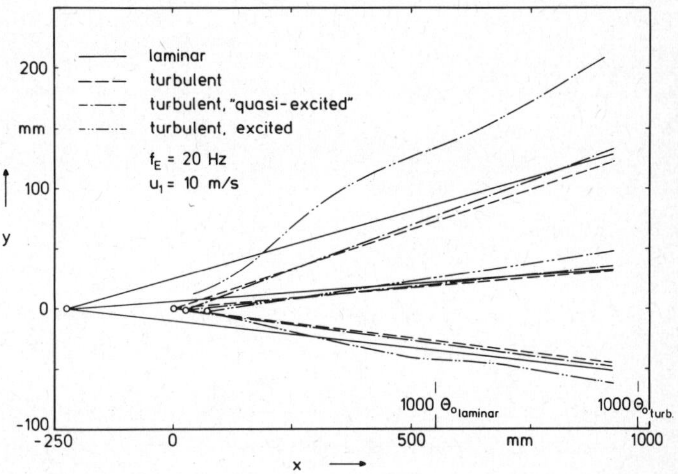

Fig. 2: Shear layer spread for different initial conditions

Fig. 2 shows isotachs for the same conditions. All profiles except for the excited condition attain selfsimilarity. For the excited case we observe regions of different spread rates where, however, the mean velocity-profiles themselves remain - unlike the scalar profiles - unchanged.

In the following we addressed ourselves to the question as to what is the detailed mechanism that causes this dramatic yet in x limited increase of spread and at the same time strong changes in the turbulence distributions? For this study 3 methods were employed:

1) Flow visualization
2) Spectrum analysis
3) Computer-aided phase-averaged measurements

4.21 Flow visualization

Fig. 3 shows three side view smoke photographs for different phase positions together with a top view. From these together with a great number of others not shwon here the following information was obtained:

1) Excitation causes 2-dimensional undulation of the separated boundary layer, followed by formation of packages of concentrated vorticity (foldings). These "nascent" vortices then roll up, thereby ingesting a large amount of entrainment fluid.

2) In the process of roll-up the elliptical cross-section of the vortex turns such that its originally horizontal main axis is then in the direction of maximum strain ($\sim 45^0$). This deformation may be responsible for the sudden increase in spread as well as an apparent increase of convection velocity as obtained from the vortex-spacings ($U_c/U_1 = 0.5 \rightarrow 0.9$).

3) Rolled-up vortices exist only over a limited length with passage frequency = f_E.

4) Pairing of vortices in the downstream region is scarcely observed; in most cases the vortices only seem to loose their identities.

5) The internal structure in a rolled-up vortex is largely lost by diffusion.

6) Excitation causes the strongest spread on the low-velocity, unstable side yet causes the strongest determinism on the high velocity side.

7) Secondary instabilities in the rolled-up vortex sheet are frequently observed.

8) Boundary layer separation at side walls occurs between vortices.

side views

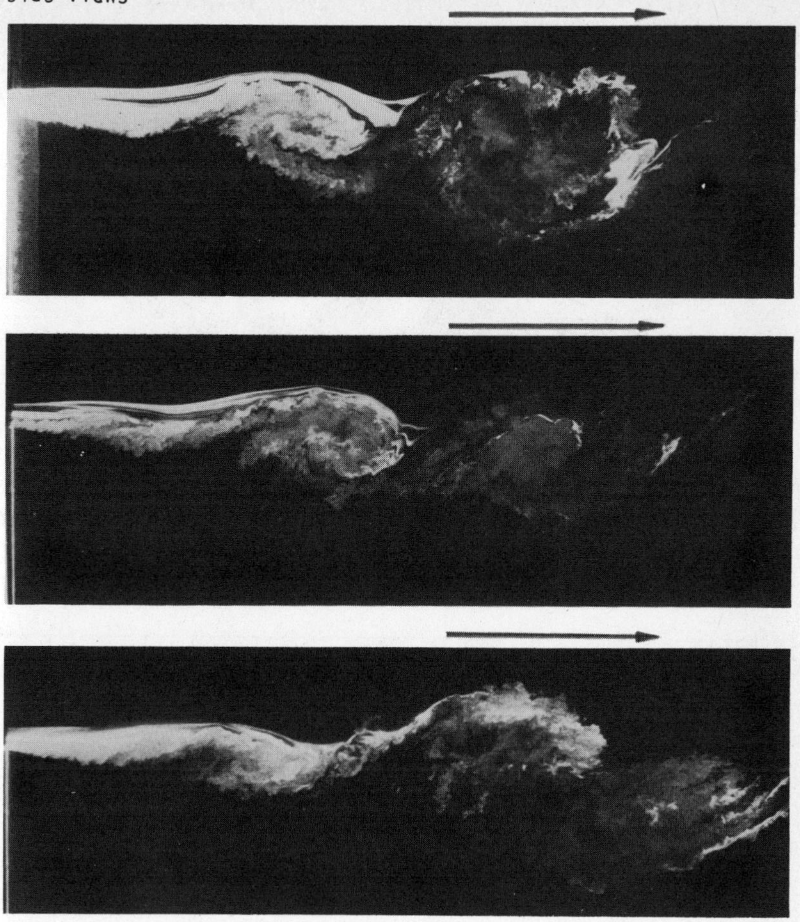

top view

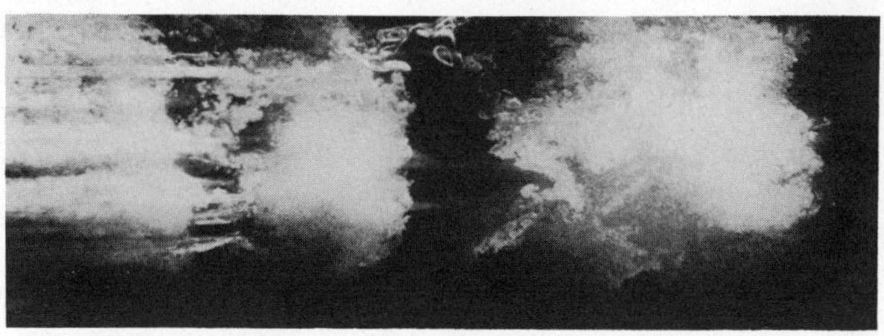

Fig. 3: Smoke photographs of excited shear layer

By comparison of mean velocity isotachs with a smoke photograph three regions are distinguished (Fig. 4):

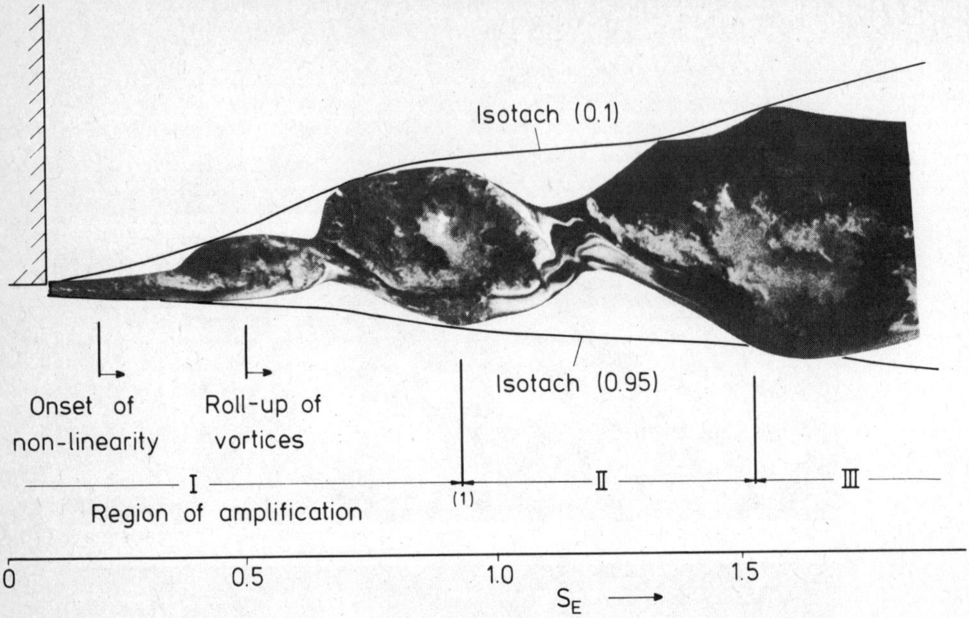

Fig. 4: Regions of excited flow

I) Amplification of imposed periodic disturbance f_E.
Formation of vortices, accompanied by non-linear, strongest increase in width of layer.
Position (1): Full amplification reached (saturation), vortices are fully developed and tilted.

II) Vortices decay with <u>occasional</u> pairing. Width increase is strongly reduced (may even become negative with stronger excitation!) Vortices loose coherence before pairing.

III) Shear layer slowly assumes neutral characteristics.
Dominating frequency $f \sim 1/x$ becomes prominent.

Some of these observations are in good agreement with the results of the extensive numerical study by Acton [28].

4.22 Spectrum analysis

Spectral analysis was then employed to investigate the amplification of the basic frequency f_E in region I:

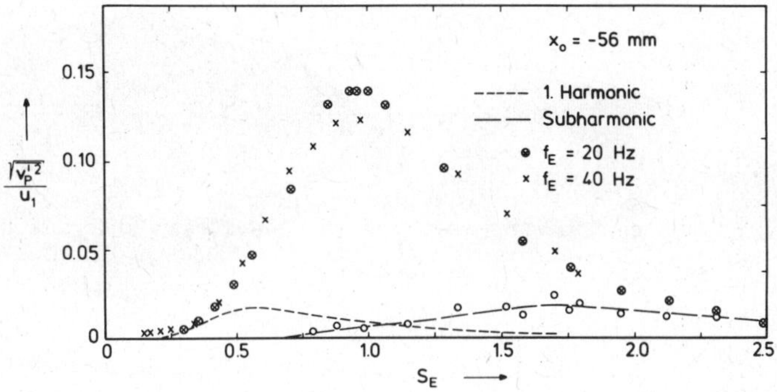

Fig. 5: Longitudinal development of basic frequency and harmonics in excited shear layer

In Fig. 5 rms-values of the periodic content of the lateral fluctuation (v') at excitation frequency f_E = 20 Hz, its first harmonic $2f_E$ and subharmonic $f_E/2$ as obtained from spectral measurements along the x-axis (y = 0) are presented. We observe a total amplification by approximately two orders of magnitude, whereas the harmonics remain typically an order of magnitude below the basic frequency's rms-value.
Only at quite large distances x the subharmonic constituent, reflecting the amalgamation of vortices, seems to reach the values of the basic frequency. The first harmonic appears at approximately 1/5 of the position of the maximum of f_E, indicating the formation of vorticity concentration at the wave crests.

Because of the apparent similarity with the behaviour in the exited laminar shear layer as for instance observed by Freymuth [29], it seems worthwhile to compare the amplification characteristic of the basic frequency with the theoretical consideration by Michalke [30].

A simple scheme for the extended application of Michalke's theory for invisced spacial wave-amplification to non-parallel flow is proposed:
For parallel flow (θ_x = const) we put $A_x/A_0 = e^{-\alpha_i x}$
where A_0 = initial saturation amplitude of disturbance.
Consideration of a stepwise diverging flow and generalization to an infinite number of steps and $\Delta x \to 0$ leads to

$$A_x/A_o = e^{\int_0^x -\alpha_i \, dx^*}$$

Defining $S_{\Theta_x} = \Theta_x f_E \quad u_1$ where $\Theta_x = \int_{-\infty}^{\infty} \frac{\bar{u}}{u_1}\left(1-\frac{\bar{u}}{u_1}\right) dy$

we write $$A_x / A_o = e^{\left(\frac{x}{\Theta_x}\right) \int_0^{S_{\Theta_x}} \frac{-\alpha_i \Theta_x}{S_{\Theta_x}} dS_{\Theta_x}^*}$$

and by inversion we finally obtain: $$-\alpha_i \Theta_x = \left(\frac{S_{\Theta_x}}{S_E}\right) \cdot S_E \cdot \frac{d}{dS_E}\left(\ln \frac{A_x}{A_o}\right)$$

which can be evaluated from Fig. 5 considering that for the one-stream shear layer $\Theta_x/x \approx 0.35$ and thus $S_E/S_{\Theta_x} \approx 28.6$

The evaluation led to a most surprising result which is shown in Fig. 6 compared with Michalke's theoretical curves: It clearly demonstrates the potential of the inviscid theory to describe a turbulent instability process. Obviously and expectedly the measurements fit the spacial amplification curve best. The still close fit for high Strouhal number i.e. strong amplitudes should, however, not be overrated. This agreement is particularly surprising in view of Crighton & Gaster's [54] work.

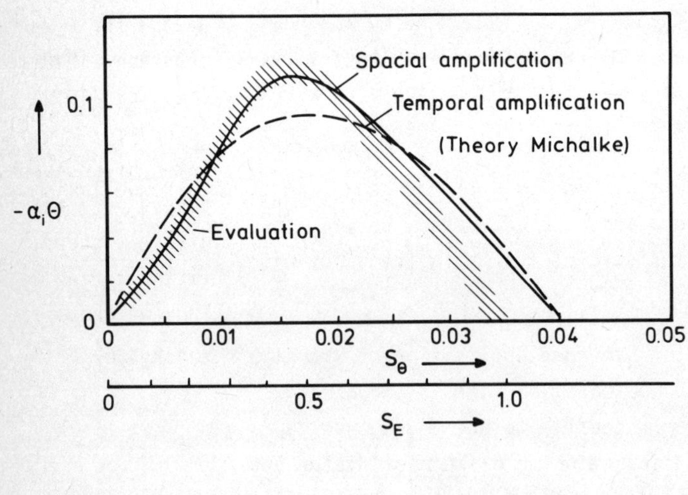

Fig. 6: Comparison of theoretical and measured amplification of excitation frequency amplitude

4.23 Flow-field measurements

This part of the investigation was carried out in a different yet essentially identical test section with u_1 = 10 m/s. A total number of 400 measurement station was processed, forming a 20 x 20 grid with equidistant spacing in x (Δx = 30 mm) and the Δy spacing varying from 1 mm (at x_{min} = 40 mm) to 12 mm (at x_{max} = 610 mm). For each grid point, 1500 cycles of the excitation signal were averaged and the values stored on the disc of a minicomputer.

A special hot wire was designed to reduce errors caused by strong vertical components in the low velocity side entrainment flow. It consisted of a normal wire and a wire under 45 degrees, thus allowing measurements with an angle of resolution of about 135 degrees.

In the processing of the averaged two-dimensional set of data, a smoothing/interpolation algorithm was used which computed the spacial derivatives appearing in the vorticity and energy production terms.

The method of phase averaging was chosen especially because of its independence on detection criteria for the signatures of structures in the time signal.

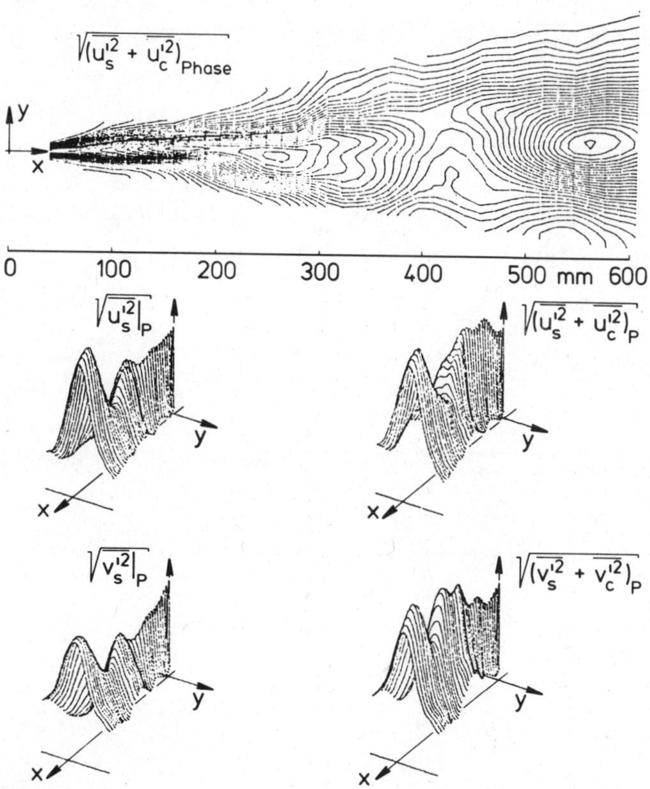

Of the many results obtained only a few are presented:

Fig. 7 shows in its upper part - for a certain phase position - lines of constant total u'-fluctuation intensity, which constitutes of a stoachstic component u_s' and a deterministic (coherent) u_c' due to the vortex motion;
Thus we decompose:
$u = \overline{u} + u'_c + u'_s$ etc.

Fig. 7: Phase averaged $\sqrt{u'^2}$ contours and $\sqrt{u'^2}$ and $\sqrt{v'^2}$ distributions

Below are the distributions of

$$\sqrt{\overline{u'^2_s}} \quad \text{and} \quad \sqrt{\overline{u'^2_s} + \overline{u'^2_c}} = \sqrt{\overline{u'^2}}$$

and of

$$\sqrt{\overline{v'^2_s}} \quad \text{and} \quad \sqrt{\overline{v'^2_s} + \overline{v'^2_c}} = \sqrt{\overline{v'^2}}$$

It is of particular interest that in the vortex cores, particularly in the second (i.e. the fully developed) vortex - which in this phase position is practically located as apparent from the isotachs - the stochastic constituent is strongly amplified, indicating strong mixing i.e. equalized scalar properties. It was ascertained by inspection of a great number of individual velocity "signatures" of vortices that this high level of random turbulence in the vortex core is real and not a consequence of phase jitter.

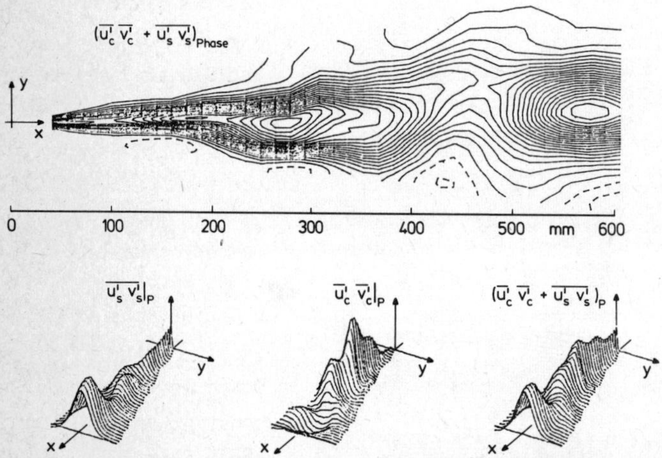

Fig. 8: Phase averaged $\overline{u'v'}$ contours and distributions

Figure 8 shows the same representation for the shear stress $\overline{u'v'}$. Again the strongest contribution of the random component is in second vortex. The deterministic contribution to the total $\overline{u'v'}$ is, however, largest in the amplification range and reaches negative values at the location of the second vortex.

The time-averaged representations are easier comparable to the overall behaviour of the shear layer.

Time-averaged $\sqrt{\overline{u'^2}}$ distributions are shown in Fig. 9. The random constituent remains almost constant in the region of amplification, yet again goes to a maximum in the fully developed vortex.

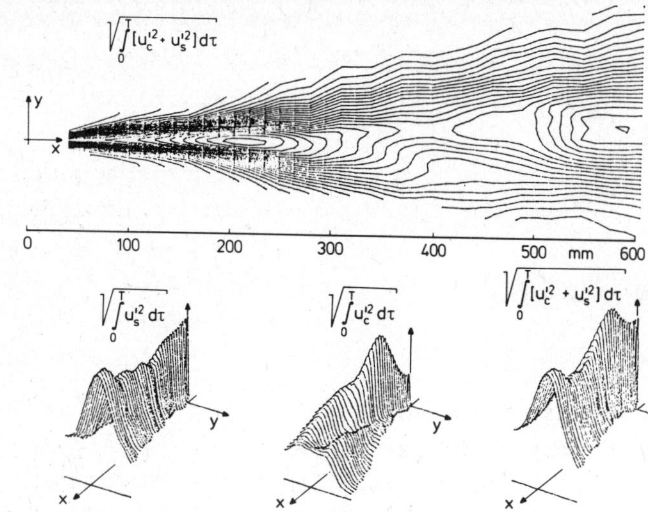

Fig. 9: Time averaged $\sqrt{\overline{u'^2}}$ contours and distributions

The deterministic constituent reaches a maximum at 220 mm ($S \approx 0.5$), i.e. before the point of maximum amplification, approximately at the beginning of the roll-up. At the same time a weak secondary maximum develops on the low velocity side due to the vortex action. The total u'-rms-distribution then has two maxima: One at $x \approx 200$ mm ($S \approx 0.5$) and one at $x \gtrapprox 600$ mm ($S \approx 1.3$): The first maximum corresponds to the position of maximum spreadrate due to vortex roll-up, and the second one to maximum spread due to decay of vortex coherence and randomization.

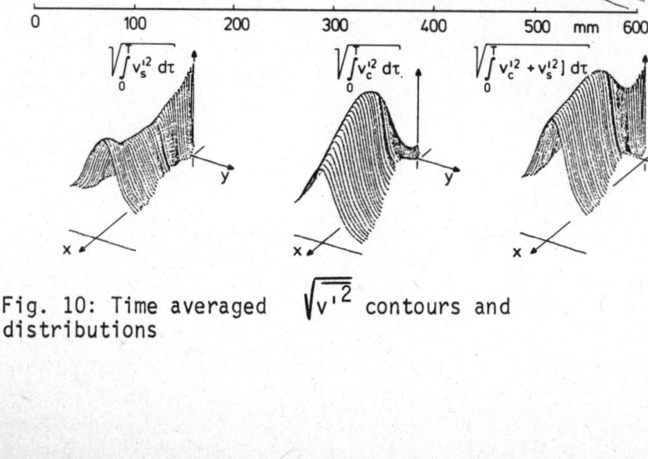

Fig. 10: Time averaged $\sqrt{\overline{v'^2}}$ contours and distributions

Time-averaged $\sqrt{\overline{v'^2}}$ -distributions are shown in Fig. 10:
Again we find almost constant values along the x-axis for the stochastic component, yet a weak maximum around $x \gtrapprox 600$ ($S \approx 1.3$). Expectedly the maximum of the deterministic part is around $x \approx 400$ ($S \approx 0.8$).

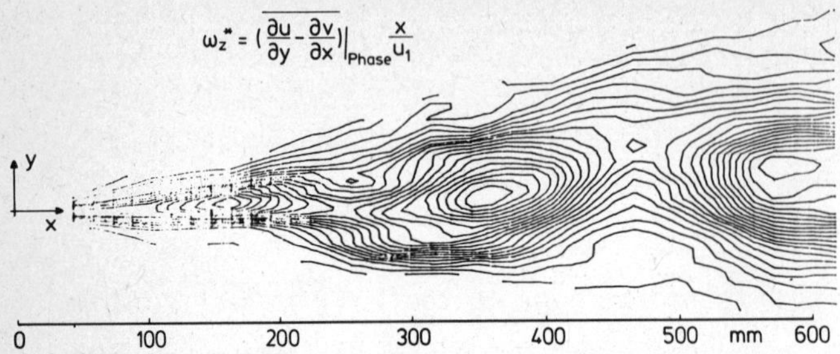

Fig. 11: Phase averaged vorticity contours

Fig. 11 has a typical vorticity contour-plot for the same phase position as above, where the locations of the vortices are distinctly located (and thus helpful for interpretation of the preceding plots) yet does not admit for a direct comparison with Michalke's [30] vorticity contour plots for the unstable parallel shear layer.

Finally the shear stress distribution is considered in Fig. 12:
Again the maximum of the stochastic constituent is at $x \gtrapprox 600$ mm ($S \approx 1.3$). The periodic part, however, reaches negative values in region II. This is where the vortices have aligned with the principal direction of strain. Consequently the complete $\overline{u'v'}$-term reaches relatively low values at the beginning of region II at $x \approx 500$ ($S \approx 1.0$).

Since the spread rate of the flow is closely related to the shear stress distribution we should find maximum spread rates at the location of the two $\overline{u'v'}$-maxima. This is indeed the case, where the first maximum then is a conseauence of the vortex-roll-up and the second caused by the strong stochastic motion in the decaying vortex. The mechanisms in region II, leading to III are not fully understood.

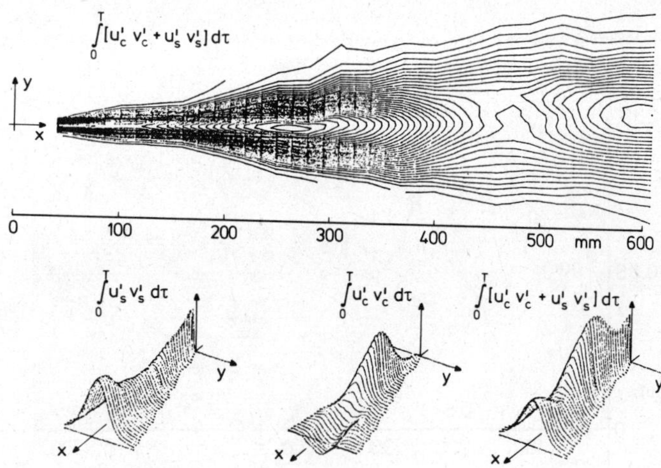

Fig. 12: Time averaged $\overline{u'v'}$ contours and distributions

There are two hypotheses concerning the vortex decay:

- according to Winant & Browand [31] and Acton [28] the vortices cease to exist as single units by amalgamation (pairing).

- according to Moore & Saffman [32] vortices are getting too tightly packed and consequently are torn. (Condition: $l_{min} = 3.5 \Big/ \left(\frac{\partial(\bar{u}/\Delta u)}{\partial y} \right)_{max}$ where l = distance between vortices).

According to our observations both mechanisms seem to coexist with pairing being apparently of minor importance - at least in the one-stream layer -, since the lifetime of a vortex is very limited. On the other hand if - what is strongly assumed - the coherent structures in region III are no longer two-dimensional, observation of the structure in only one plane x-y should reduce the probability of observing an amalgamation considerably.

Regarding the life cycle of a vortex we tried to obtain information about its energy balance. Generally speaking the vortex energy increases in region I, while in region II part of the vortex energy is returned to the mean flow, resulting in a reduced rate of spread. At the same time the vortex is continually loosing energy to the stochastic "medium" scale of the turbulent motion within, the scale of which is typically one order of magnitude below the scale of the coherent motion.

According to Starr [33] a condition for negative production of energy (return to the mean) would be the tilting of the vortices after roll-up, i.e. at the beginning of region II.

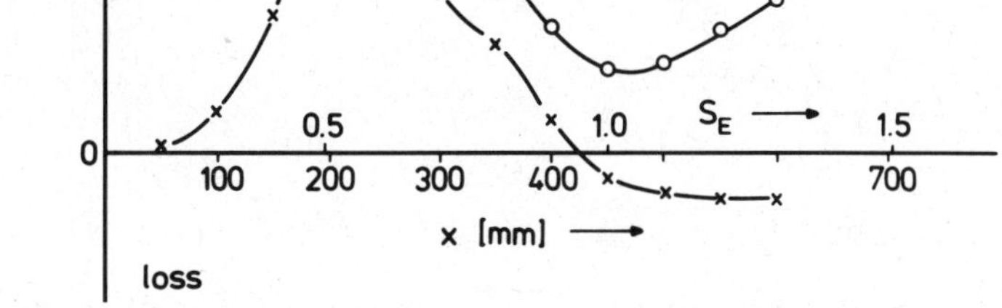

Fig. 13: Energy production in excited shear layer

From an evaluation of the energy flux from the mean motion to the vortex and from there to the stochastic turbulence Fig. 13 was obtained.
In the diagram the energy production from the shear stresses alone is shown for the complete turbulent motion and - separately - for the coherent, i.e. periodic vortex-motion, as evaluated along the line of maximum flux (i.e. $y \approx y_{u/2}$).

Obviously the energy gain of the vortex becomes zero at the position of maximum amplification (1), and is negative in region II, i.e. energy is returned to the mean flow. The overall flux has a minimum at the beginning of region II (see also [18] and [23]).

Recently Takaki [34] has proposed a "phenomenological model of the organized vortex in the two-dimensional mixing layer" where at least for the part of the deterministic stress he finds comparable results. Very similar results from a theoretical analysis were also reported by Knight [35].

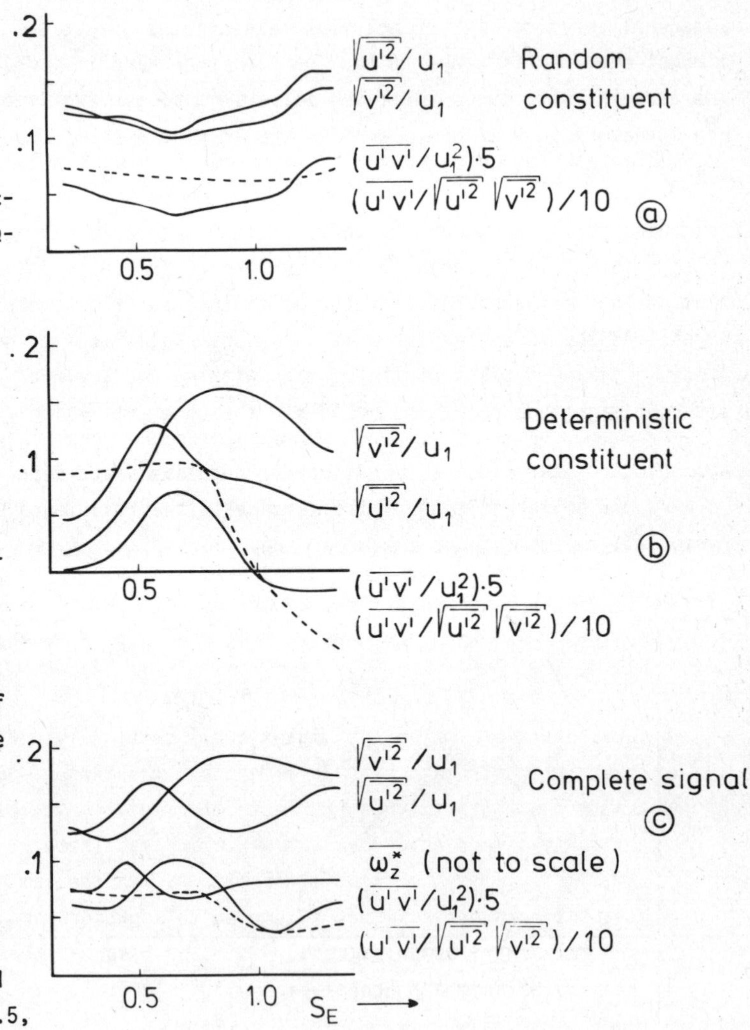

Fig. 14: Structural development of signal constituents

A synopsis of the main outcome from the time averaged flow measurements is presented in Fig. 14, showing the evolution of some characteristic turbulence quantities along the x-axis. We find that the random signal constituents $\sqrt{\overline{u'^2}}$ and $\sqrt{\overline{v'^2}}$ reach minimum values at $S \approx 0.6$, corresponding approximately to the locus of vortex formation (maximum amplification rate), while being well correlated and almost of equal magnitude over the whole length of observation; their phase correlation is approximately constant: $\overline{u'v'} / \sqrt{\overline{v'^2}} \sqrt{\overline{u'^2}} \approx 0.65$ (a).

The deterministic constituents are again well correlated up to $S \approx 0.5$, i.e. for the amplified wave. At vortex formation the phase correlation between u'_D and v'_D

decreases, is zero at $S = 0$ and becomes negative for $S > 1$. Expectedly the v'_D - component becomes considerably larger than u'_D in the vortex domain (b). The development of the complete signal is shown in (c). Clearly it is less indicative of the structural evolution. The asymptotic values obtained in the non-excited (natural) flow are - for comparison - :

$$\sqrt{\overline{u'^2}} / u_1 = 0.166$$

$$\sqrt{\overline{v'^2}} / u_1 = 0.121$$

$$\overline{u'v'} / u_1^2 = 0.009 \quad (\overline{u'v'} / \sqrt{\overline{u'^2}} \sqrt{\overline{v'^2}} \approx 0.45).$$

4.3 Summary of observations

The observations concerning the vortex dynamics can be summarized as follows:

1) Excitation causes small periodic deflections of the separating boundary layer. These disturbances are amplified exponentially.

2) Maximum amplifications of the basic mode is reached for $S \approx 1$. This amplification is well described by inviscid stability theory. At S_o constituents of the first harmonic $2f_o$ appear. The subharmonic $f_o/2$ comes into existence at $S \approx 0.5$. Compared to the amplification in a laminar flow the agreement is best for $S \sim 0.5$. For $S > 0.5$ the main differences are in a smoother maximum for the turbulent flow with subsequent decay of the basic frequency until its disappearance in the turbulence spectrum and in only the appearance of the modes $3g_E$, $2f_E$ and $f_E/2$ (see Miksad [36]).

3) In the roll-up process entrainment fluid is incorporated; the vortex volume - and thus the spread - are strongly increased - particularly on the low velocity side.

4) During its evolution the vortex cross section is deformed such that eventually its main axis is aligned to the axis of principal strain. In this position it has its most deterministic state until decay and/or three-dimensional amalgamtion dominates. There - considering the similarity of the excitation behaviour with the laminar-turbulent transition process - one is tempted to look for similarities with the relaminarization process. A similar idea has been touched upon by Narasimha & Sreenivasan [37].

5) After the vortices eventually loose their identity the shear layer seems to approach the "neutral" state regarding structure and spread, yet with another virtual origin.

In all cases irrespective of whether the flow is excited or not, the fluctuation spectrum displays a long $f^{-5/3}$-region for $f > f_E$ (three-dimensionality).

Excited shear layers with $u_2/u_1 > 0$ (e.g. 0.4, 0.6) display comparatively long regions of vortex existence. There the x-axis is effectively stretched by the factor $\Sigma\, u_{1,2}/\Delta u_{1,2}$, thus also the range of vortex existence i.e. the number of vortices in a row, is increased which then causes real structural differences, as for example a strong minimum in the $\sqrt{\overline{u'^2}}$-distribution.

5. Effects of boundary conditions on flow development

The structure of the shear layer and thus its consequences, i.e. developmemt, spread rate etc. may be influenced by a number of boundary conditions, some of which may - unnoticed - cause erroneous results and give rise to wrong if generalized interpretations.

The following table lists some of the more obvious parameters which may modify the flow to some extent:

Static:	
initial:	Condition of the separating boundary layer Geometry (e.g. angularity) of trailing edge
side walls:	Vortex-induction Boundary layer (separation) Compression / stretching of vortices Curvature Pressure gradients ($\partial p/\partial x$, $\partial p/\partial z$)
bottom walls:	Vortex-induction (distance) Influence on stability behaviour Pressure gradients ($\partial p/\partial x$, $\partial p/\partial y$) Curvature
exit:	Stagnation Suction Obstacles, asymmetries
flow:	Inhomogeneities, shear, stratification

Dynamic random:

Turbulence in main streams
Random vibrations
Unstationary flow

Dynamic periodic:

Excitation - initial / boundary
Periodic vibrations (mechanical)
Organ pipe frequency
Blower frequency
Initial laminar / turbulent transition
Feedback mechanisms
"Exit - frequency"

A number of those have been discussed in various publications (e.g. Hussain & Zedan [46]), at large, however, their effects on the flow are still not fully documented nor understood and it is believed that a number of discrepancies in published results might be traced back to one or more of those effects, which is particularly troublesome if a reliable model of the flow is to be established.

The influence of the condition of the separating boundary layer (initial condition) on the shear layer structure was studied in a number of investigations (e.g. Batt [38], Foss [39], Oster et al [23]). General agreement is found for a stronger spread as a consequence of a turbulent separating boundary layer in the one-stream shear layer as compared to the case where the boundary layer is laminar.
Oster's observation of the (weak) inverse effect in the two stream layer is difficult to understand and is not undisputed.
These observations consequently lead to the question of the existence of a "final" universal i.e. initial-condition-independent state.
Investigations by Fiedler & Thies [41], Birch & Eggers [42], and Hussain & Kleis [43] confirm that notion.
Some disagreement was found concerning the influence of random disturbances on the two-dimensional nature of the coherent structures (Chandrsuda et al [44], Wygnanski et al [22]).
Organ-pipe effects, blower-frequency-excitation, excitation by mechanical vibrations etc. were observed by Fiedler & Thies [41]; Fiedler, Korschelt & Mensing [18], Pui & Gartshore [45].

Recently the authors re-examined a number of external effects on the "natural" and the excited two-stream shear layer.

Concerning the spread rate the following was observed:

1) Vibrations or resonance frequencies in the test arrangement result in a "quasi-excitation" of the shear layer flow where the spread behaviour is similar as in case of controlled excitation:

   a) The spread is larger than in the natural (i.e. undisturbed) shear layer. Defining the"maximum slope thickness" as $\delta' = k(u_1-u_2)/(u_1+u_2)$ we find for laminar boundary layer:

quasi excited ("organ pipe"):	k = 0.184
natural:	k = 0.170
Liepmann & Laufer [48]:	k = 0.162
for comparison:	
Brown & Roshko [1]:	k = 0.181
Browand & Latigo [47]:	k = 0.170

   b) The varying rate of spread in the quasi-excited shear layer is responsible for ambiguous data for the spread rate, depending on the range considered (Fig. 15).

   Thus spread rates obtained for $u_2/u_1 = 0.4$ were:

$db/dx = (y_{0.1} - y_{0.95})/(x - x_o)$	x - range
0.071	800 - 1200 mm
0.082	800 - 1700 mm

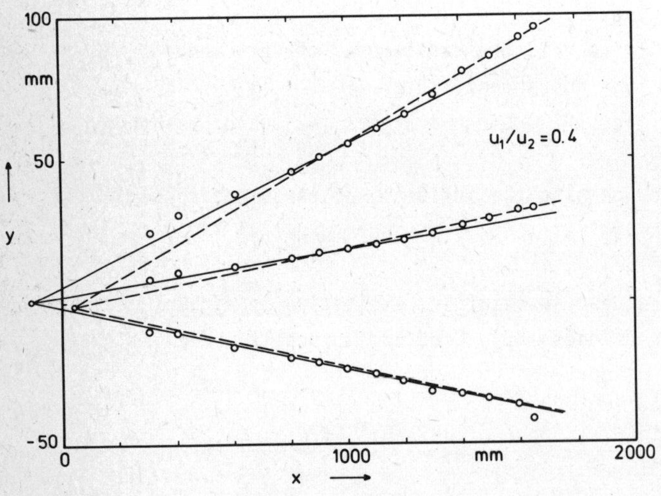

Fig. 15: Spread rates db/dx

It may therefore be misleading to extrapolate an initial momentum thickness from the shear layer data. This is by principle wrong if this extrapolated momentum thickness is used to obtain a measure x/θ_o for the flow development, i.e. the approach to self-preservation (Bradshaw [49]).

c) Transverse correlation is increased at the position where $f_{resonance}/f_{natural} \approx 1$.

2) All profiles measured with trip over a length of 1700 mm showed an increase of the spread rate as compared to the untripped case. The spread itself is, however, smaller over most of the observed range. A tendency for a universal range seems obvious, but was not reached for $x < 2000$ mm ($x/\theta_{BL} < 1000/1500$) - Fig. 16.

3) Longitudinal pressure-gradient $\partial p/\partial x$ as a consequence of side wall adjustments has stronger influence on the spread for increasing velocity ratio and generally exceeds the area ratio.

4) No influence of the test section length, which was extended from 2000 mm to 3500 mm, was observed. This question is of particular interest for the shear layer extending to the bottom - or top-wall.

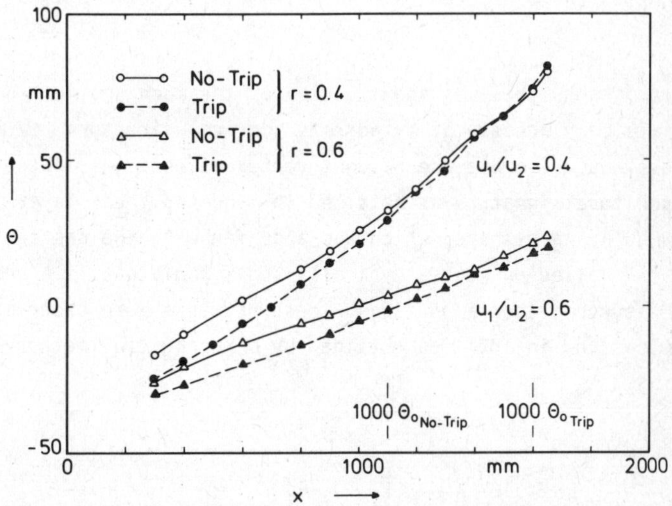

Fig. 16: Comparison of momentum thickness θ with and without trip wire

6. Summary and re-examination of natural shear layer

The investigation and experiments described have shed some light on the dynamic behaviour of the artificially "stabilized" shear flows, in particular the plane shear layer.

How then does the knowledge obtained fit into the observations of the natural layer and further: How can the natural layer be described in an overall way?

One has to answer first a fundamental question: Is the truly undisturbed natural shear layer of more then academic interest, since its strong receptivity to many disturbances is clearly established?
In our opinion the answer is yes: To obtain true understanding of the observed phenomena and finally to derive a model for this flow its unperturbed characteristics have first to be known and the external influences valued and understood separately.
Of the natural flow we do, unfortunately, not have as clear a picture as for the excited one, particularly because of

- the many (partly unknown) disturbing influences, which are not easily eliminated
- the need for sufficiently large Re-numbers and
- the experimental difficulties in extracting true structural information from an essentially random flow.

In the region of flow development of the neutral one-stream shear layer ($x/\Theta_0 \lesssim 1000$) developing from a turbulent boundary layer the flow exhibits certain characteristic features, some of which bear strong similarities with certain stages of the excited flow:

- Smoke photographs, besides showing clearly the much smaller yet constant spread rate only occasionally indicate vortical structures of generally short lifetime. Those resemble the nascent vortex of region I in the excited case. Its temperature signature as obtained in a heated shear layer is typically of the triangular ramp-shape, which was also found in the nascent excited vortex, whereas the rolled-up vortex typically shows a plateau, i.e. equalized properties in its core, (see also [19]). These patterns were observed as individual realizations and in addition obtained by a recognition method (Figs. 17 and 18).

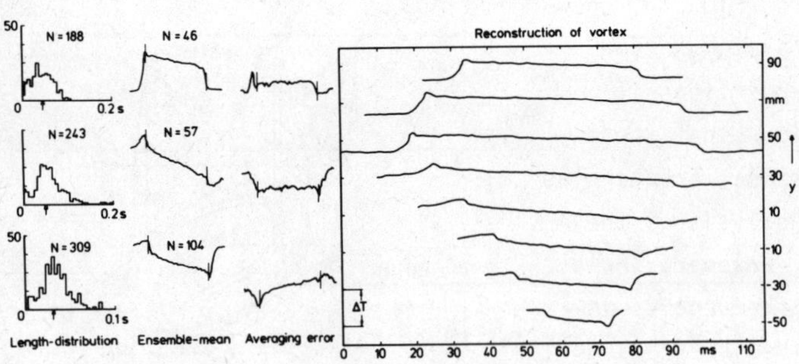

Fig. 17: Mean temperature fluctuation pattern (pattern recognition) and reconstructed vortex - region I

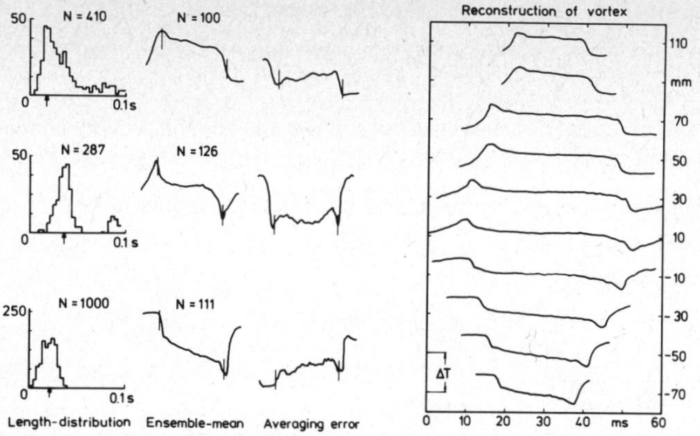

Fig. 18: Mean temperature fluctuation (pattern recognition) and reconstructed vortex - region II

They are easily understood from the vortex geometry as explained in Fig. 19. Mean turbulence properties are also very similar to those in the "nascent vortex region" of the excited flow. Thus the large scale structure in the natural shear layer may best be characterized as "folded."

In searching for deterministic constituents spectra were measured along the x-axis. From Fig. 20, showing spectra of u' for various x-positions it seems obvious that there are no distinct frequencies present. Analysis of the v'-signal, however, shows preferred frequencies in this flow region (Fig. 21).

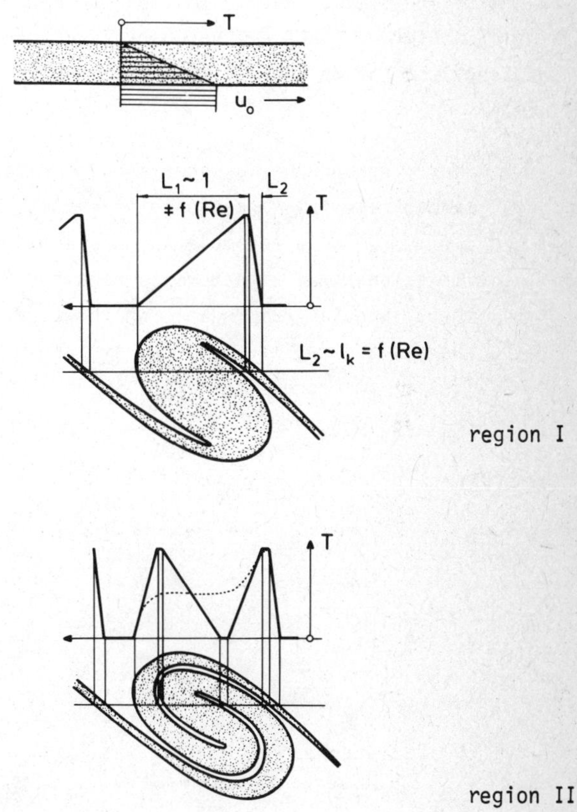

Fig. 19: Model for temperature signature (ramp) from vortex geometry

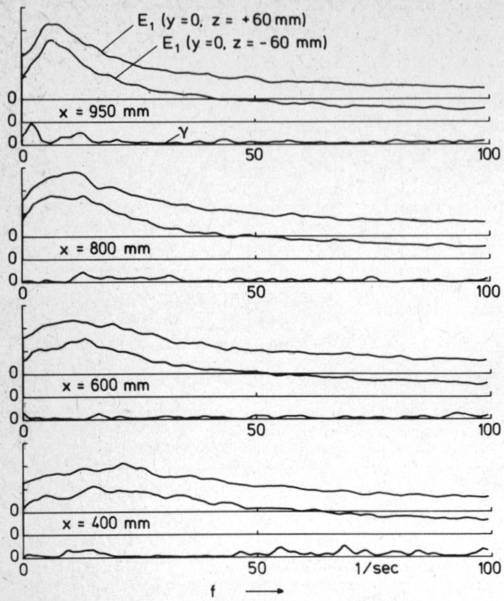

Fig. 20: Spectra and coherence-function γ of u'-signal in natural flow. (Coordinate shift corresponds to γ = 0.1.)

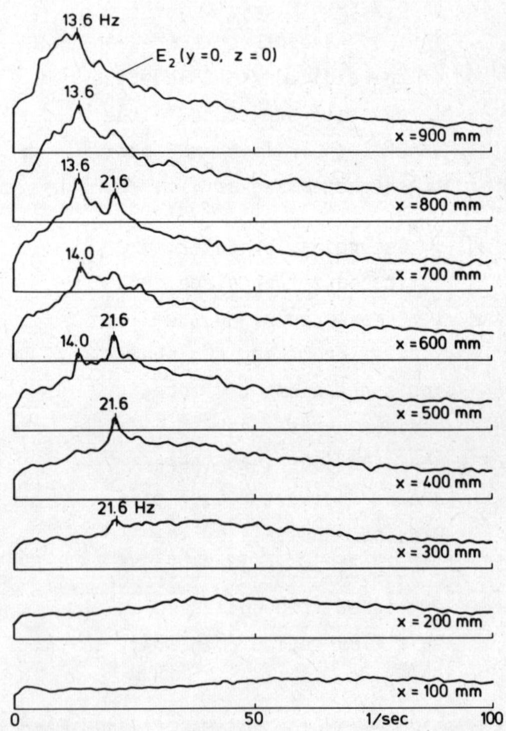

Fig. 21: Spectra of v'-signal in natural flow.

- Coherence functions taken from two points 120 mm apart in z at $y = 0$ and various x up to $x = 950$ mm remained essentially zero (< 0.1), indicating negligible or at least very weak two-dimensionality in the sense that - as was observed from visualization as well as from temperature-sensor rake observation - the doubtlessly existing coherent "lumps" seem to be elongated in z-direction yet with sufficient irregularity in slant-angle so that simple statistical methods will fail to indicate their structure.

  It is felt that this structure is ultimately becoming universal and independent of initial conditions. This idea is strongly supported by measurements of Browand & Troutt [50]. It suggests an amalgamation process, which is three-dimensional in the sense that it is initiated at one position in z and then moves across the flow. It is also understood that the degree of coherence is weakest, i.e. the three-dimensional effects strongest for the one-stream layer.

  The initiated roll-up on the other hand reflects to some extent an amalgamation process with a high degree of two-dimensionality. It exhibits all characteristics of the natural amalgamation, e.g. entrainment increase, spread increase, shear stress increase.

Based on the information available we may propose a general picture for the shear layer development according to four initial conditions for which the observations by Jimenez [51] and Clark & Kit [52] provide an important link (although it seems questionable, whether the three-dimensionalities in the mean as reported by Jimenez are a feature of the flow or of the apparatus) and which provides also an explanation for the interesting observation of turbulence suppression in the excited shear layer as observed by Zaman & Hussain [53]:

In any case the initial region of the flow is dominated by a strong tendency to forming large two-dimensional structures as a consequence of flow instability.

In a separated laminar boundary layer the characteristic frequency ($S\theta_0 \approx 0.017$) is amplified resulting in a roll-up of two-dimensional vortices of strong periodicity. Formation of these structures is followed by laminar-turbulent transition with well preserved determinism leading eventually to strong three-dimensional interactions and increase of fluctuation-energy. The flow then attains a high degree of randomization that is stronger than in the originally turbulent case. Gradual development toward the final state is characterized by slow recovery of coherence and determinism.

Excitation may delay the three-dimensional decay and may thus effectively reduce the fluctuation strength of the longitudinal component in that region.

In the natural flow with turbulent separation the tendency to formation of two-dimensional structures is also present yet by virtue of its high perturbation level as well as by its considerably stronger non-parallelism these structures are weaker than in the laminar case. They are gradually approaching an asymptotic value, while loosing again some of its determinism (see Fig. 18). Thus the developing turbulent shear layer with turbulent initial condition may exhibit a region of stronger determinism than the shear layer with laminar initial condition after transition.

Excitation of the turbulent flow, necessitating a certain threshold amplitude, then causes strong determinism accompanied by an overshoot of turbulent energy (comparable to the laminar case), eventual decay of the periodic constituent and increase of the random energy.

In light of these considerations the pending question as to whether the coherence in the shear layer is a major or a weak constituent may partly be answered by knowing the initial conditions as well as the flow regime considered. Still a definitive and quantifying statement regarding the strength of the coherent constituents in a neutral shear layer and their influence on the flow- and scalar field cannot yet be given.

Acknowledgement

A substantial part of the work presented was supported by Deutsche Forschungsgemeinschaft.

References

1 Brown, G.L.; Roshko, A.: "The effect of density differences on the turbulent mixing layer." "Turbulent shear flows", AGARD - CP-93 (1971) 23 (1-12).

2 Roshko,A.: "Structure of turbulent shear flows: A new look." AIAA J. 14 10 (1976) 1349.

3 Liepmann, H.W.: "The rise and fall of ideas in turbulence." American Scientist, 67, March/April (1979).

4 Laufer,J.: "New trends in experimental turbulence research." Annual Rev. Fluid Mech., 7 (1975) 307-326.

5 Townsend, A.A.: "The structure of turbulent shear flow." Cambridge: Cambridge Univ. Press 1959.

6 Fuchs, H.V.; Mercker, E. and Michel, U.: "Large-scale coherent structures in the wake of axisymmetric bodies." JFM, 93 1 (1979) 185-207.

7 Mensing, P.; Fiedler,H.: "Eine Methode zur Sichtbarmachung von hochturbulenten Luftströmungen mit großen Reynoldszahlen." To appear in ZFW.

8 Yule,A.J.: "Phase scrambling effects and turbulence data analysis". Proc. 2nd Symp. on turbulent shear flows, July 2-4. 1979, London.

9 Rösgen, Th.: "Digitale Signalerkennung in quasikohärenten turbulenten Strukturen". Diplomarbeit, HFI, TU Berlin 1978.

10 Fiedler, H.: "On data acquisition in heated turbulent flows." Proceedings Dynamic Flows Conference, Marseille 1978.

11 Wygnanski, I.: "Turbulent spots in boundary layers." Intern. Conf. on the role of coherent structures in modelling turbulence and mixing, Madrid 1980.

12 Eckelmann, H.; Wallace, J.M.: "A comparison of the characteristic features of coherent turbulent structures found using the variable interval time average (VITA) technique and using the pattern recognition technique". Intern. Conf. on the role of coherent structures in modelling turbulence and mixing, Madrid 1980.

13 Coles. D.: "Useful strategies in turbulence experiments and modelling". Intern. Conf. on the role of coherent structures in modelling turbulence and mixing, Madrid 1980.

14 Wille, R.: "Beiträge zur Phänomenologie der Freistrahlen." ZfW, 11 (1963) 222-233.

15 Hussain, A.K.M. F.: "Perturbed and unperturbed turbulent jets." Intern. Conf. on the role of coherent structures in modelling turbulence and mixing,Madrid 1980.

16 Korschelt, D.: "Experimentelle Untersuchung zum Wärme- und Stofftransport im turbulenten ebenen Freistrahl mit periodischer Anregung am Düsenaustritt." Dissertation, HFI, TU Berlin 1980.

17 Murthy, S.N.B.: "Turbulent mixing in non-reactive and reactive flows" A Review, Project SQUID, Purdue University 1974.

18 Fiedler, H.; Korschelt, D.; Mensing, P.: "On transport mechanisms and structure of a scalar field in a heated plane shear layer." Lecture Notes in Physics 76 Berlin: Springer 1977.

19 Wygnanski, I.; Oster, D.; Fiedler, H.: "A forced, plane, turbulent mixing-layer; A challenge for the predictor." Proc. 2nd Symp. on Turbulent Shear Flows, July 2-4, 1979 London.

20 Münch, F.: "Strukturuntersuchungen in einer turbulenten Scherschicht." Diplomarbeit HFI, TU Berlin 1978.

21 Mening, P.; et al: "Einfluß kontrollierter Störungen auf eine ebene turbulente Scherschicht." DGLR Symp. "Strömungen mit Ablösung", München 1979.

22 Wygnanski, I.; Oster, D.; Fiedler, H.; Dziomba, B.: "On the perseverance of a quasi two-dimensional eddy-structure in a turbulent mixing layer." JFM, 93 2 (1979) 325-335.

23 Oster, D.; Wygnanski, I.; Fiedler, H.: "Some preliminary observations on the effect of initial conditions on the structure of the two-dimensional turbulent mixing layer." Proc. of SQID Conf., 1976.

24 Hussain, A.K.M.F.; Clark, A.R.: "On the coherent structure of the axisymmetric mixing layer." Report FM 6, University of Houston 1979.

25 Oster, D.: "The effect of an active disturbance on the development of the two-dimensional turbulent mixing layer." Ph.D.-Thesis, Tel Aviv University 1980.

26 Fiedler, H.; Rösgen, Th.: "On the anatomy of a rolled-up vortex in an excited plane turbulent shear layer." 32nd Meeting of the APS, Nov. 18-20 1979.

27 Ho, Chi-Ming: private communication

28 Acton, E.: "The modelling of large eddies in a two-dimensional shear layer." JFM, 76 3 (1976) 561-592.

29 Freymuth, P.: "On transition in a separated laminar boundary layer." JFM, 25 4 (1966) 683-704.

30 Michalke, A.: "On spatially growing disturbances in an inviscid shear layer". JFM, 23 3 (1965) 521-544.

31 Winant, C.D.; Browand, F.K.: "Vortex pairing: the mechanism of turbulent mixing-layer growth at moderate Reynolds number." JFM, 63 2 (1974) 237-255.

32 Moore, D.W.; Saffmann, P.G.: "The density of organized vortices in a turbulent mixing layer." JFM, 69 (1975) 465-473.

33 Starr, V.P.: "Physics of negative viscosity phenomena." McGraw Hill 1968.

34 Takaki, R.: Private communication.

35 Knight, D.: 32ndMeeting of the APS. Nov. 18-20 1979.

36 Miksad, R.W.: "Experiments on the nonlinear stages of free-shear-layer transition." JFM, 56 4 (1972) 595-719.

37 Narasimha, R.; Sreenivasan, K.R.: "Relaminarization of fluid flows." Advances in applied mechanics, 19 Academic Press 1979.

38 Batt, R.G.: "Some measurements on the effect of tripping the two-dimensional shear layer." AIAA J., 13 2 (1975) 245-246.

39 Foss, J.F.: "The effect of the laminar/turbulent boundary states on the development of a plane mixing layer." Proc. 1st Intern. Symp. on turbulent shear flows. The Penn. State Univ. 1977.

40 Oster, D.: "An experimental investigation of the two-dimensional turbulent mixing layer." M.Sc.-Thesis, Tel Aviv University 1976.

41 Fiedler, H.; Thies, H.J.: "Some observations in a large two-dimensional shear layer." Lecture Notes in Physics 75 Berlin: Springer 1978 108-117.

42 Birch, S.F.; Eggers, J.M.: "A critical review of the experimental data for developed free turbulent shear layers." Free Turbulent Shear Flows, Vol. 1, Conf. Proc., NASA Rep. No. SP-321 (1973).

43 Hussain, A.K.M.F.; et al: Private communication.

44 Chandrsuda, C.; et al: "Effect of free-stream turbulence on large structures in turbulent mixing layers." Dept. Aerodynamics, Imperial College, London (1976).

45 Pui, N.K.; Gartshore, I.S.: "Measurements of the growth rate and structure in plane turbulent mixing layers." JFM, 91 1 (1979) 111-130.

46a Hussain, A.K.M.F.; Zedan, M.F.: "Effects of the initial condition on the axisymmetric free shear layer: Effects of the initial momentum thickness." Phys. Fluids, 21 12 (1978) 1100-12.

46b Hussain, A.K.M.F.; Zedan, M.F.: "Effects of the initial condition on the axisymmetric free shear layer: Effect of the initial fluctuation level." Phys. Fluids, 21 12 (1978) 1475-81.

47 Browand, F.K.; Latigo, B.O.: "The growth of the two-dimensional mixing layer from a turbulent and non-turbulent boundary layer." Physics of Fluids, 22 6 (1979) 1011-1019.

48 Liepmann, H.W.; Laufer, J.: "Investigation of free turbulent mixing". N.A.C.A. Techn. Note No. 1257, (1947).

49 Bradshaw, P.: "The effect of initial conditions on the development of a free shear layer". JFM, 26 2 (1966) 225-236.

50 Browand, F.K.; Troutt, T.: A note on spanwise structure in the two-dimensional mixing layer." JFM, 97 4 (1980) 771.

51 Jimenez, J.: "On the origin and evolution of three-dimensional effects in the mixing layer." Fin. Techn. Rep., DAERO-78-G-079, Dec. 1979 School of Aeronautics, Univ. Politechnica, Madrid.

52 Clark, J.A.; Kit, L.: "Shear layer transition and the sharp-edge orifice." ASME, Fluids Eng. Conf., New Orleans, LA., March 1980.

53 Zaman, K.M.B.Q.; Hussain, A.K.M.F.: "Turbulence suppression in free shear flows by controlled excitation." Report FM-7, Univ. of Houston 1979.

54 Crighton, D.G.; Gaster, M.: "Stability of slowly diverging jet flow." JFM, 77 2 (1976) 397-413.

COHERENT STRUCTURES AND STUDIES OF PERTURBED AND UNPERTURBED JETS

A.K.M.F. Hussain
University of Houston, Texas 77004

Abstract: The following aspects of large-scale coherent structures are addressed: definition and identification of the coherent structure, difficulties in the eduction of naturally occurring structures, need for studies of structures induced via controlled excitations, relevance of induced structures to natural structures, the effects of initial and boundary conditions including characterization of the initial condition, and some analytical and experimental considerations in investigations of coherent structures. Progress made in the understanding of large-scale coherent structures in the following experimental studies of perturbed and unperturbed axisymmetric turbulent jets are then summarized: coherent structures in the jet near field induced by impulsive and sinusoidal excitations; coherent structure distortion due to the Taylor hypothesis and applicability of the hypothesis to turbulent shear flows; the sensitivity of the jet 'preferred mode' coherent structure to the Reynolds number and initial condition; and coherent structures in the jet near and far fields in the absence of any excitation. The signigicance of coherent structures vis-a-vis the background turbulence is discussed, and on the basis of available evidence it is suggested that although coherent structures are most probably inherent to all turbulent shear flows, their predominance in fully developed flows may have been exaggerated.

I. Introduction

The relatively recent discovery that flows, which have been historically regarded as fully (random) turbulent, indeed consist of quasi-deterministic large-scale structures[1-5] have engendered an upsurge of research activity. The intense current interest in large-scale coherent structures stems from the expectation that these structures play the key roles in transports of heat, mass, and momentum and noise production in turbulent shear flows.

The presence of transitional large-scale structures in the near fields of free shear flows has been well known (probably for centuries), even though there are doubts regarding persistence of these structures or of their two-dimensionality at higher Reynolds numbers.[6-9] Large-scale coherent structures have been found in the fully developed turbulent boundary layers as well as in wakes[10-15]. The existence of large-scale coherent structures in the far field of axisymmetric jets is still an open question (our efforts[16] in this direction will be discussed).

Details of coherent structures in free shear flows have been investigated by us with three goals in mind: (i) to understand the dynamics of the structures and thus the physics of shear flow turbulence, (ii) to explore the control of transport phenomena and noise production via control of structures and their interactions, and (iii) to provide data for model functions that can be used as building blocks for the formulation of a realistic theory of turbulence. Since large-scale coherent structures appear to be inherent to perhaps all turbulent shear flows,[17,18] it is clear that realistic theories of turbulent shear flows must include these structures as well as their phase information which is discarded by Reynolds averaging. It is hoped that a theory incorporating these structures and their interactions will capture enough essential physics of turbulence such that the theory might be viable and fairly universal.

IIa. What is a coherent structure?

In spite of widespread use of this term, quite evident also from the title of this conference, there is neither a definition of, nor a general agreement on, what is precisely meant by "coherent structure". It is not likely that a simple definition will be accepted without disagreements, or even strong protests.[17] Although it is perhaps premature now to do so, I will attempt to briefly define here what I mean by coherent structure, as otherwise it will be fruitless to discuss an undefined entity.

A coherent structure is a turbulent fluid mass with coherent vorticity and of a size of the order of the width of the shear flow. Even though the dynamics of these structures is essentially inviscid, they are turbulent, i.e. they have random three-dimensional vorticity. However, there must be an underlying organized (i.e. non-random) vorticity connecting different parts of the structure; this is the coherent vorticity. Coherent structures perform significant large-scale mixing and are also likely to be dominant in heat and momentum transfer. Especially, the coherent Reynolds stress (defined in Sec. III) must be significant. Coherent structures do not necessarily have to be highly energetic. Thus, conventional correlation measurements of linear momenta cannot uniquely detect a coherent structure. Otherwise, a non-turbulent blob of fluid 'inserted' in a turbulent flow would appear to be a coherent structure, even though this fluid neither has vorticity nor performs transport.

A coherent structure is not an eddy as classically understood.[19] Eddies co-exist in spatial superposition and can be extracted through some form of signal decomposition, say, Lumley's orthognal decomposition.[20] In contrast, coherent structures are non-overlapping. At the location of a structure, the entire signal is associated with a structure. In some sense, this non-superposition makes coherent structure interactions inherently nonlinear.

Flow visualization can be used to identify coherent structures and infer their qualitative topology -- more definitively so in their pristine states as in transitional flows than in fully developed turbulent regions. However, for either flow situation, any meaningful description of the coherent structure must include quantitative data.

IIb. Naturally occurring coherent structures and their eduction

In an unexcited turbulent shear flow, detection as well as detailed investigation of the coherent structures is complicated by the large dispersion in the shape, size, orientation, strength, and convection velocity of the structures. The structures typically undergo cycles of formation, growth and decay. Thus, even if evolutions of successive structures were identical in each cycle, the locations of initiation and termination of each cycle are random; the trajectory of each structure is also random in each cycle. Thus, phase scrambling[21,22] is only a small part of the complicacy in structure detection.

The passage of a structure carries with it both phase-coherent and phase-incoherent (i.e. random) turbulent motions. Sifting out the coherent part, which alone characterizes the structure, requires spatial filtering; this is typically achieved through ensemble-averaging a large number of signals, since the phase-incoherent motions would cancel out in the average. Furthermore, even if the shape, size, orientation, strength and convection velocity of successive structures were identical, a single probe does not necessarily intercept all successive structures along the same line. That is, because of random formation point in space as well as random trajectory, a sensor or a rake of sensors traverses different cuts of successive structures. In order to educe the coherent structure, the signal containing footprints of successive structures, even if they were of identical measures, must be obtained with a multi-sensor probe in a three-dimensional array. In that case, only those structures with identical configuration with respect to the array should be ensemble averaged.

Thus, in an unexcited flow, structures buried in signals obtained with a multi-sensor probe in a three-dimensional array must be classified on the basis of their characteristic measures viz. size, shape, orientation, strength, convection velocity, 'age' etc., and only structures of identical measures should be ensemble averaged. For each narrow range of the measures, contours of various properties over the extent of the structure should be determined in convenient planes from the ensemble-averaged signals. Furthermore, since the structures are in general three-dimensional, data should include all three velocity and vorticity components (as well as temperature, concentration, etc.) at each sensor. Clearly, in addition to the costs of multi-sensor measurements, probe interference and experimentation and signal analysis times make such studies prohibitive. One must thus look for simpler approaches and be content with limited data and even partial information.

A major simplification in experimentation is possible if the multi-dimensional probability density distribution as a function of various structure measures has sharp peaks representing 'preferred modes'. One then can limit investigation to only these modes. If the probability density distribution has a sharp isolated peak, study of this mode alone may be adequate since this mode can be expected to represent the essential physics of the flow. Such a mode exists in the axisymmetric jet.[1,23,24] However, most other shear flows are probably not blessed as such.

Even if such drastic simplifications were possible, the structure eduction is still not simple. A practical solution is to obtain data with a limited few sensors (single or x-wires) typically in a linear array. From simultaneous traces of the sensor array, passage of a structure can be inferred only after some careful signal conditioning. The signal may sometimes indicate a characteristic front common to all. In free turbulent shear flows, however, the signal is typically buried in relatively large-amplitude random fluctuations. The first step is to smooth these signals. Even then the structure signature may not be apparent. The signature can be enhanced by cross-correlating the signals, taking a suitable one from the array as a common reference. Clearly, the integration time in the (double or multiple) cross-correlation should be large enough to suppress random turbulence but small enough to capture the structure details and is optimum when it is about equal to the structure time scale. Low-pass filtering the signals before obtaining cross-correlation improves signal detection. It is important to emphasize that a common front or feature in simultaneous signals from a sensor array does not necessarily unambiguously indicate passage of a structure, as external disturbances inducing 'puffing' or 'flapping' motion of the free shear flow can produce simultaneous 'bumps' in the signal traces. Since coherent vorticity is the key feature of coherent structures, the passage of the structure must be inferred from the distribution of at least one vorticity component, derivable from the smoothed signals from the sensor array.

In a turbulent shear flow, the structure typically spans the width, and hence most of the "jitter" in structure detection can be assumed to be in the streamwise direction. Thus, if the structures are essentially identical (especially when driven), the structure properties can be obtained by ensemble-averaging after suppressing phase scrambling, i.e., through phase-alignment of successive signatures.[25-27] Since typically the signal would not have a sharp feature for phase reference, shifting of each realization in time should be by an amount that maximizes the cross-correlation of each realization with the ensemble average. This, like many other aspects of eduction, of course, must be iterative. Since the coherent structure is large-scale, the alignment process can be accelarated by low-pass filtering the signals first. Of course, after having determined the necessary time shift, the unfiltered signal should be used for eduction of structure properties. Realizations requiring excessive time-shifts must be regarded as 'freak' because the corresponding structures have been subjected to unusual turbulence activity, and thus should be discarded.[26,27] Further refinement of signal eduction is possible by discarding realizations which are atypical, identified by weak peak cross-correlation values. However, care should be taken to assure that the acceptance/rejection ratio is not too low and the educed structure itself is not 'freak' or too select to serve any useful purpose.

IIc Perturbation-induced coherent structures

In an unexcited flow, the large dispersion in the structure characteristics makes eduction extremely difficult, and alignment in time can only partly alleviate the problem. One must align only structures of identical characteristics before obtaining phase average (such an effort is underway in our laboratory). The experimental effort can be significantly reduced through induction of structures artificially and eduction of structure details phase-locked to the excitation. For simplicity, the disturbance can be periodic. If isolated structure details are sought, the period can be large compared to the time-scale of the structure.[26] If controlled structure interactions are desired, the excitation period can be comparable to the structure time-scale.[28] Note that periodic excitation at the time-scale of the structure is quite different from the case when the excitation period is much larger than the structure time-scale: in the former case, all structures in the flow are induced, while in the latter case induced structures are separated by and interact with naturally occurring structures. Because of the controlled phase and strength of successive induced structures in the former case, the structure interactions and flow evolution are likely to be different from the latter case. The induced structure in the latter case is more likely to resemble a natural one than in the former case.

Controlled excitation can simplify measurement also because a single sensor can be used to obtain spatial distributions of structure properties by repeating the measurement at the same phase by moving the sensor in space. Because of this simplicity, this approach has found the widest use. A significant, further simplification in experimentation can be achieved by inferring spatial distribution from temporal data by applying the Taylor hypothesis. The validity of the Taylor hypothesis has been quantitatively examined and will be discussed.

IId. <u>Relevance of the artificially-induced structure</u>

An inquiry of the relevance of the 'artificially' induced structures to the naturally occurring ones is both valid and expected. It is important to recognize that no flow is unperturbed: all flows have some residual oscillation arising from a number of factors, including rotating stall, fan blade wake, tunnel settling chamber cavity resonance, laboratory standing waves, feedback from downstream obstructions or structures, etc. It is impossible to completely eliminate these disturbances which manifest themselves as "free-stream" turbulence. These residual 'random' disturbances typically trigger the initiation of coherent structures. Thus, all "natural" coherent structures are in some sense "driven".

Small-amplitude controlled excitation can produce large changes in the time-average measures in the near fields of mixing layers and jets. This does not necessarily mean that the induced structures are different from natural structures. Typically, a flow is unstable to a particular range of disturbance frequencies. Thus when the natural disturbance contains a Fourier mode in this range an instability is induced. In the presence of a broadband 'white-noise' type disturbance, the most unstable frequency i.e. the preferred mode dominates. A small-amplitude excitation at the same frequency, therefore, should produce the same result. If the excitation is periodic, the preferred unstable mode is sustained. However, when not 'artificially' excited, the preferred mode occurs in a random manner along with other modes in the unstable band. Thus, controlled excitation merely paces the formation of the structures and affects their interactions through the control of the phase. In fact, modification of the time-average measures of a jet when excited at its preferred mode[24] is much smaller than that when excited for stable vortex pairing.[23] Of course, if the amplitude of excitation is large, the structure evolution is a strong function of the excitation. Furthermore, control of period or phase in nonlinear instability can produce interactions quite different from those occurring naturally.[29]

Controlled excitation, through the availability of phase reference, allows eduction of coherent structure properties with relatively simple experimental techniques. It is not likely that even following the most sophisticated techniques these properties can be determined to such detail for the natural structures. There is also some evidence that individual structures induced by controlled excitation resemble closely the naturally occurring ones.[24,28,30]

IIe. Initial and Boundary Conditions

The characteristics of the exit boundary layer, i.e. the initial condition, have a profound effect on the free shear flow in not only the near field but also perhaps the far field.[31-33] For sufficiently high Reynolds numbers, the incompressible flow is only a function of the initial and boundary conditions. The boundary condition in a free shear flow is felt primarily through the pressure field. When the boundaries are sufficiently removed from the flow, the flow then principally depends on the initial condition. Different characteristics of the exit boundary layer can be used as identifiers of the initial condition.[33] Since the boundary layer at the lip is typically thin, measurements of V or $\overline{uv}$ are not likely to be accurate. Thus, the initial condition characterization is limited to single-wire measures only. It should be noted parenthetically that invasive measurements near the lip of an initially laminar shear layer are almost sure to be strongly in error due to the probe-induced shear layer tone phenomenon,[34] unless special care is taken to avoid it.

The initial condition may be classified as: laminar, nominally laminar, highly disturbed, and fully turbulent. Most experiments involve the second and third which are intended to be the two asymptotic limiting cases, i.e. the first and the fourth. The first is defined as the case with a velocity profile U(y) matching the Blasius profile with a shape factor of 2.59 and a sufficiently small fluctuation intensity u'/U_e everywhere in the boundary layer (say below 0.5%). [An initial Blasius profile is typical with a straight lip following the contraction. All of our facilities have straight lips parallel to the exit flows, and we recommend use of a straight lip for all free shear flows. However, if a straight lip is not used, the exit profile can be different from the Blasius profile (D. Bechert, private communication). The initial state then is still laminar if u'/U_e is small and the u-spectrum ϕ_u has no sharp peaks.] In the second (nominally laminar) case, the mean profile agrees with the Blasius profile but the fluctuation intensity is large. In the third case, the mean profile noticeably differs from the Blasius profile and ϕ_u is not as broadband as in the fourth (fully turbulent) case and typically contains sharp spectral peaks; the fluctuation intensity in this case is always large. The fourth case must satisfy the following criteria: (i) a clear logarithmic region of the mean velocity profile extending at least one decade in the universal (U^+,y^+) coordinates; (ii) a "wake strength" consistent with the equilibrium flat plate boundary layer[35] at the corresponding Re_θ, (iii) a monotonically increasing u' from the free-stream value to the maximum at about $y^+ = 15$; (iv) broadband ϕ_u without any sharp spectral peak. A consequence of (i) and (ii) is that the shape factor is about 1.4. [A straight lip is assumed in all the four cases]. A large number of previous experiments used tripping devices to obtain the fourth case, but provided no documentation of the exit boundary layer. It is very unlikely that unless the trip is properly designed and placed sufficiently upstream of the lip, the exit boundary layer can become fully turbulent. Thus, there are sufficient reasons to believe that previous tripped shear layers were not initially fully turbulent. Experience shows that a trip of a height equal to the local displacement thickness, with spanwise notches of length, width, depth and spacing equal to this thickness and placed at least 100 displacement thicknesses upstream from the lip is adequate to produce the fully turbulent initial state.

Note that in addition to the mean and rms fluctuation profiles U(y), u'(y), the spectral content of the boundary layer fluctuations is a key identifier of the initial condition.[36] The second case above (i.e. nominally laminar) is associated with distinct spectral peaks representing free-stream oscillations which produce the peak u' value typically at $y/\delta^* \cong 1$.[31] The third case (i.e. highly disturbed) typically involves trips producing separated shear layers which fail either to reattach or to relax to an equilibrium turbulent boundary layer; this should emphasize the criticality of the trip regarding its three-dimensionality and location upstream from the lip. Note that the second and third cases also typically arise due to separation in the nozzle,[37] unless properly designed.

In view of the probably lingering effect of the initial condition, all future studies, including those in the far field, must carefully document the initial condition, i.e. U(y), u'(y), ϕ_u etc., taking especial care to avoid the probe-induced shear tone.

The instability and roll-up of a free shear layer depends on the initial condition;[38,39] an initially fully turbulent layer can also become unstable and roll up depending on the initial condition.[40] Since structure interaction at each stage depends on the preceding structures, the initial condition can continue to retain control of a free shear flow.[41] The axisymmetric free shear layer has been found to be strongly dependent on the initial condition[33,36,42] and increase of the jet diameter up to 27 cm in our laboratory ($Re_D \simeq 2X10^6$) has failed to indicate independence of the shear layer integral measures from the initial condition. Since both asymptotic and local invariances suggest independence from the initial condition,[33,41] it was suggested that a sufficiently large mixing layer must achieve a unique asymptotic state.[41-45] Our data in a large plane mixing layer (span 96 cm) tend to suggest that all integral measures of the layer eventually become identical, independent of the initial condition.[46] It thus appears that while the plane mixing layer has a unique asymptotic state, the axisymmetric mixing layer is persistently dependent on the initial condition.

Evolution of the coherent structures is often complicated by geometric constraints. In a round jet, the spanwise curvature starts becoming significant unless $\theta/D << 1$, where θ is the shear layer momentum thickness. Typically the curvature effect should be non-negligible for $x/D \gtrsim 1$. [It is surprising that time-average measures of the axisymmetric mixing layer for $x/D \lesssim 4.5$ agree very closely with those of the plane mixing layer.[42,43]] On the other hand, there is possibly some unavoidable influence of the end bounding walls, which define the span of the plane mixing layer, on the structure of the layer. In this sense, a large axisymmetric jet near field is the ideal two-dimensional mixing layer, because it is free from any end effects. The presence of side walls (parallel to a plane mixing layer) is almost sure to enhance the spanwise organization of the coherent structure of the layer, unless the layer is sufficiently away from the walls. The relatively cleaner large-scale structures in the two-stream mixing layer as compared to the one-stream layer could be partly due to such an influence from the two bounding parallel walls in the two-stream case. Furthermore, termination of the test section by an expansion at the downstream end of the two-stream layer, may further organize the structures upstream through feedback, similar to that in the whistler nozzle.[47] The presence of pressure gradient is also likely to alter the structure. A favorable gradient may stabilize the structures while an adverse gradient may hasten downstream evolution and breakdown.

III. Some Analytical Considerations

For the purpose of analysis, an instantaneous flow signal f in the presence of a structure could be regarded as a superposition of the time-mean $\bar{f}$, the coherent motion $\tilde{f}_p$ and the phase-incoherent motion f_r.[48,49] That is, $f = \bar{f} + \tilde{f}_p + f_r$. The corresponding continuity and momentum equations are:

$$\frac{\partial \bar{u}_i}{\partial x_i} = \frac{\partial \tilde{u}_{pi}}{\partial x_i} = \frac{\partial u_{ri}}{\partial x_i} = 0 \tag{1a}$$

$$\frac{\bar{D}}{Dt}\bar{u}_i = -\frac{1}{\rho}\frac{\partial \bar{p}}{\partial x_i} + \nu\frac{\partial^2 \bar{u}_i}{\partial x_j \partial x_j} + \frac{\partial}{\partial x_j}(-\overline{\tilde{u}_{pi}\tilde{u}_{pj}} - \overline{u_{ri}u_{rj}}) \tag{1b}$$

$$\frac{\bar{D}}{Dt}\tilde{u}_{pi} = -\frac{1}{\rho}\frac{\partial \tilde{p}_p}{\partial x_i} + \nu\frac{\partial^2 \tilde{u}_{pi}}{\partial x_j \partial x_j} - \frac{\partial}{\partial x_j}(\bar{u}_i\tilde{u}_{pj})$$

$$-\frac{\partial}{\partial x_j}(\tilde{u}_{pi}\tilde{u}_{pj} - \overline{\tilde{u}_{pi}\tilde{u}_{pj}}) - \frac{\partial}{\partial x_j}(\langle u_{ri}u_{rj}\rangle - \overline{u_{ri}u_{rj}}) \tag{1c}$$

where $\bar{D}/Dt \equiv \partial/\partial t + \bar{u}_j\partial/\partial x_j$ is the material derivative following the time-mean motion of a fluid element; the overbar is the time-average and $\langle f \rangle$ is the phase-average of f. Note that summation is not implied for subscripts p and r. For a two-dimensional flow, $-\tilde{u}_p\tilde{v}_p$ is the coherent Reynolds stress, and $-\langle u_r v_r \rangle$ is the phase-average of the background turbulence Reynolds stress. Note that in addition to $\tilde{u}_p\tilde{v}_p$, the turbulent transport by the structure also includes $U\tilde{v}_p + \tilde{u}_p V$.

Some further understanding of the dynamics can be obtained from the study of the energetics. The total mean kinetic energy is the sum of the mean kinetic energy in each field. That is,

$$\frac{1}{2}\overline{u_i u_i} = \underset{E}{\frac{1}{2}\bar{u}_i\bar{u}_i} + \underset{\varepsilon}{\frac{1}{2}\overline{\tilde{u}_{pi}\tilde{u}_{pi}}} + \underset{e}{\frac{1}{2}\overline{u_{ri}u_{ri}}} \tag{2a}$$

$$\frac{\bar{D}E}{Dt} = -(-\overline{\tilde{u}_{pi}\tilde{u}_{pj}})\frac{\partial\bar{u}_i}{\partial x_j} - (-\overline{u_{ri}u_{rj}})\frac{\partial\bar{u}_i}{\partial x_j} + \ldots\ldots \tag{2b}$$

$$\frac{\bar{D}\varepsilon}{Dt} = -\overline{\tilde{u}_{pi}\tilde{u}_{pj}}\frac{\partial\bar{u}_i}{\partial x_j} - \overline{(-\langle u_{ri}u_{rj}\rangle)\frac{\partial\tilde{u}_{pi}}{\partial x_j}} + \ldots\ldots \tag{2c}$$

$$\frac{\bar{D}e}{Dt} = -\overline{u_{ri}u_{rj}}\frac{\partial\bar{u}_i}{\partial x_j} - \overline{\langle u_{ri}u_{rj}\rangle\frac{\partial\tilde{u}_{pi}}{\partial x_j}} + \ldots\ldots \tag{2d}$$

The mean field provides energy to the coherent structure and random turbulence by the works of the coherent and background Reynolds stresses ($-\overline{\tilde{u}_{pi}\tilde{u}_{pj}}$ and $-\overline{u_{ri}u_{rj}}$) against the mean strain rate. Note that these appear as source terms in the corresponding equations with signs opposite to those in eqn (2b). The coherent structure also produces random turbulence by the work of the phase-average background Reynolds stress $-\langle u_{ri}u_{rj}\rangle$ against the coherent motion strain rate; the same term appears as a source term in eqn (2d) with the opposite sign.

The preceding derivations assume that $\langle\tilde{f}_p g_r\rangle = \overline{\tilde{f}_p g_r} = 0$. That is, the coherent and background turbulence are uncorrelated. This does not mean that the two fields are independent. In fact, the background turbulence is both produced and organized by the coherent structures. These two are not, however, in strong mutual interaction because of the largely differing time-scales: the phase-incoherent motion is mostly of a time-scale considerably smaller than the coherent structure time-scale.

The above triple decomposition provides a formalism for exchange of kinetic energy between the time-mean, coherent and phase-incoherent motions, but is somewhat artificial. The coherent structure is not a perturbation from the time-mean motion; the latter is the result of combinations (and interactions) of many coherent structures. At the time of passage of a structure, the entire motion (minus the phase-random part) is due to the structure. That is, the structure is not riding on the time-mean profile; the structure is the flow.

Perhaps the double decomposition[49] $f = f_p + f_r$ is more appropriate. The corresponding continuity, momentum and vorticity equations are:

$$\frac{\partial u_{pi}}{\partial x_i} = \frac{\partial u_{ri}}{\partial x_i} = 0, \tag{3a}$$

$$\frac{\hat{D}}{Dt}u_{pi} = -\frac{1}{\rho}\frac{\partial p_p}{\partial x_i} + \frac{\partial}{\partial x_j}\left[\nu\frac{\partial u_{pi}}{\partial x_j} - \langle u_{ri}u_{rj}\rangle\right], \tag{3b}$$

$$\frac{\hat{D}}{Dt} u_{ri} = -\frac{1}{\rho}\frac{\partial p_r}{\partial x_i} - u_{rj}\frac{\partial u_{pi}}{\partial x_j} + \frac{\partial}{\partial x_j}\left[\nu\frac{\partial u_{ri}}{\partial x_j} - (u_{ri}u_{rj} - \langle u_{ri}u_{rj}\rangle)\right], \tag{3c}$$

$$\frac{\hat{D}}{Dt}\omega_{pi} = \omega_{pj}\frac{\partial u_{pi}}{\partial x_j} + \nu\frac{\partial^2 \omega_{pi}}{\partial x_j \partial x_j} + \frac{\partial}{\partial x_j}\left[\langle u_{ri}\omega_{rj}\rangle - \langle u_{rj}\omega_{ri}\rangle\right], \tag{3d}$$

$$\frac{\hat{D}}{Dt}\omega_{ri} = \omega_{pj}\frac{\partial u_{ri}}{\partial x_j} + \omega_{rj}\frac{\partial u_{pi}}{\partial x_j} + \nu\frac{\partial^2}{\partial x_j \partial x_j}\omega_{ri} - u_{rj}\frac{\partial \omega_{pi}}{\partial x_j}$$
$$- \frac{\partial}{\partial x_j}(\omega_{ri}u_{rj} - \langle\omega_{ri}u_{rj}\rangle) + \frac{\partial}{\partial x_j}(\omega_{rj}u_{ri} - \langle\omega_{rj}u_{ri}\rangle). \tag{3e}$$

where $\hat{D}/Dt = \partial/\partial t + u_{pj}\partial/\partial x_j$ is the material derivative following the phase-average motion of a fluid element; no summation is implied by subscripts p and r. The corresponding kinetic energy, Reynolds stress equations are:

$$\frac{\hat{D}}{Dt}\frac{u_{pi}u_{pi}}{2} = \underset{\text{I}}{-\frac{\partial}{\partial x_j}\frac{p_p u_{pj}}{\rho}} + \underset{\text{II}}{\langle u_{ri}u_{rj}\rangle\frac{\partial u_{pi}}{\partial x_j}} - \underset{\text{III}}{\frac{\partial}{\partial x_j}(u_{pi}\langle u_{ri}u_{rj}\rangle)}$$
$$+ \underset{\text{IV}}{\nu\frac{\partial}{\partial x_j}\left\{u_{pi}\left(\frac{\partial u_{pi}}{\partial x_j} + \frac{\partial u_{pj}}{\partial x_i}\right)\right\}} - \underset{\text{V}}{\frac{\nu}{2}\left\{\frac{\partial u_{pi}}{\partial x_j} + \frac{\partial u_{pj}}{\partial x_i}\right\}\left\{\frac{\partial u_{pi}}{\partial x_j} + \frac{\partial u_{pj}}{\partial x_i}\right\}}, \tag{4a}$$

$$\frac{\hat{D}}{Dt}\frac{\langle u_{ri}u_{ri}\rangle}{2} = -\frac{\partial}{\partial x_j}\langle u_{rj}\left(\frac{p_r}{\rho} + \frac{1}{2}u_{ri}u_{ri}\right)\rangle - \langle u_{ri}u_{rj}\rangle\frac{\partial u_{pi}}{\partial x_j}$$
$$+ \nu\frac{\partial}{\partial x_j}\langle u_{ri}\left(\frac{\partial u_{ri}}{\partial x_j} + \frac{\partial u_{rj}}{\partial x_i}\right)\rangle - \frac{\nu}{2}\langle\left(\frac{\partial u_{ri}}{\partial x_j} + \frac{\partial u_{rj}}{\partial x_i}\right)\left(\frac{\partial u_{ri}}{\partial x_j} + \frac{\partial u_{rj}}{\partial x_i}\right)\rangle, \tag{4b}$$

$$\frac{\hat{D}}{Dt}\langle u_{ri}u_{rj}\rangle = -\frac{1}{\rho}\langle\left(u_{ri}\frac{\partial p_r}{\partial x_j} + u_{rj}\frac{\partial p_r}{\partial x_i}\right)\rangle - \frac{\partial}{\partial x_k}\langle u_{ri}u_{rj}u_{rk}\rangle - \langle u_{rj}u_{rk}\rangle\frac{\partial u_{pi}}{\partial x_k}$$
$$- \langle u_{ri}u_{rk}\rangle\frac{\partial u_{pj}}{\partial x_k} + \nu\frac{\partial^2\langle u_{ri}u_{rj}\rangle}{\partial x_k \partial x_k} - 2\nu\langle\frac{\partial u_{ri}}{\partial x_k}\frac{\partial u_{rj}}{\partial x_k}\rangle. \tag{4c}$$

In eqn. (4a), the terms denote: I - coherent pressure work, II - coherent production of background turbulence, III - coherent momentum transport by background turbulence, IV - reversible viscous work by coherent motion, V - irreversible viscous work by the coherent motion (i. e. coherent dissipation). The terms in eqn (4b), which results from contraction of (4c), represent the corresponding terms for the background turbulence field, the last term denoting the phase-average background turbulent dissipation. These equations are written down for completeness of discussion and further details are available elsewhere.[49]

For two-dimensional coherent structures, the equation for the coherent vorticity $\Omega_p = \partial v_p/\partial x - \partial u_p/\partial y$ becomes, after an order of magnitude analysis,

$$\frac{\hat{D}}{Dt}\ \Omega_p = -\frac{\partial^2}{\partial x^2}\langle u_r v_r\rangle + \frac{\partial^2}{\partial y^2}\langle u_r v_r\rangle \qquad (5)$$

That is, the maximum decrease of coherent vorticity is associated with the 'saddle point' (i.e.minimax) of the background turbulence Reynolds stress distribution. Physically, the continual advection of turbulent fluid from the braids into the structure cores induced by the structures dilutes the peak vorticity at the core centers.

The double decomposition allows analysis of the phase-coherent fluctuations but does not directly address the evolution of the coherent structures themselves. Further decomposition of the coherent motion into mean and time-dependent parts merely brings back the problem of triple decomposition.[50] A finite-difference scheme with sufficiently small spatial resolution should be able to predict the evolution in space and time of coherent structures of specified initial characteristics. This is likely to be an extremely time-consuming undertaking. Computation costs so far have dictated pursuit of temporal evolution alone[50] or simplification of the flow by point or discrete vortices.[51-56] The analysis or discussion of the coherent structures may be more useful in a quasi-Lagrangian form in a frame convected with a velocity U_c equal to that of the structure center. However this approach is also not free from problems. Flow visualization has revealed that a structure survives only for a distance about equal to its length scale.[9] Consequently, the convection velocity is not a meaningful property. Secondly, the structure convection velocity is neither uniquely defined nor easily measureable[57-59] and varies considerably.

IIIa. Coherent Structure Properties

From the point of view of measurements as well as analytical formulations, it is important to agree on the measures which characterize the coherent structures. Since coherent vorticity is the key characteristic of a structure, it must be one of its primary properties. Thus, measurements should include $\Omega_p = \partial v_p/\partial x - \partial u_p/\partial y$ and $\tilde{\Omega}_p = \partial\tilde{v}_p/\partial x - \partial\tilde{u}_p/\partial y$. In addition, other important structure properties are: coherent Reynolds stresses $u_p v_p$ and $\tilde{u}_p\tilde{v}_p$; time-average and phase-average background turbulent Reynolds stresses $\overline{u_r v_r}$ and $\langle u_r v_r\rangle$; coherent strain rates $S_p = \partial u_p/\partial y + \partial v_p/\partial x$, $\tilde{S}_p = \partial\tilde{u}_p/\partial y + \partial\tilde{v}_p/\partial x$; coherent structure production $\langle P\rangle = u_p v_p(\partial U/\partial y + \partial V/\partial x)$; coherent trubulence productions $\langle\tilde{P}\rangle = -\langle u_r v_r\rangle(\partial\tilde{u}_p/\partial y + \partial\tilde{v}_p/\partial x)$ and $\langle P\rangle = -\langle u_r v_r\rangle(\partial u_p/\partial y + \partial v_p/\partial x)$. The last two should be contrasted with the time-average production $\bar{P} = -\overline{u_r v_r}(\partial U/\partial y + \partial V/\partial x)$. Note that these production terms include only "shear production" and do not include "normal production" terms like $\langle u_r^2\rangle\partial u_p/\partial x$, $\langle v_r^2\rangle\partial v_p/\partial y$, $\langle u_r^2\rangle\partial U/\partial x$, $\langle v_r^2\rangle\partial V/\partial y$, etc.

Additional properties of interest are: vector patterns $\hat{\Psi}_p=(u_p,v_p)$ and $(\tilde{u}_p,\tilde{v}_p)$, instantaneous streamlines ψ_p, pseudo-streamfunctions $\langle\psi\rangle = \int_0^r r^*(u_p - U_c)dr^*$ and coherent intermittency γ_p. Note that both $\hat{\Psi}_p$ and $\langle\Psi\rangle$ are not invariant under Galilean transformation and can change by a great extent depending on the reference frame of the observer. In the computation of $\langle\Psi\rangle$, the most appropriate frame has been found to be that moving downstream with the structure center convection velocity U_c.[26-28] The lines drawn through the vector patterns at any phase over the extent of the structure give the corresponding instantaneous streamlines with respect to the laboratory frame. The function $\langle\psi\rangle$ will correspond to streamfunction if the flow were steady and two-dimensional; hence the name pseudo-stream function. The coherent intermittency γ_p is an ensemble average of the phase-aligned instantaneous intermittency functions.

The above are some properties which can quantitatively characterize the structure properties (and perhaps its boundary), and thus should be measured if possible. The properties discussed above address coherent structures which are essentially two-dimensional. When these are strongly three-dimensional, the velocities, vorticities, and transport terms in the other two directions must also be considered.

In a flow with temperature and concentration denoted by $\theta = \theta_p + \theta_r$ and $s = s_p + s_r$, the coherent heat and mass fluxes will be $\langle v_p\theta_p\rangle$ and $\langle v_p s_p\rangle$ in the transverse direction and $\langle u_p\theta_p\rangle$ and $\langle u_p s_p\rangle$ in the longitudinal direction. The corresponding phase-average background turbulent heat and mass fluxes will be $\langle v_r\theta_r\rangle$, $\langle v_r s_r\rangle$, $\langle u_r\theta_r\rangle$, $\langle u_r s_r\rangle$, etc.

In the following, we summarize our measurements involving some of the above quantities and describe the structure characteristics based on these properties. For further discussion of measurement details and constraints as well as of other properties not reported here, see cited references, especially [27, 28].

IV. EXPERIMENTAL RESULTS

IVa. Coherent Structures and Vortex Pairing in a Circular Jet

The coherent structures and some details of vortex pairing dynamics in the near field of a 7.62 cm circular air jet at a jet Reynolds number of 3.2×10^4 have been explored through controlled excitation. It was found that under certain excitation conditions the coherent vortical structures could be induced to undergo 'stable' pairing, which occurred at the identical physical location at successive intervals of excitation.[23] This enabled measurement of flow properties at a location by sampling the signal from that location at periodic intervals, the sampling being locked to the same phase of occurrence of the structure, and then obtaining an ensemble average at that phase; this is called the phase average. By moving the probe to different spatial locations, the phase average of any property over the physical extent of the structure was obtained for the selected phase. Thus, the structure properties are 'actual' within the measurement uncertainty and do not involve use of the Taylor hypothesis.

In order to educe the structure details, a particular phase of the coherent structures must be selected. Fig. 1 schematizes the structures at the selected phase in the four Regions I-IV as well as the on-line computer-controlled measurement system. To reduce detection jitter, the phase for signal eduction was obtained from a reference hot-wire in the flow rather than from the excitation signal; the correspondence between the reference signal and structure phase was established through simultaneous stroboscopic flow visualization. Fig. 2 shows contours of

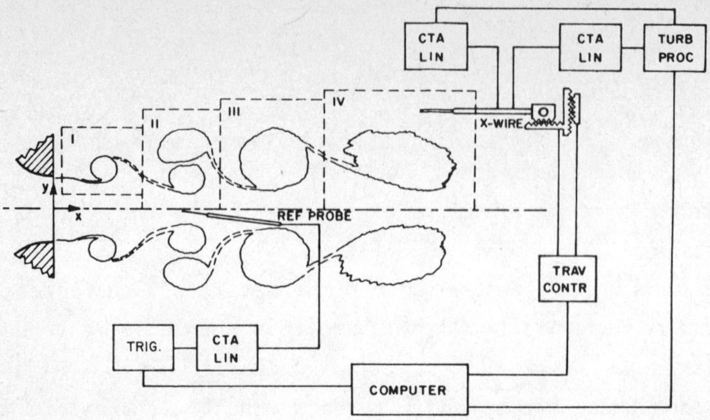

Fig. 1. Schematics of structures in Regions I-IV and instrumentation.

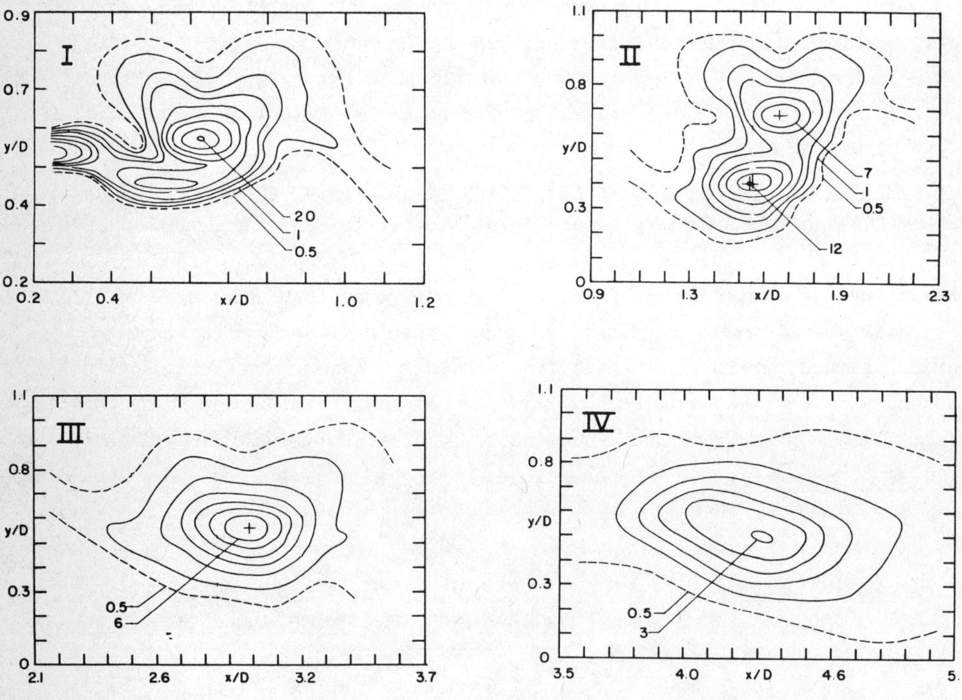

Fig. 2. Contours of Ω_p for Regions I-IV.

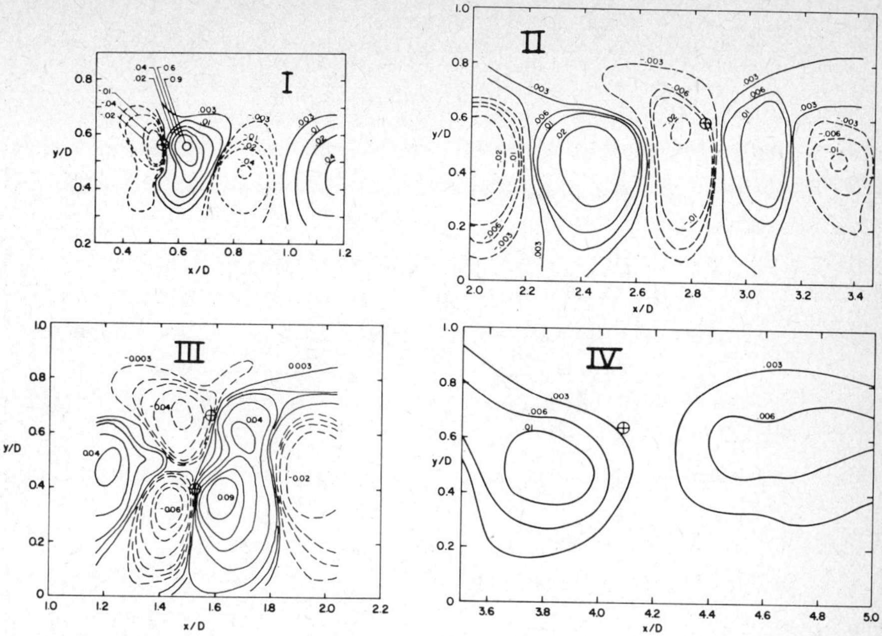

Fig. 3. Contours of coherent Reynolds stress $u_p v_p$ in Regions I-IV.

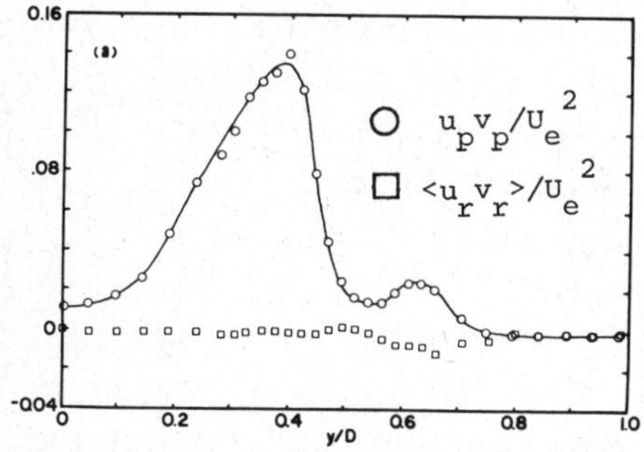

Fig. 4a. $\langle u_p v_p \rangle$ and $\langle u_r v_r \rangle$ in Region II.

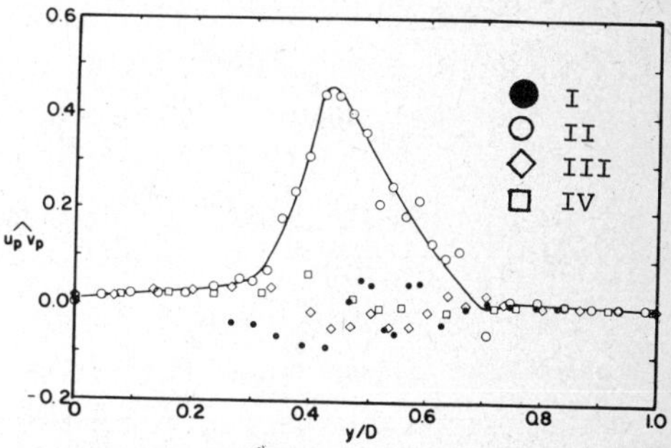

Fig. 4b. $\widehat{u_p v_p}$ in Regions I-IV

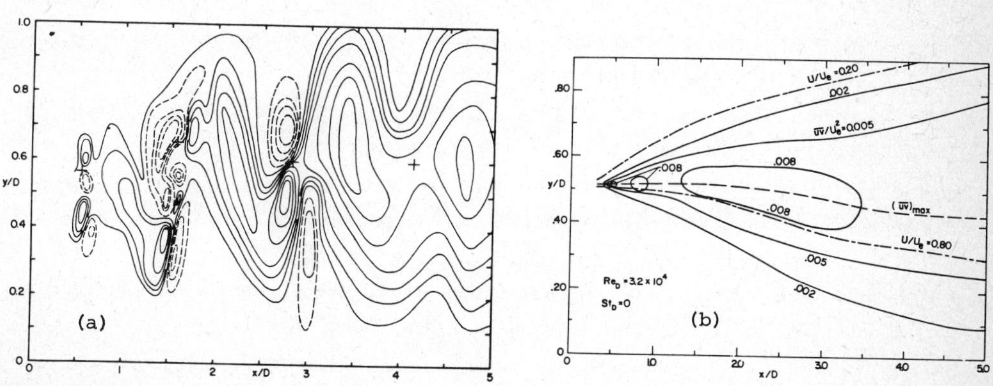

Fig. 5. (a) $\langle u_r v_r \rangle$; (b) unexcited time-average Reynolds stress $\overline{uv}$.

phase-average azimuthal vorticitiy Ω_p for the four Regions I-IV. Note that at the selected phase, the Region I contains a rolled up vortex, Region II contains two vortical structures in nearly radial configuration in the middle of pairing, Region III contains a structure immediately following pairing, and Region IV the same paired structure farther downstream. In each region, (and elsewhere in this paper) vorticity is nondimensionalized by the roll-up frequency f_p, which is the frequency of controlled excitation. Three interesting observations can be made. First, the peak vorticity decreases with time; amalgamation of structures of similarly directed vorticity does not result in its increase. Second, the vorticity in the braids is considerably lower than the structure peak vorticity; this is to be expected because of continual (structure-induced) migration of vortical fluid from the braids to the structure centers[60]. Third, the low-speed side distortion (i.e. 'dip') in the vorticity contours is not real but a measurement artefact resulting from two independent effects: jitter in the structure arrival at the measurement probe and flow reversal.[28,61]

Figure 3 shows contours of the coherent Reynolds stress $u_p v_p$ in the Regions I-IV. The corresponding structure centers identified by peak vorticity are shown by the $\oplus$ symbols. Note that the Regions are marked by alternate positive and negative contours with a periodicity in x twice that of the structure, as to be expected. It should be clear from these spatial patterns of $u_p v_p$ that coherent Reynolds stress, even when averaged in space or time, can be of either sign depending on the relative strengths of positive and negative contributions. Thus, counter-gradient average momentum transport should not appear surprising[62,63] and can be explained through coherent structure interactions.[62,23,28] The Region IV shows no negative peak in $u_p v_p$ because this is both very low and perhaps smeared out due to jitter. The transverse variation of $u_p v_p$ through its peak location in Region II is shown in Fig. 4a, which shows that the coherent Reynolds stress is significantly larger than the background turbulence Reynolds stress. However, because $u_p v_p$ contours consist of both positive and negative contributions, comparison with the time average is meaningful only when an average is obtained over the spatial extent of the structure. Figure 4b shows the transverse variation of the zone-average coherent Reynolds stress $\widehat{u_p v_p}$, which is the average at each y of $u_p v_p$ in each Region. It is clear that although Region I consists of large positive and negative peaks, the zone average is no larger than those in Regions III and IV. Thus, significant Reynolds stress production is associated with the pairing event.[3,28,62] When the coherent structures are not in intense interaction like pairing, the zone-average coherent Reynolds stress is not necessarily larger than the background turbulence Reynolds stress.

Figure 5a shows a composite of the background turbulence Reynolds stress $\langle u_r v_r \rangle$ for the four Regions I-IV combined. Note that the vorticity peaks occur at the locations (marked by +) of minimaxes of $\langle u_r v_r \rangle$, as predicted by eqn. (5). This figure suggests that the background turbulence is organized by the coherent structure and produces maximum decrease of coherent vorticity at its 'saddle points'.

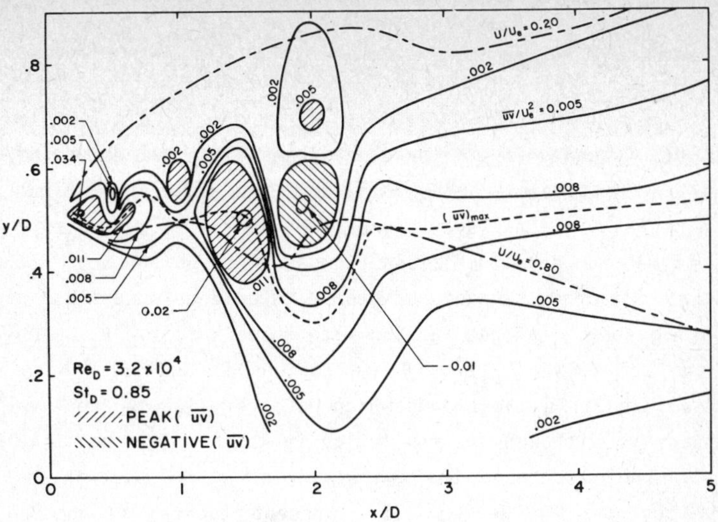

Fig. 5c. Time-average Reynolds stress $\overline{uv}$ under excited condition.

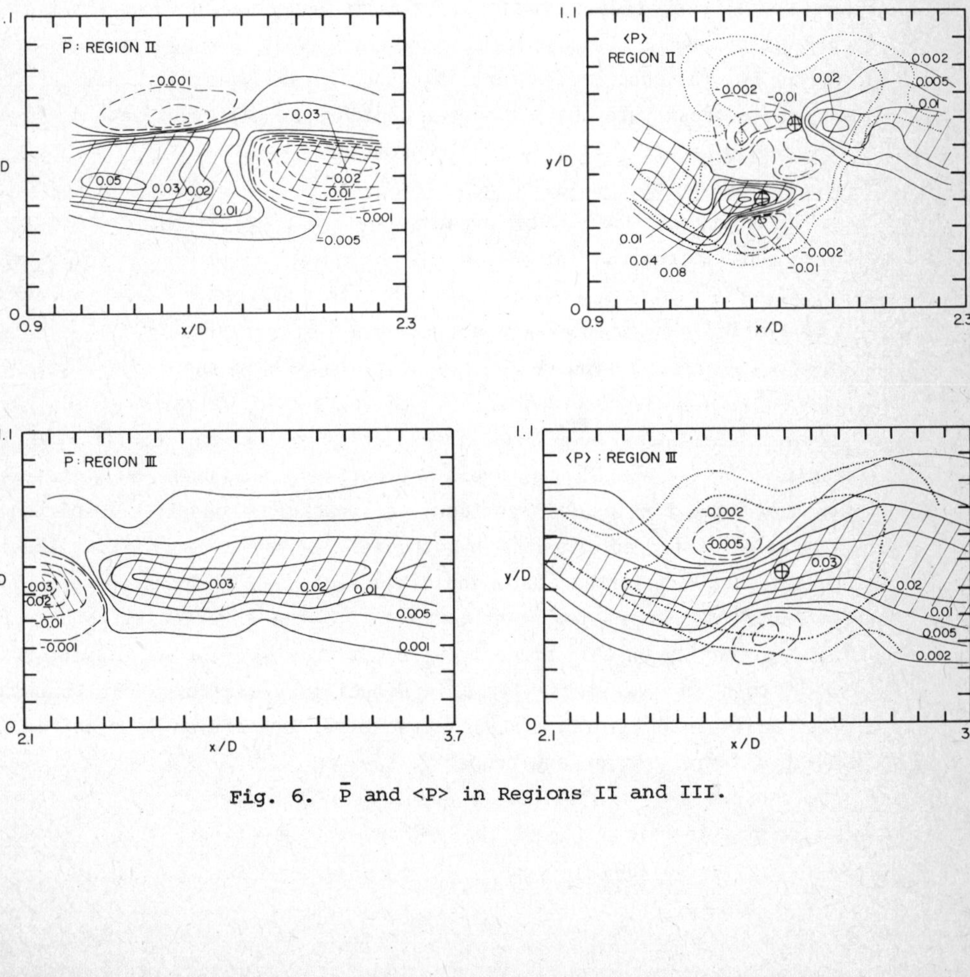

Fig. 6. $\bar{P}$ and <P> in Regions II and III.

The coherent structures induce the braid turbulent fluid to migrate to the structure centers and dilute the peak vorticity. Figures 5b,c show the corresponding time-average Reynolds stress distributions without and with excitation, respectively. Note that pairing can induce large variations of the Reynolds stress and also of a sign opposite to that suggested by the gradient transport hypothesis. Note also the large distortion of the time-average measures of the shear layer induced by the stable pairing, especially in the layer boundaries identified by U/U_e=0.2 and 0.8 as well as the line of maximum Reynolds stress (compare Figs 5b and 5c). It is reasonable to assume that the production of background turbulence occurs not uniformly in space or time but in some organized fashion with respect to the structure boundary. Figure 6 shows the contours of $\bar{P}$ and $\langle P \rangle$ in Regions II and III. Note that large production is associated with the braids which have very low coherent vorticity but large strain rates. Even though the resolution in Region II is not fine enough, it is clear that the region between the two vortices is not associated with large production. The contours in Region III represent typical production around a large-scale structure in a passive state i.e. when neither pairing nor tearing.

In Regions III and IV, the zone-average and time-average background turbulence productions are comparable, thus suggesting that the induced structures are typical of the natural structures. It should be mentioned that since azimuthal lobes appear[28] in Region IV, the educed structure is not quite axisymmetric. However, the azimuthal alignment necessary to obtain sharper contours at different lateral planes of the structure would be prohibitively time-consuming.

IVb. <u>On the Applicability of the Taylor hypothesis</u>

The inapplicability of the Taylor hypothesis (i.e., $\partial/\partial t = -U_T\partial/\partial x$) to flows with large shear and fluctuation intensities has been well known.[21,64-68] However, in most experimental situations, use of the hypothesis is unavoidable, sometimes necessitated by limited experimental times and number of sensors.[26,27,69-73] Even though the extent of error introduced by the hypothesis for the large wavenumber components appear to be well-documented,[65-70] the extent of validity of the hypothesis for large scales has never been investigated. The applicability of the Taylor hypothesis to large-scale coherent structures has been quantitatively investigated by comparing the 'acutal' contours of structure properties (reported in Sec. IVa) with those inferred via the hypothesis. A review of the hypothesis, experimental procedures and further data and discussions can be found in ref. [74]. Since coherent vorticity is an important property of the coherent structure, our study was focused primarily on this property, i.e., $\omega = \partial v/\partial x - \partial u/\partial y$. Taylor hypothesis enters in this in two steps: first in deducing $\partial v/\partial x$ from $\partial v/\partial t$ to obtain $\omega(t,y)$ and then again in obtaining $\omega(x,y)$. The latter conversion is required for inferring spatial distribution of any property, say, $\langle u_r^2\rangle^{1/2}$, $\langle u_r v_r\rangle$, u_p, v_p, etc. from stationary sensors and is intuitively quite clearly perceptible. The former, more subtle, step has therefore formed the primary thrust of our study. Thus, we first obtained $\omega(t,y)$ as follows:

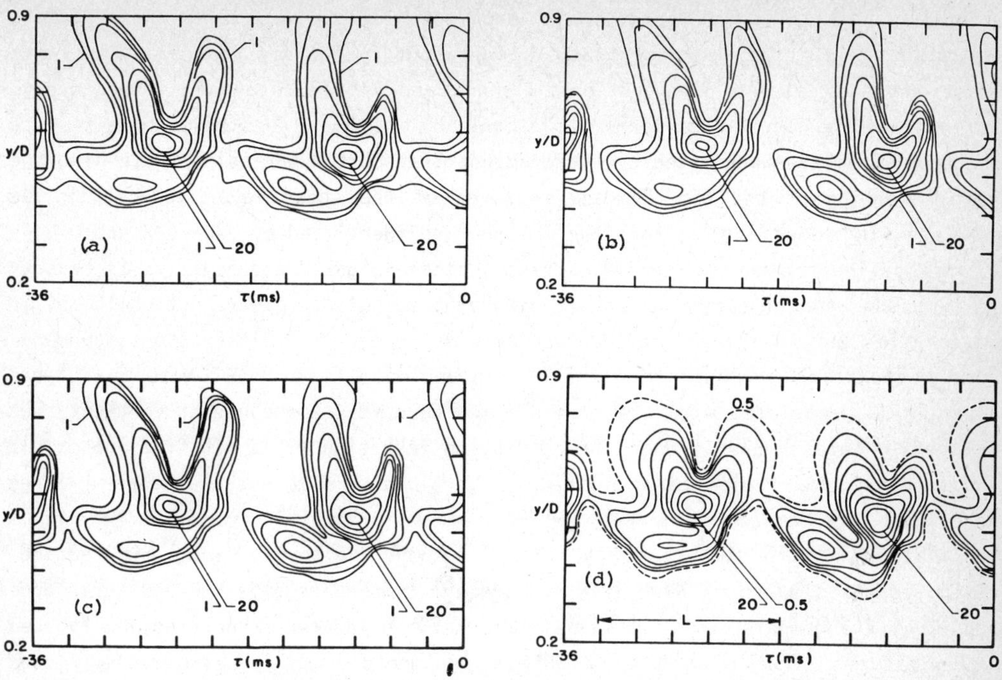

Fig. 7. $\Omega_p(t,y)$ contours for Region I.

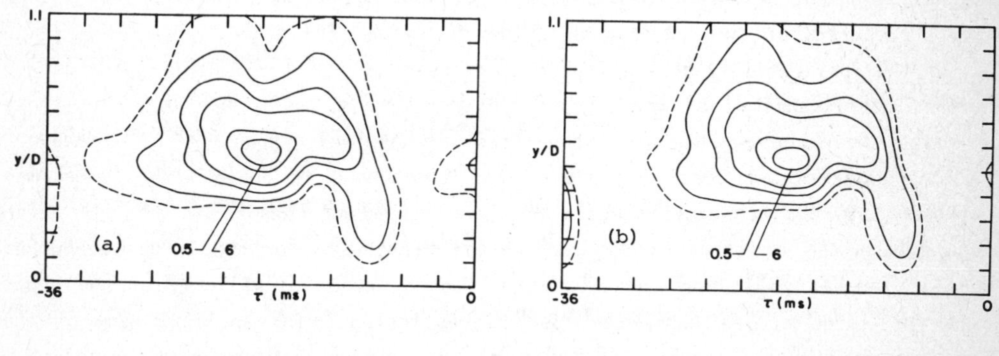

Fig. 8. $\Omega_p(t,y)$ contours for Region III.

$$\tilde{\Omega} = -\frac{\partial\tilde{u}}{\partial y} - \frac{1}{U_T}\frac{\partial\tilde{v}}{\partial t}\,; \qquad \Omega_p \equiv \langle\tilde{\Omega}\rangle \tag{6}$$

where Ω_p represents phase-average of ω, and U_T is the convection velocity in the Taylor hypothesis. Four choices of U_T appear worth investigating:

a) $U_T = U(y)$, the time-mean velocity;

b) $U_T = \langle u\rangle(y,t)$, the coherent or phase-average velocity;

c) $U_T = u(x,t)$, the instantaneous velocity;

d) $U_T = U_c$, the structure center velocity.

Use of (a) is quite widespread, especially in the conversion of frequency-spectrum to wavenumber-spectrum. Choice of (b) or (c) is a logical improvement over (a) because it is with this velocity that the fluid is advected past the sensor instantaneously[75] or on a phase-average basis. Since the convection velocity of the same structure varies greatly with time as well as from structure to structure, U_c is neither unique nor easily measurable.[76,77] Furthermore, structures usually undergo continual deformation in cycles of regeneration, growth and decay[9,15] and thus there is no apriori reason to prefer any of the choices (a)-(d). As an example, Figs. 7a-d show the $\Omega_p(t,y)$ contours computed by schemes (a)-(d), respectively for Region I (see Fig. 1). These two vortices which will pair in Region II are not expected to be identical. Note that (a)-(c) provide unacceptably large distortions. Only (d) produces structures with the least distortion. Figure 8 shows the $\Omega_p(t,y)$ contours for Region III. Note that distortions for the four schemes (a)-(d) are similar. As we will see, this is because $\partial v_p/\partial x$ is considerably smaller than $\partial u_p/\partial y$. Figure 9 shows the $\Omega_p(x,y)$ contours in the four Regions through conversion (b) for Regions I-III and (a) for Region IV. Conversions (a)-(c) all produce comparable and unacceptably large distortions. In Region IV, $\Omega_p(t,y)$ corresponding to the four schemes are essentially similar, yet $\Omega(x,y)$ obtained through the $t \rightarrow x$ transformation using schemes (a)-(c) produce large distortions like those shown (scheme a). Thus, contour distortion due to application of the Taylor hypothesis in obtaining $\Omega_p(t,y)$ is considerably smaller than that introduced by the hypothesis in the $t\rightarrow x$ conversion. For either step, scheme (d) produces the least distortion.

Considering negligible viscous effects, the equation for the phase-average vorticity may be written as

$$-\Omega = \frac{\partial u_p}{\partial y} + \frac{1}{u_p}\frac{\partial v_p}{\partial t} + \frac{v_p}{u_p}\frac{\partial v_p}{\partial y} + \frac{1}{u_p}\frac{\partial\langle v_r^2\rangle}{\partial y} + \frac{1}{u_p}\frac{\partial\langle u_r v_r\rangle}{\partial x} + \frac{1}{\rho u_p}\frac{\partial p_p}{\partial y} \tag{7}$$

For the Region III, spatial distributions of these quantities are shown in Fig. 10(a)-(e)as an example; (f) shows actual $\partial v_p/\partial x$ without using the hypothesis. Note that of the last four terms neglected by the Taylor hypothesis, only the turbulence Reynolds stress term can be neglected. Note that $(1/u_p)\partial v_p/\partial t$ is qualitatively equal and of opposite sign to that of actual $\partial v_p/\partial x$, although they disagree in details. The pressure term in eqn. (7) has been estimated to be large and hence non-negligible.

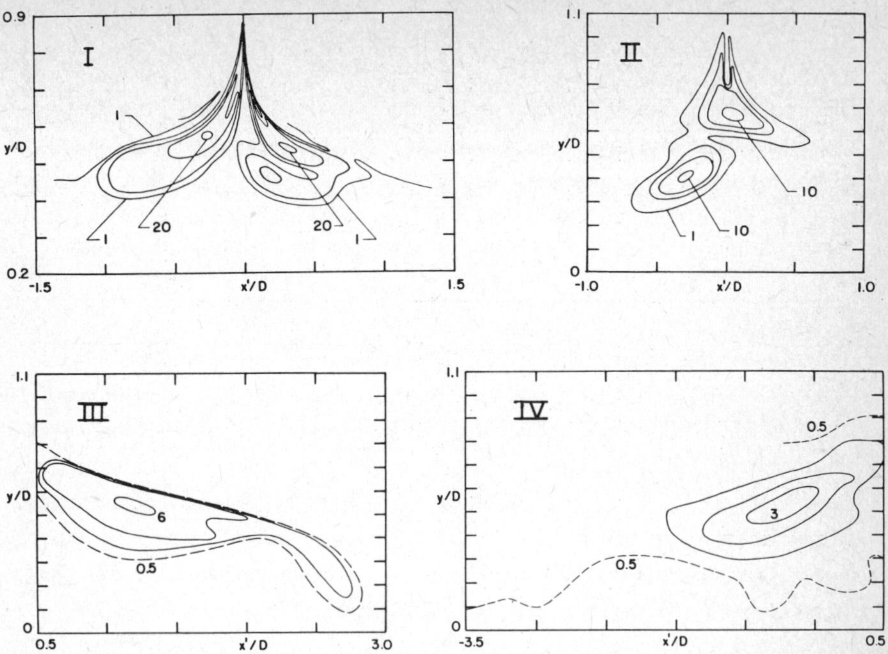

Fig. 9. $\Omega_p(x,y)$ contours for Regions I-IV.

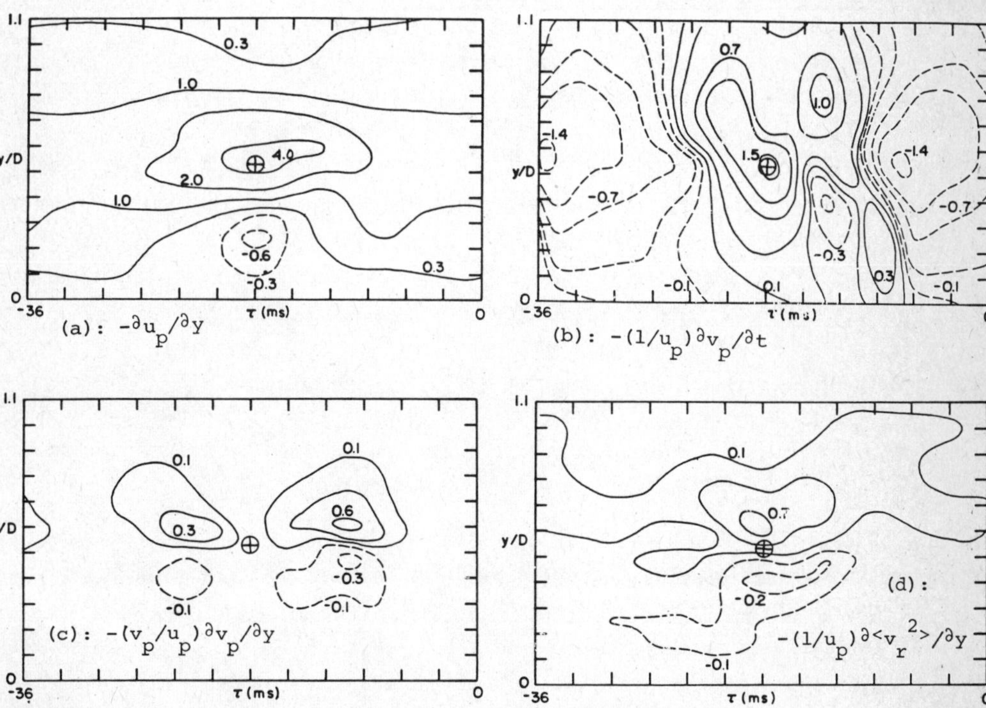

Fig. 10a-d. Distribution of terms in Ω_p/f_p.

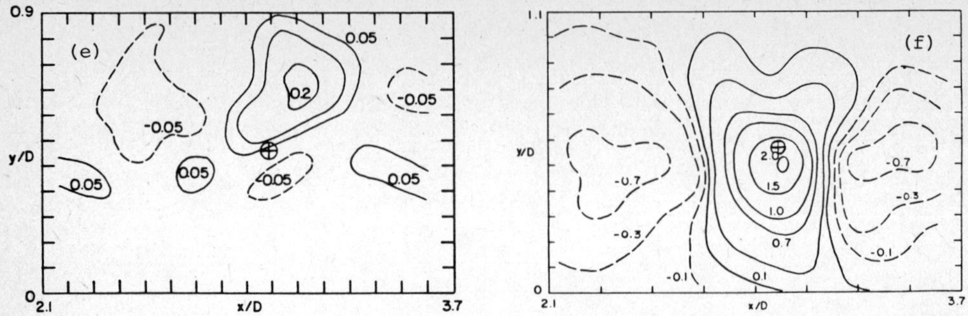

Fig. 10(cont'd): (e) $(1/u_p)\ \partial\langle u_r v_r\rangle/\partial x$; (f) actual $\partial v_p/\partial x$.

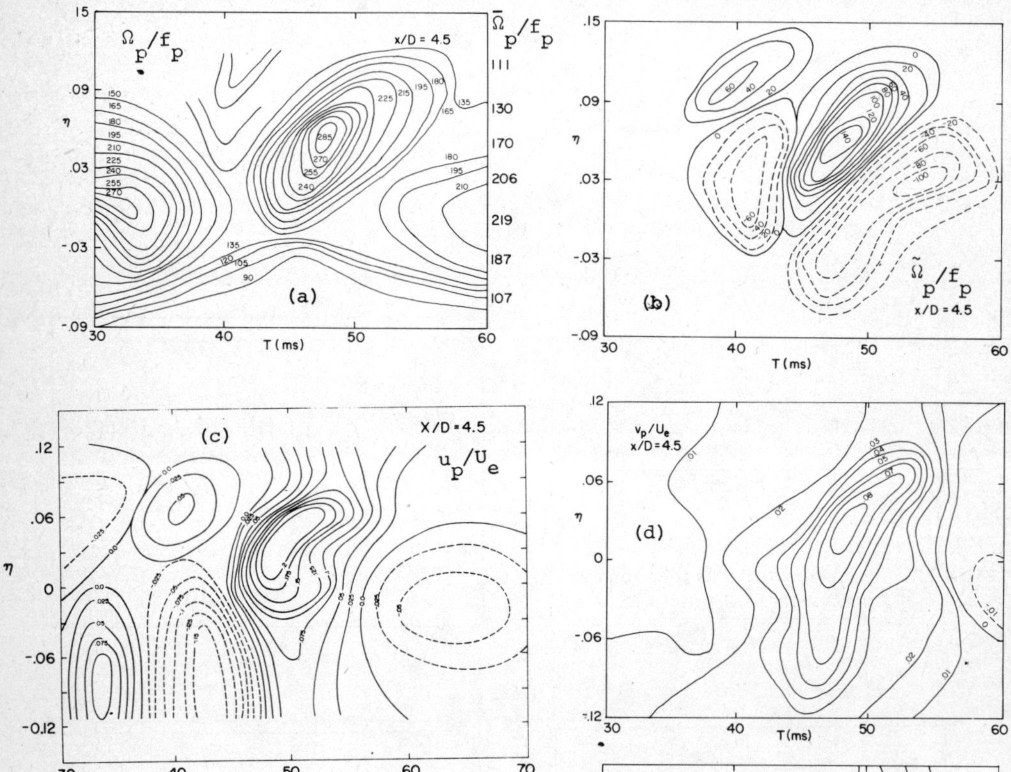

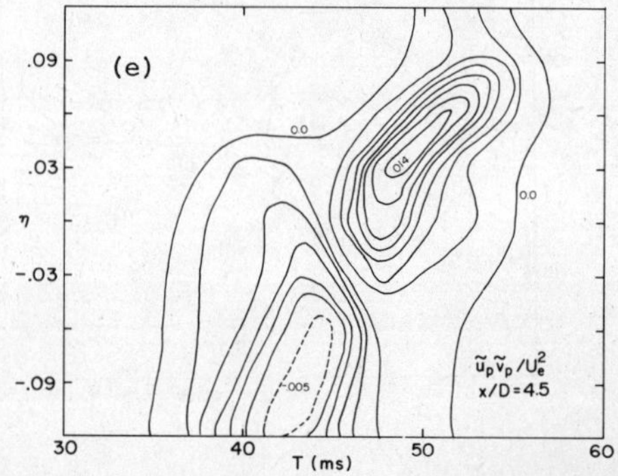

Fig. 11. "Spot" contours: (a) Ω_p; (b) $\tilde{\Omega}_p$; (c) u_p; (d) v_p; (e) $\tilde{u}_p\tilde{v}_p$

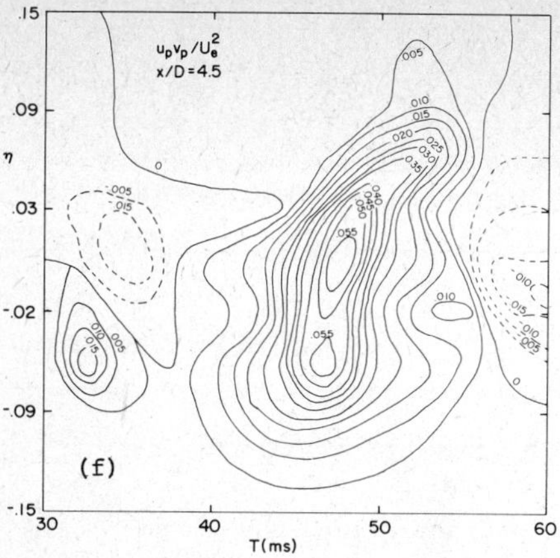

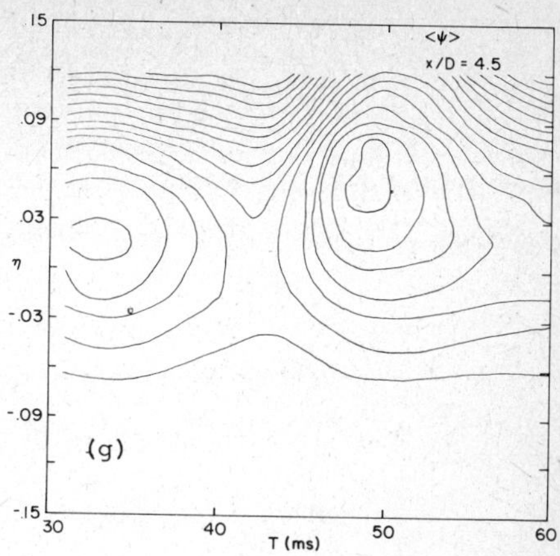

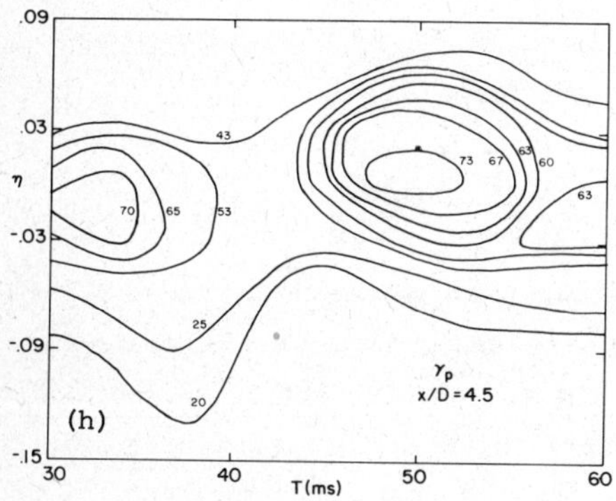

Fig. 11 (cont'd): (f) $u_p v_p$; (g) <ψ>; (h) γ_p.

Fig. 12. Contours of phase-average production <P> at x/D = 1.5,3.0 and 4.5.

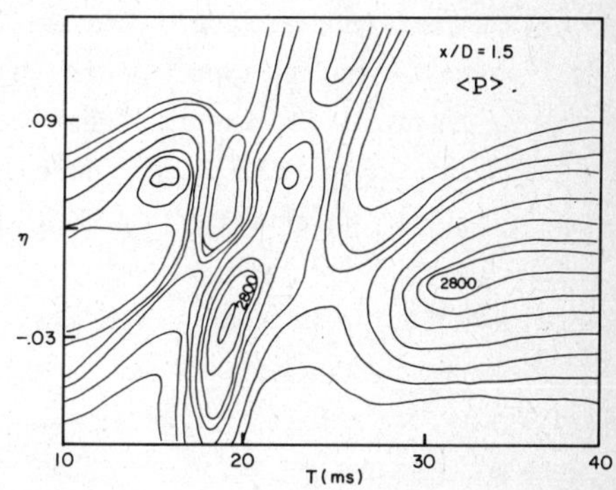

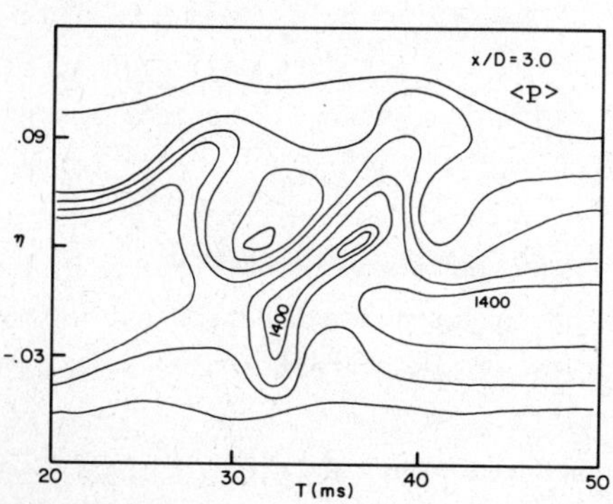

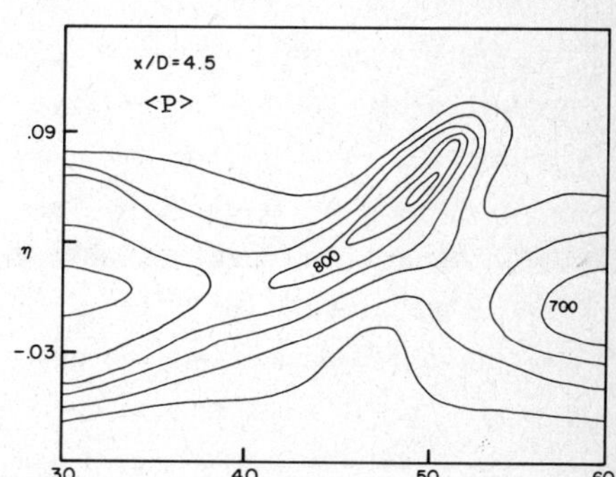

To summarize, there is no choice of U_T for which the Taylor hypothesis is acceptable for the large-scale coherent structures. The popular use of the local time-mean (or even instantaneous streamwise) velocity produces unacceptably large distortions. Thus, all reported wavenumber-spectra in shear flows at low k should be viewed with suspicion. Neither is the phase-average streamwise velocity a solution. The least objectionable choice is the use everywhere across the shear flow of a constant convection velocity U_T in the hypothesis equal to the structure passage velocity. However, this is tolerable only for structures which are essentially passive. For structures undergoing intense interaction like pairing or tearing no choice of the convection velocity renders the Taylor hypothesis acceptable. It has been shown that terms associated with the coherent motion field but typically neglected by the hypothesis should be retained. In particular, the pressure term is large and thus cannot be neglected.

In turbulent shear flows which appear to be dominated by coherent structures, use of the Taylor hypothesis is unacceptable. The least objectionable form of the Taylor hypothesis will be that with use, everywhere across the flow, of a velocity equal to the average of the velocities across the shear region.

IVc. A 'Turbulent Spot' in a Turbulent mixing layer

The results for 'stable pairing', though obtained rather painstakingly, are subject to jitter which increases with increasing separation of the measurement probe from the reference probe. Consequently, the contour properties in Region IV, say, are somewhat smeared out. This is an inevitable fate of all phase-average measurements, because of the inherent jitter primarily due to dispersion in the structure convection velocity. To increase the sharpness of the structure features, the phase scrambling effect[21,22] must be reduced, as complete elimination is impossible. This may be done via an alignment procedure,[25-27] which involves relative time-shifting of successive signal realizations with respect to one another before ensemble averaging. Since a distinct feature or 'front' is typically unavailable in free flows, alignment was done through a method of optimization of cross correlation. By applying this technique, a spark-induced large-scale coherent structure -- called a 'spot' for the lack of a better name -- has been educed at three stations in the mixing layer of a 12.7 cm jet at an exit speed of 21 m/s. Realizations requiring shifts larger than three times the standard deviation of all shifts were discarded but the additional step of rejecting weak realizations was not taken. Figures 11 (a)-(h) show the contours of Ω, $\tilde{\Omega}$, u_p, v_p, $\tilde{u}_p\tilde{v}_p$, $u_p u_p$, $\langle\psi\rangle$, and γ_p for the third station i.e. x/D = 4.5.

Contours of γ_p are most reliable for structure boundary identification. The $\langle\psi\rangle$ contours agree closely with the γ_p contours and thus provide the simplest method of identification of structures, because $\langle\Psi\rangle$ can be obtained from single-wire data only. However, all the other contours are distinctly different from one another. This is in sharp contrast with the boundary layer spot where $\tilde{u}_p$ contours have been used to identify structure boundary.[25,73]

The contours of background turbulence production for the three stations are shown in Fig. 12. It is clear that production associated with the structures is not significantly larger than the time-average production, indicated by the contours near the right hand boundary. Since sufficiently away from the spot the flow must contain naturally occurring structures, these data suggest that the 'spot' is probably a naturally occurring structure.

In an attempt to understand coherent structure interactions, two sparks were fired simultaneously at two diamterically opposite points inside the nozzle lip, and the resulting coherent structure properties were educed at different azimuthal planes. The two sparks together trigger a single axisymmetric structure, suggesting that perhaps one spark also induces an axisymmetric structure.

Further detailed study[78] indeed show that the boundary layer spot induced by a single spark decays soon after separation from the lip due to the lack of a sustaining mechanism. The mixing layer spot orginates at a point downstream from the lip as a result of local instability of the shear layer caused by the spark-induced acoustic field and grows downstream. Consequently, the mixing layer spot is initially axisymmetric, as confirmed by detailed azimuthal data. The structure undergoes intense interaction and develops three-dimensionality and breakdown further downstream. The structure average convection velocity is about 60% of core fluid velocity, a result consistent with that for naturally occurring structures.[9] Simultaneous signal traces from an azimuthally spaced rake of hot-wires also reveals that the natural instability of the jet shear layer occurs axisymmetrically on an instantaneous basis even though the jet diameter is considerably larger than the instability wavelength. The natural instability is amplitude-modulated in a random manner; the modulation is also axisymmetric. It is indeed surprising that the large-scale coherent structure induced by the localized spark is not only axisymmetric on a phase-average basis but is also individually axisymmetric.

Evolution of a spark-induced 'spot' has also been carried out in a plane mixing layer (of 96 cm span). Educed properties show strong spanwise variations.[79] The same technique, however, failed to educe any structure when the mixing layer was initially fully turbulent.

IVd. The 'Preferred mode' of the axisymmetric jet

The "preferred mode" coherent structure in the axisymmetric jet has been investigated for four flow cases [24]. In order to enhance the structure, the jet has been excited in each case through settling chamber resonance at $St_D=0.3$ at the exit excitation amplitude of 2%. The values of jet diameter (D), Reynolds number (Re_D), excitation frequency (f_p) and the initial condition (laminar or turbulent) for the four cases are: (a) D = 2.54 cm, Re_D = 25,000, f_p = 312 Hz, turbulent; (b) D = 7.62 cm, Re_D = 55,000, f_p = 44 Hz, turbulent; (c) D = 7.62 cm, Re_D = 110,000, f_p = 87 Hz, turbulent; (d) D = 7.62 cm, Re_D =55,000, f_p = 44 Hz, laminar. Note that these cover distinctly different jet speeds, excitation frequencies, and initial conditions in two different jets. Cases (a), (b), (c) give sensitivity to Re_D for the same initial condition, and (b) and (d) to the initial condition at the same Re_D.

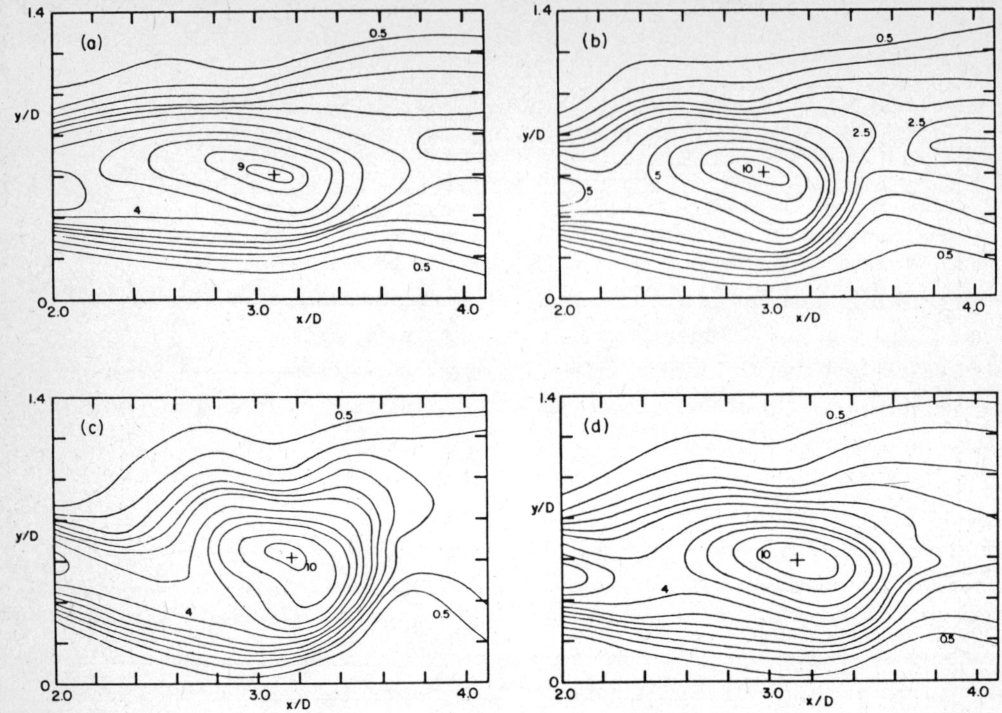

Fig. 13. Coherent vorticity Ω_p for the preferred mode.

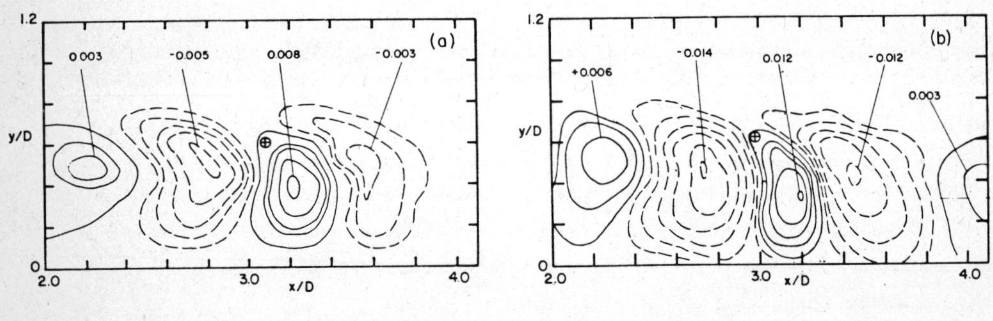

Fig. 14. Coherent Reynolds stress $u_p v_p$ for the preferred mode.

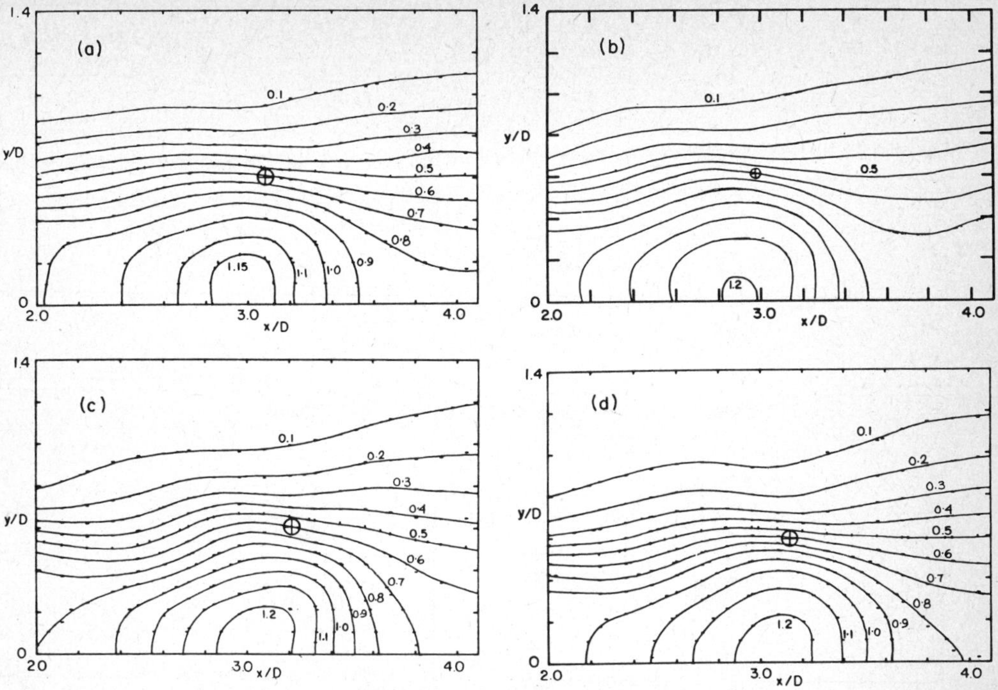

Fig. 15. Contours of u_p for the preferred mode.

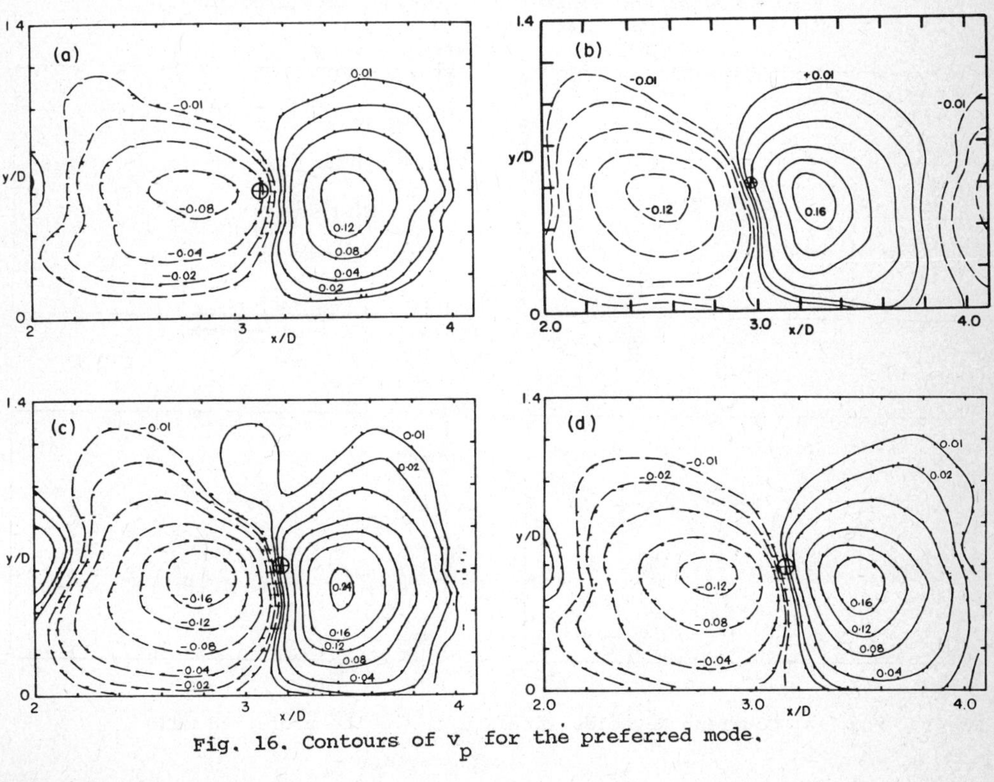

Fig. 16. Contours of v_p for the preferred mode.

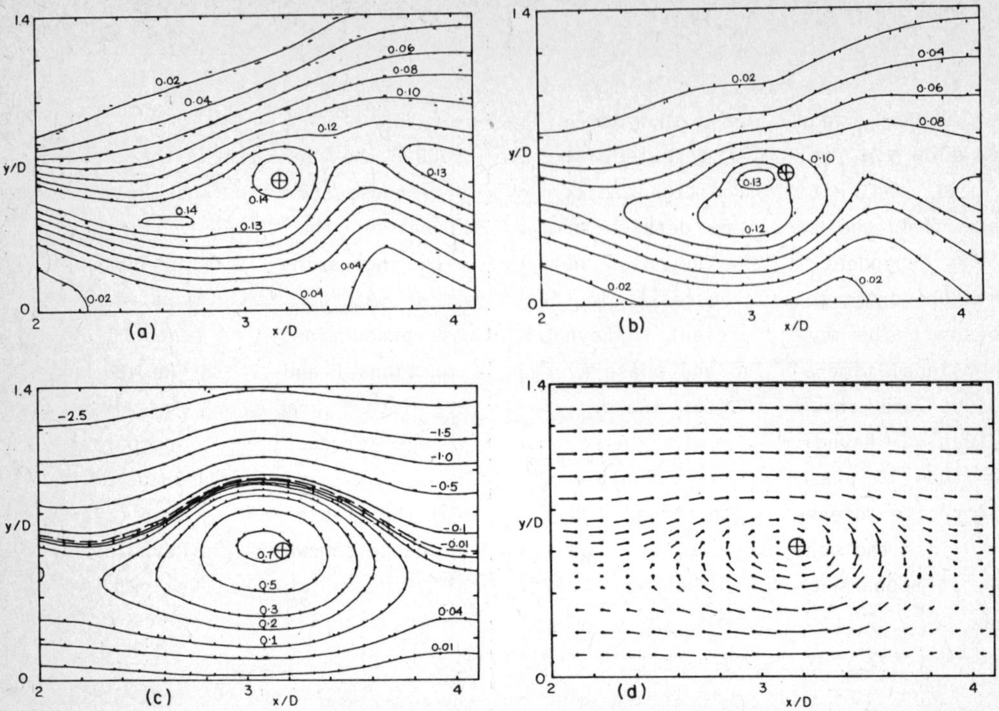

Fig. 17. Preferred mode at Re_D = 110,000: (a) $\langle u_r^2\rangle^{\frac{1}{2}}$; $\langle v_r^2\rangle^{\frac{1}{2}}$; (c) $\langle\psi\rangle$; (d) $\hat{\psi}_p$.

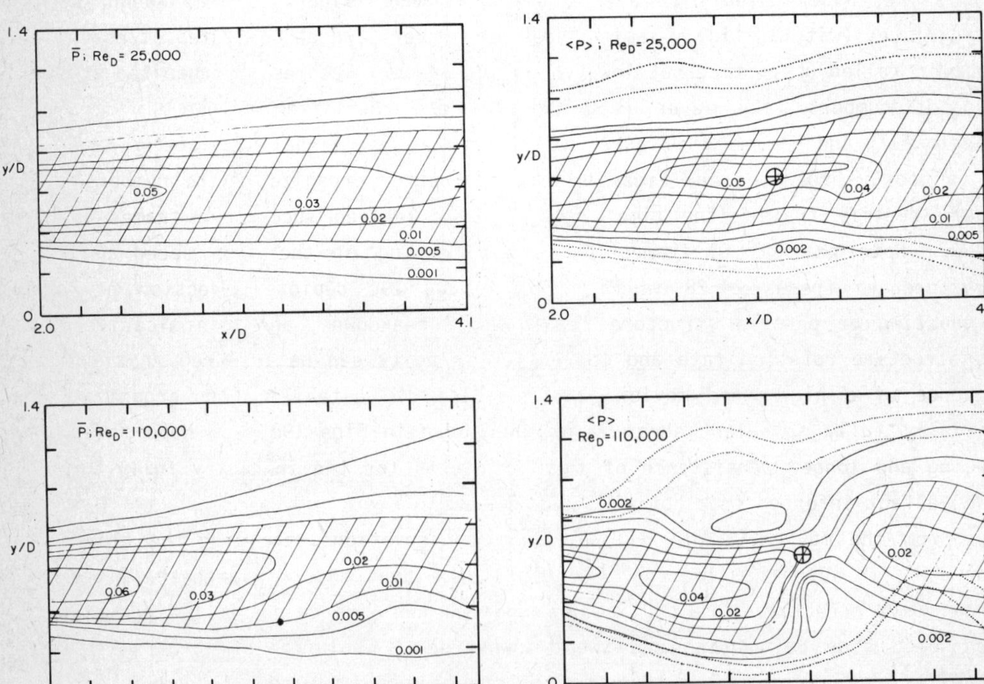

Fig. 18. $\bar{P}$ and $\langle P\rangle$ for the preferred mode.

The preferred mode coherent vorticity Ω_p/f_p, coherent Reynolds stress $u_p v_p/U_e^2$, longitudinal velocity u_p/U_e, and transverse velocity v_p/U_e are shown in Figs. 13-16, respectively. The cases (a)-(d) in each figure correspond to the four cases identified above. Figures 17a-d show the contours of $\langle u_r^2\rangle^{\frac{1}{2}}$, $\langle v_r^2\rangle^{\frac{1}{2}}$, $\langle\Psi\rangle$, and $\hat{\Psi}_p$ for the same flow i.e. initially turbulent at Re_D=110,000. The variation of these contours with Re_D and initial condition are comparable to those shown in Figs 13-16. It is clear that the structure details are independent of the initial condition, but mildly dependent on the Reynolds number. With increasing Reynolds number, the structure cross-section decreases in the streamwise length, becomes more rounded and appears to be more efficient in Reynolds stress production. The contours of non-dimensional time-average and phase-average productions $\bar{P}$ and $\langle P\rangle$ for Re_D=25,000 and 110,000 are shown in Figs 18a-d. It is clear that production increases with increasing Reynolds number. Also, time-average production is comparable in magnitude to phase-average production. Note that the production contours are in qualitative agreement with those in Region III of the stable pairing case (see IVa).

Since the coherent structures are found to change somewhat with Reynolds number Re_D, it appears that information obtained with laboratory jets of Re_D sufficiently lower than 10^5 cannot convey the physics of high Re_D practical jets.

IVe. Naturally occurring structures in the axisymmetric mixing layer

In an attempt to understand the nature of organization of the near field of a jet, the mixing layer of a 18 cm jet at an exit speed of 30 m/s was visualized with TiO_2 smoke, illuminated with a thin sheet of laser light. It was found that the mixing layer (with initially laminar boundary layer), which consisted of essentially periodic rolled up vortices at lower Reynolds number, becomes disorganized at higher Reynolds numbers; the layer is organized into identifiable coherent structures occasionally and randomly. Figure 19a shows a cross-sectional view of the layer when it is poorly organized and Figs 19b, c when better organized. The numbers on the sides identify frames, the time separation δt between successive frames is indicated. Note pairing in frames 17-23 and tearing of the high speed part of a structure in frames 24-28 in Fig 18b. Figure 19c depicts advection of a non-interacting or passive structure leading to 'breakdown', and is indicative of how the structure rotation rate and convection velocity can be inferred from a picture sequence of a high-speed movie. Two examples of instability and organization of initially fully turbulent shear layer are shown in Figs 19d, e. Note the smaller spacing and longer persistence of the structures for the initially fully turbulent case, a fact inferred from hot-wire measurments.[40]

From the movies, a set of qualitative observations regarding the flow can be made. It is seen that tearing[51] is also a dominant mode of large-scale behavior. Large-scale structures are continually sheared and divided due to a fragment on the high-speed side being torn and swept downstream. Evolution of large-scale structures occurs not primarily through complete pairing as widely believed, but frequently through 'fractional pairing' when substructures torn from larger structures pair up, or 'partial pairing' when one structure captures only a part of

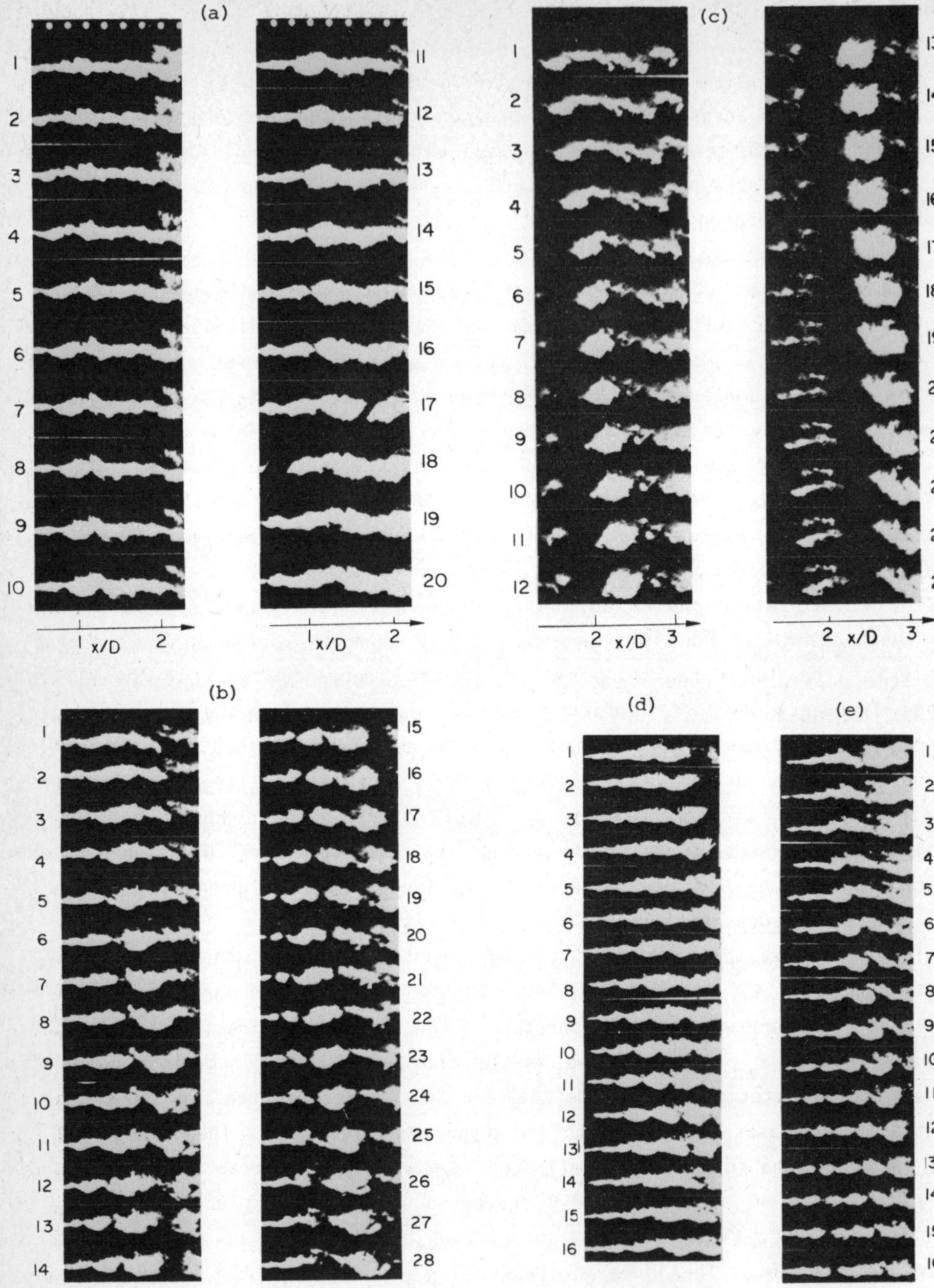

Fig. 19. Visualization pictures of the unexcited mixing layer: (a) laminar, δt = 0.56ms; (b) laminar, δt = 1.43ms; (c) laminar, δt = 0.56ms; (d) turbulent, δt = 1.67ms; (e) turbulent, δt = 1.61ms.

another. The survival distance of the large-scale structure is typically about the streamwise length of the structure. The rotation rate of the structures is such that the 'turnover' time equals the survival time of the structures. The convection velocity variation across the layer obtained from space-time trajectories of the smoke-laden fluid parcels agrees quite well with that inferred from the wavenumber-celerity spectra, obtained from double-Fourier transformation of signals from two longitudinally separated hot-wires.[40,77]

It must be recognized that the thin-slit light technique has the constraint that longitudinal motion can be confused by azimuthal motion of marked fluid. While azimuthal motion is quite apparent from the appearance and disappearance of smoke-laden fluid from the illuminated plane, it is also clear, however, that observations based on the same smoke-marked fluids regarding structure interaction and advection are not confused by the azimuthal motion. The movies show that along with entrainment of nonvortical ambient and core fluids, radially outward ejection of turbulent fluid into the ambient is an important aspect of jet mixing. The visualization also revealed that flow reversal[28,61] on the low-speed side is associated with the passage of large-scale coherent structures.

The visualization movies do not bear out the claim of Lau[80] that the mixing layer vortices bifurcate into two streets of coaxial vortex rings starting at $x/D \cong 1.5$. The large radial variation of the frequency count of the structure passage frequency[40,77,80,81] which prompted Lau to propose the two street model (presumably experimentally verified by him) can be explained by the continual tearing, fractional and partial pairings as well as relative slippage of structures. These events will also explain why different structures were educed by Bruun[82] on the high- and low-speed sides of the mixing layer and why Yule[30] was unable to educe any structure on the low-speed side while triggering on the high-speed side u-signal.

IVf. <u>Coherent structure in the farfield of a jet</u>

The self-preserving region of the axisymmetric jet was investigated in a 2.54 cm jet with $Re_D = 4.6\ 10^4$ for $x/D > 40$. Figure 20a shows the normalized space-time correlation between two sensors separated in the streamwise direction and located at the half radius; τ_{op} in each case is the time separation for maximum correlation between the two stations and is defined in the insert. Figure 20b shows that τ_{op} essentially increases linearly with the probe separation Δx. The contours of constant values of the correlation coefficient R_{AB} with the upstream probe A located at $y = y_{0.5}$, $x/D = 40$ and the probe B traversed in the axial plane at $x/D = 50$ are shown in Fig. 20c. Note that the correlation peaks off-center and is significant within one quadrant. The linear increase of τ_{op} with x and the invariant shape of $R_{AB}(\tau,\Delta x)$ suggests the existence of large-scale structures. In fact, the wavenumber-celerity spectra data in the nearly self-preserving regions $(x/D \geq 30)$ of circular and plane jets showed[57] that the most energetic eddy size in a jet increases in proportion with the local jet width: the streamwise length scale of these eddies is about 3 times the local jet width in the circular jet and 3.5 times

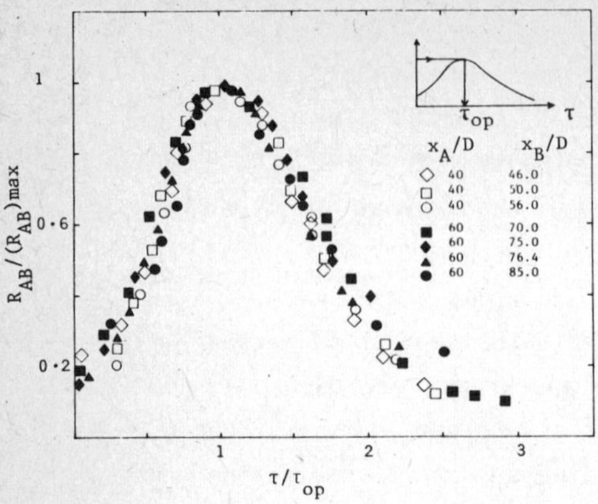

Fig. 20a. $R_{AB}(\tau)$ for probes located at x_A and x_B.

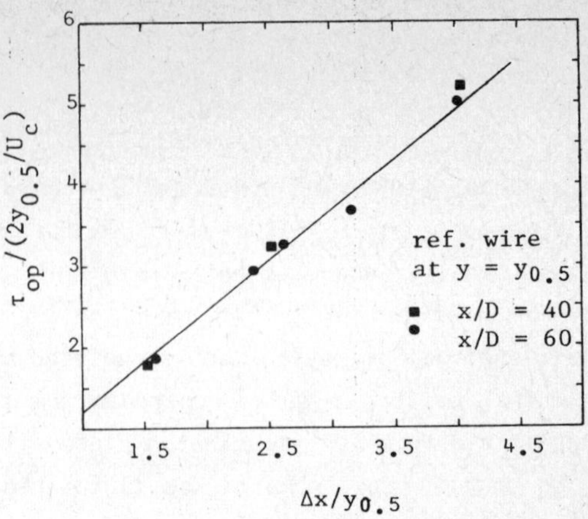

Fig. 20b. Optimum time delay τ_{op} as a function of Δx.

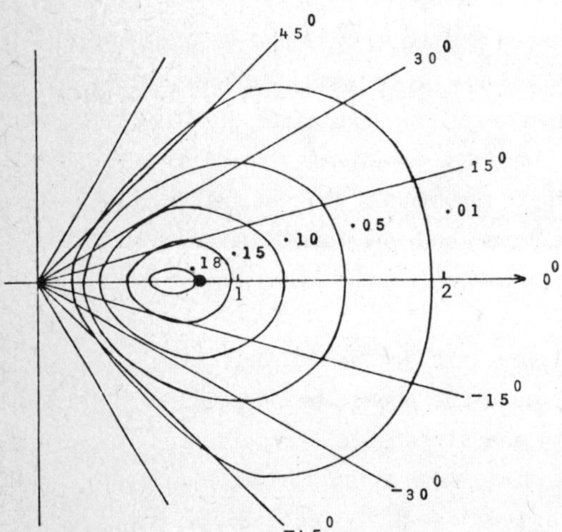

Fig. 20c. Contours of constant correlation in a transverse plane.

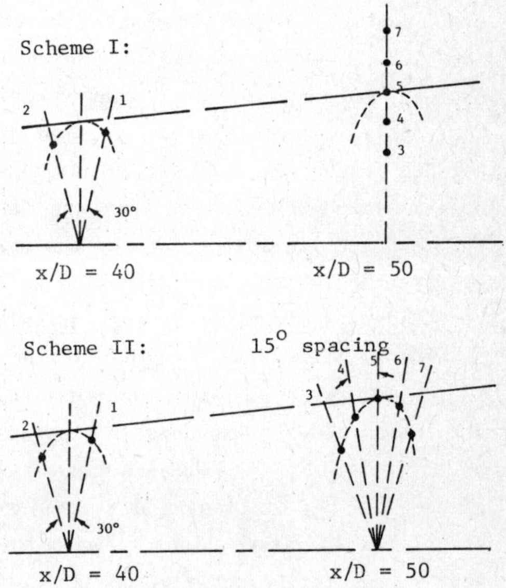

Fig. 20d. Schemes of hot-wire arrangements for $\hat{R}_{ij}$ and $\hat{C}_{ij}$.

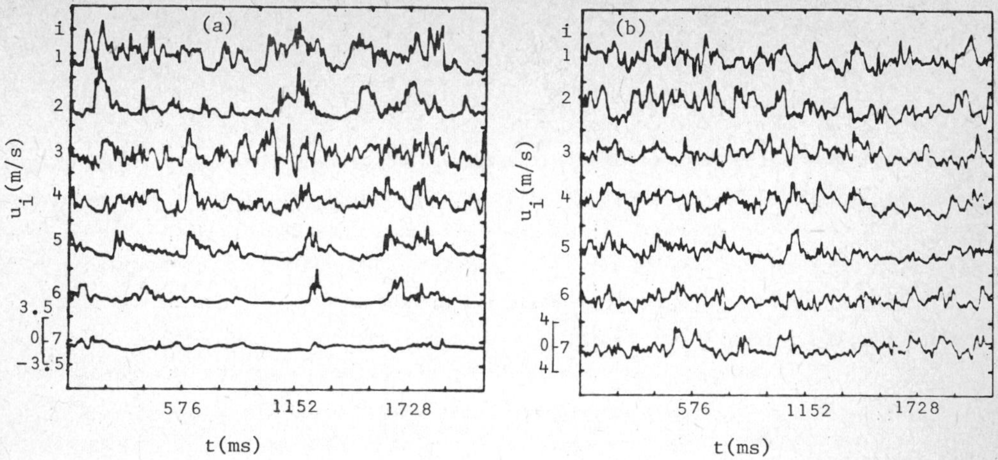

Fig. 21. Sample traces of u(t); (a) Scheme I; (b) Scheme II.

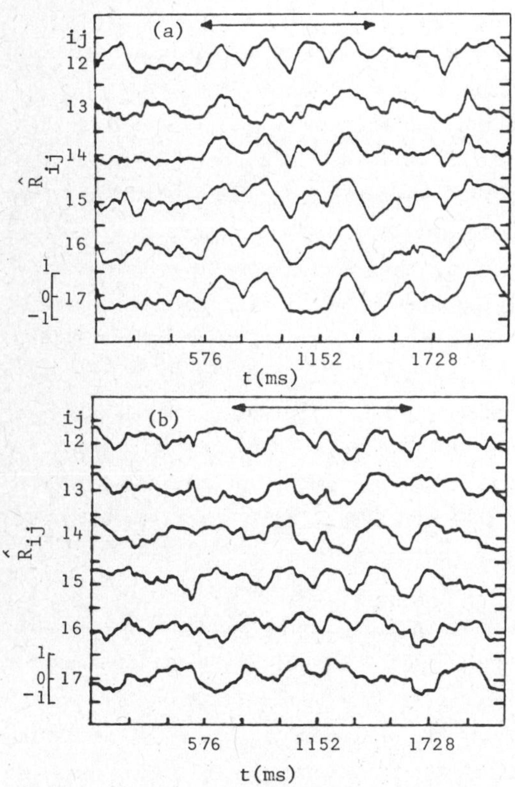

Fig. 22. Traces of $\hat{R}_{ij}(t)$:

(a) Scheme I; (b) Scheme II.

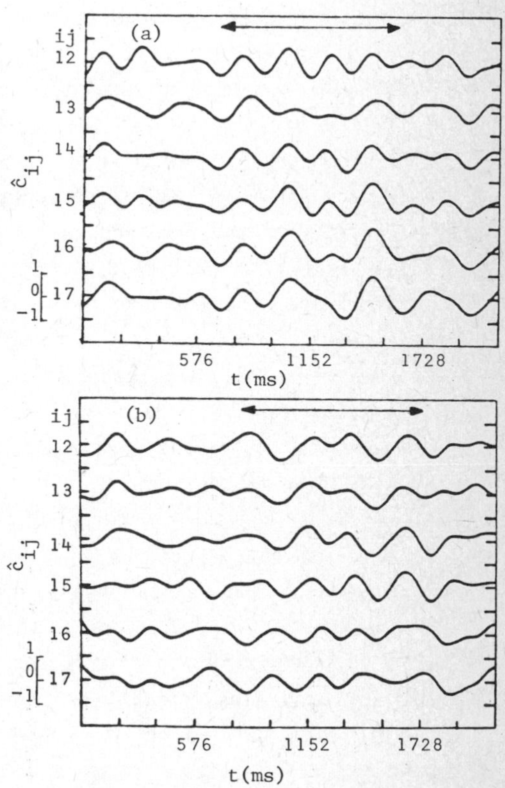

Fig. 23. Traces of $\hat{C}_{ij}(t)$:

(a) Scheme I; (b) Scheme II.

the local jet width in the plane jet. Based on the correlation data in Figs. 20a,b, it appears that the length of the structure is about $3\tau_{op}$. This corresponds to roughly about $6y_{0.5}$ or three local jet diameters, in agreement with the data in ref. [57].

Figure 20d shows two schemes of hot-wire array for simultaneous recording of u-signals. In both, two wires are placed upstream. In scheme I five downstream wires are aligned radially while in scheme II these are spaced azimuthally.

Figure 21a shows a sample time trace of u(t) fluctuations from hot-wires in the radial configuration (scheme I). Figure 21b shows the corresponding traces for scheme II. Note that these traces reveal very little on their own. However, in order to uncover the possible existance of structures, short time-average cross-correlations $\hat{R}_{ij}(t)$ of these traces have been obtained and are shown in Figs. 22a,b, corresponding to Figs 21a,b, respectively. The top trace in each case is the short-time cross-correlation of signals from the two upstream probes; the remainders are between the signals from one upstream probe and one downsteam probe. The top trace in each case has been time-shifted by $\hat{\tau}_{op}$ in order to provide a common time base. There are phase coincidences in $\hat{R}_{ij}(t)$, suggesting passage over the hot-wire array of large-scale coherent structures.

In order to further refine the detection scheme, the $\hat{R}_{ij}(t)$ traces in Figs. 22a,b have been correlated with a sine wave and the resulting correlation functions $\hat{C}_{ij}(t)$ are shown in Figs. 23a,b, respectively. The period T of the sine wave for all $\hat{C}_{ij}(t)$ traces is the same and is equal to the estimated average period of the large undulations in Figs. 22a,b. Note that by this short-time cross-correlation with the sine wave, the instants of underlying structures still hidden in $\hat{R}_{ij}$ have been emphasized. Furthermore, even though this has resulted in a smoothing of the signals, there is an amazing resemblance between the $\hat{R}_{ij}(t)$ and the $\hat{C}_{ij}(t)$ traces. Two observations can be made from long records of $\hat{C}_{ij}(t)$ (not shown). Even at instants when the coincidence intervals are not large, a characteristic front is apparent in all the traces. The $\hat{C}_{ij}(t)$ traces also show ranges of multiple periodic undulations, thus suggesting occurrence of the structures in groups. However, the possible existence of structures in groups, like those found by Townsend[15] in the boundary layer and the wake, appears quite surprising and needs to be confirmed through more careful investigations. These will include sample functions of adjustable periods, maximization of the matching with data as well as distributions of these periods, and classification of structures by size and strength. Continuing studies are being carried out to educe contours of coherent vorticity, Reynolds stress and other properties over the extents of the structures.

V. Concluding Remarks

Even though the large-scale coherent structures have been studied for quite some time, our knowledge of these structures is far from adequate. Understanding of the naturally occurring structures is especially limited because the large dispersion in the formation time, strength, shape, size, orientation, convection velocity, and 'age' of these structrues present formidable eduction problems. Clearly, these must be studied with a three-dimensional array of sensors and the structures must be classified first on the basis of the characteristic measures like shape, size, orientation, strength, age, etc.; that is, the probability density distribution of the structures as a function of these measures must be determined in each class of flows. The spatial distributions of structure details then should be educed by ensemble averaging only structures of identical measures.

Computation and experimental limitations will force use of minimal number of sensors, typically in a linear array. Since the structure is buried in large-amplitude random fluctuations, visual check of the signals -- after the signals are smoothed by filtering or short-time cross-correlation -- may indicate structures via coincidences of oscillations. However, I feel that only coherent vorticity can identify a structure; vorticity contours can be obtained from an array of signals through spatial filtering, most conveniently achieved through ensemble averaging of appropriately aligned signals containing structure footprints.

Although structures should be aligned in all three orthogonal coordinates, brevity of efforts forces alignment in x only as dispersion in y and z is comparatively smaller. Since there is no apriori knowledge of the structure property that will result from ensemble averaging, and typically no sharp front or feature is apparent from the signatures dominated by background turbulence, neither detection nor alignment is in general possible on the basis of the instantaneous signal. The best alignment must rely on optimization of cross-correlation. Individual realizations producing extremely low peaks of cross-correlation or requiring excessive time shifts for alignment must be regarded as 'freak' and thus be discarded. Here one faces two conflicting requirements. If the selection criterion is too strict, stable ensemble average would require prohibitively large time and cost, and the structure, though sharply defined, may itself be regarded as 'freak' or too special to be of general relevance. However, on the other hand, if the criteria are not sharp, the structure characteristics may be too smeared to represent the physics. In the latter case, computed phase-average turbulence intensities will be high due to pseudo-turbulence in regions of steep gradients of phase average velocity.

A problem associated with inferring spatial distributions from time traces is the Taylor hypothesis. It is shown that the hypothesis is inapplicable to a turbulent shear flow in general and especially to large-scale coherent structures, but is least objectionable for passive structures when the structure center velocity is used everywhere across the shear flow as the convection velocity in the hypothesis. In situations involving structure pairing or tearing, no choice of the convection velocity is acceptable. The popular use of the local time-mean velocity, or even the phase-average or instantaneous velocity, produces unacceptably large structure distortions, and thus cannot be used in any shear flow, except at the center of the shear width.

Educed properties of coherent structures at larger x for the case of stable pairing in the circular jet, namely, in Regions III and IV suggest existence of weaker structures (see IVa). This is probably due to two reasons. The initial toroidal structure develops azimuthal lobes[28,30,83,84] and undergoes "breakdown" near the end of the potential core. The second possibility is that with increasing jitter with respect to the structure detection at x/D=1.5, the educed structure properties would be smoothed out somewhat even if distinct structures existed beyond the potential core. In fact, the voritcity contours in the preferred mode suggest presence of structures - though probably weak - as far downstream as 8 diameters from the exit[24]. It is possible that coherent structures with sharply defined boundaries could be educed there through alignment. A possible approach could be to educe a structure at small x-separations and fill in the data between the stations using the Taylor hypothesis (after the ensemble average at each station has been obtained through time-alignment procedures); Taylor hypothesis for time intervals short compared with the structure time-scale is tolerably acceptable. Since the large-scale structures near the end of the potential core are three-dimensional, the structures could be educed if aligned azimuthally. This will be hard because of random azimuthal locations of successive structures. Otherwise, only structures having the same azimuthal location with respect to a sensor array should be accepted for ensemble averaging.

It is interesting to note that the thin vortical braids connecting the coherent structures contain very little vorticity but are the locations of large turbulence production because of associated large strain rates. Thus, smaller scale turbulence produced at the braids are advected towards the centers of structures by the structure-induced motion; these in turn produce peak turbulence intensities and dilution of peak coherent vorticity at the structure centers while increasing the structure size through migration of turbulent braid fluid into the structure cores. Increase of flow width is also contributed by complete, fractional and partial amalgamations, while tearing[9,51] counters rapid growth. A delicate balance between fluid migration into the cores, amalgamation, and tearing produces the resulting growth of the shear layer. Hernan & Himenez[85] suggest that about 80% of structure growth is due to processes not connected with pairing.

There is no evidence of vortex pairing in the preferred mode. Since this is the most frequently occurring structure in a jet beyond about two diameters downstream, it is clear that pairing activity cannot be the dominant mechanism for noise production in an unperturbed subsonic jet where noise production peaks near the end of the potential core. Thus, it is very likely that the process of rapid development of azimuthal lobes in the initial toroidal structure and the resulting catastrophic breakdown process is the major source of jet noise; it is not known to me that this breakdown process has been suspected as being dominant in jet noise production. It should be emphasized that pairing, when it occurs, is likely to produce significant noise. Thus, pairings in the 'shear layer mode',[23,86] occurring near the lip, should be expected to be major sources of noise. However, since experimental evidences suggest that the peak noise source in a subsonic jet is located at $4 \lesssim x/D \lesssim 8$, it appears that amalgamation process is not the primary source of jet noise. [Note also that peak noise location depends on the Mach number.]

We have shown that the preferred mode is independent of the initial condition and mildly dependent on the Reynolds numer. With increasing Re_D, the structures appear to be more rounded in cross-section and capable of producing comparatively larger Reynolds stress and production. This, however, is not intended to question the Reynolds number similarity or asymptotic invariance†. Namely, time-average nondimensional Reynolds stress or production is not necessarily higher with higher Re_D. One possible way to explain this apparent conflict would be that the structures occur less frequently at higher Reynolds numbers, as revealed by our visualization study.

In spite of the inherent limitation of visualization in a turbulent flow with the flow markers introduced far upstream, and in spite of the sensitivity of the thin-sheet illumination technique to azimuthal motion, our visualization movies of the high-speed mixing layer have revealed a wealth of information regarding the natural large-scale coherent structures in the self-preserving region of a mixing layer. These show that with increasing Reynolds numbers, the layer becomes progressively more intermittently and less frequently organized.

There is some evidence that the self-preserving region of an axisymmetric jet consists of coherent structures. Further details must be obtained regarding the probability density distribution of the structures as a function of the characteristic measures and the spatial distributions of the structure properties. This study is in progress. Even though coherent structures can probably be found to exist beyond the end of the potential core, it is not suggested that the same structures survive in the self-preserving region. Because of strong evidence for structure breakdown near the end of the potential core, it appears that all remants of the upstream toroidal structures are essentially broken up at around $x/D = 8$. The turbulent jet further downstream has its own instability which can produce a new breed of structures downstream. The self-preserving region coherent structures are probably these or results of evolution and interactions of these newly formed structures. Occurrence of a few successive periods in the short-time correlation functions $\hat{R}_{ij}(t)$ suggests eddy groups in the far field of a jet, similar to those

† The underlying concept in Reynolds number similarity or asymptotic invariance is that the large-scale eddies are inviscid at large Reynolds numbers which affect the small scales only. Thus, two flows with identical initial and boundary conditions must have identical large eddies independent of the Reynolds numbers provided they are large enough. How large the Reynolds number has to be is arbitrary. Since many investigations of large-scale coherent structures have been made at low Reynolds numbers, one should guard against straightforward extrapolation to higher Reynolds numbers. Our experience suggests that studies of jets at Reynolds numbers significantly below 10^5 may not be useful in understanding the physics of large Reynolds number jets. Parenthetically, it should be recognized that although experiments in general suggest Reynolds number similarity, conclusive evidence is still lacking. In spite of progressively increasing separation of the largest and smallest scales with increasing Reynolds numbers, the interaction between the two scales is still important, a fact specifically recognized in the Kolmogoroff's refined theory for the fine scales.[87-89]

reported by Townsend[15] for boundary layers and wakes. Possible existence of such groups in the far-field of a jet is surprising and needs to be carefully documented.

While I believe that coherent structures are characteristic of all turbulent shear flows, care is necessary to guard against prematurely emphasizing the significance of the structures. The significance of the coherent structures is still an open question. Our experience suggests that the structures themselves are not necessarily extremely dominant in transport except at instants of intense interaction. Thus, perhaps the dominance of the coherent structures have been exaggerated. Our results suggest that with increasing distance from the location of initial structure formation, the background turbulence becomes progressively important in momentum transport, approaching that of the coherent structures. This raises some doubt about validity of theories based only on deterministic structures, even with random phase, because background turbulence effects are not negligible.

Another unresolved question is: are all turbulent shear flows superpositions of the same coherent structure module? There is little doubt that the hairpin type vortex loop structure constituting the boundary layer[90] is unique to wall bounded flows only. It is thus inescapable to conclude that at least bounded and free flow coherent structures are inherently different. While the near fields of jets consist of vortex rings, the far field probably contains structures not too dissimilar from the Hill's spherical vortex. Perhaps each class of shear flows has its unique coherent structure.

Most of the experiments discussed above and some new experiments that are being carried out in our laboratory attempt to answer these questions.

The spatial details of structure properties educed so far (and being educed) provide the data necessary for formulating theories of turbulent shear flows by using these structures as the building blocks through some model functions fitted to the data. It is hoped that theoreticians would undertake the woefully lacking thrust in this direction by attempting to include the coherent structures in newer formulations. Since coherent structures appear to be integral parts of all turbulent shear flows, new theories of turbulence must explicitly incorporate these structures in order to capture the physics and thus be viable.

The results briefly discussed above are the outcome of collaborative efforts by K.B.M.Q. Zaman, S.J. Kleis, M. Sokolov, A. R. Clark, J. Tso, L.S.G. Kovasznay, Z.D. Hussain and others, who have contributed in a variety of ways. Their specific contributions should be evident from the cited references. I have benefitted from various discussions with these colleagues some of whom, especially D.J. Shlien, have reviewed parts of this manuscript. These research efforts have been supported over a period of time by NSF, ONR, NASA and AFOSR.

REFERENCES

1. Crow, S.C. & Champagne, F.H. 1971 J. Fluid Mech. 48, 547.
2. Brown, G.L. & Roshko, A. 1974 J. Fluid Mech. 64, 775.
3. Winant, C.D. & Browand, F.K. 1974 J. Fluid Mech., 63, 237.
4. Liepmann, H.W. 1979 American Scientist 67, 221.
5. Laufer, J. 1975 Ann. Rev. Fluid Mech. 7, 307.
6. Chandrsuda,C., Mehta, R.D., Weir, A.D. & Bradshaw, P. 1977, J. Fluid Mech., 85, 693.
7. Batt, R.G. 1978, J. Fluid Mech., 82, 53.
8. Pui, N.K. & Gartshore, I.S. 1978, J. Fluid Mech., 91, 111.
9. Hussain, A.K.M.F. & Clark, A.R. 1980 J. Fluid Mech. (to appear).
10. Kline, S.J., Reynolds, W.C., Schraub, F.A. & Rundstadler, P.W. 1967 J. Fluid Mech. 30, 741.
11. Kovasznay, L.S.G., Kibens, V. & Blackwelder, R.F. 1970 J. Fluid Mech. 41, 283.
12. Gupta, A.K., Laufer, J. & Kaplan, R.E. 1971 J. Fluid Mech. 50, 493.
13. Wallace, J.M., Brodkey, R.S., Eckelmann, H. 1977 J. Fluid Mech. 83, 673.
14. Falco, R. 1980 AIAA Paper No. 80-1356.
15. Townsend, A.A. 1979 J. Fluid Mech. 95, 515.
16. Tso, J., Kovasznay, L.S.G. & Hussain, A.K.M.F. 1980 AIAA Paper No. 80-1355.
17. Roshko, A. 1976 A.I.A.A. J. 10, 1349.
18. Hussain, A.K.M.F. 1979 Symp. Turbulence, U. of Missouri-Rolla (in print).
19. Tennekes, H. & Lumley, J.L. 1974 A First Course in Turbulence, MIT Pr.
20. Lumley, J. L. 1967 Stochastic Tools in Turbulence. Academic Pr.
21. Yule, A.J. 1979, 2nd Symp. Turb. Shear Flows, London, p. 7.1.
22. Blackwelder, R.F. 1977 Phys. Fluids 20, S232.
23. Zaman, K.B.M.Q. & Hussain, A.K.M.F. 1980 J. Fluid Mech. (to appear).
24. Hussain, A.K.M.F. & Zaman, K.B.M.Q. 1980 J. Fluid Mech. (to appear).
25. Zilberman, M., Wygnanski, I. & Kaplan, R.E. 1977 Phys. Fluids 20, S258.
26. Sokolov, M., Hussain, A.K.M.F., Kleis, S.J. & Husain, Z.D. 1980 J. Fluid Mech. 98, 65.
27. Hussain, A.K.M.F., Kleis, S.J. & Sokolov, M. 1980 J. Fluid Mech. 98, 97.
28. Hussain, A.K.M.F. & Zaman, K.B.M.Q. 1980 J. Fluid Mech. (to appear).
29. Ho, C.M., private communication, 1980.
30. Yule, A.J. 1978 J. Fluid Mech. 89, 413.
31. Hussain, A.K.M.F. & Clark, A.R. 1977 Phys. Fluids 20, 1416.
32. Gutmark, E. & Wygnanski, I. 1976 J. Fluid Mech. 73, 465.
33. Hussain, A.K.M.F. & Zedan, M.F. 1978 Phys. Fluids 21, 1100.
34. Hussain, A.K.M.F. & Zaman, K.B.M.Q. 1978 J. Fluid Mech. 87, 349.
35. Coles, D.E. 1962 Rand Corp. Rept. R-403-PR.
36. Hussain, A.K.M.F. & Zedan, M.F. 1978 Phys. Fluids 21, 1475.
37. Hussain, A.K.M.F. & Ramjee, V. 1976 J. Fluids Engrg. 98, 58.
38. Michalke, A. 1972 Prog. Aero. Sci. 12, 213.

39. Crighton, D.G. & Gaster, M. 1976 J. Fluid Mech. 77, 397.
40. Clark, A.R. & Hussain, A.K.M.F. 1979 Turb. Shear Flows, London, p. 2.30.
41. Hussain, A.K.M.F. 1978 Lecture Notes in Physics 75, Springer, p. 103.
42. Husain, Z. D. & Hussain, A.K.M.F. 1979 A.I.A.A. J. 17, 48.
43. Foss, J.F. 1977 Turb. Shear Flows, Penn. State U., p. 11.33.
44. Birch, S.F. 1977 Turb. Shear Flows, Penn State U., p. 16.10.
45. Fiedler, H. & Thies, H.J. 1978 Lecture Notes in Physics 75, Springer, p. 108.
46. Kleis, S.J. & Hussain, A.K.M.F. 1979 Bull. Am. Phys. Soc. II, 1132.
47. Hasan, M.A.Z. 1978 MS Thesis, U. of Houston.
48. Hussain, A.K.M.F. & Reynolds, W. C. 1970 J. Fluid Mech. 41, 241.
49. Hussain, A.K.M.F. 1977 Cardiovascular Flow Dynamics and Measurements, (eds. N.H.C. Hwang & N. Norman), U. Park Pr, Baltimore, p. 541.
50. Knight, D. 1979 Symp. Turbulence, U of Missouri-Rolla (in print).
51. Moore, D. W. & Saffman, P. G. 1975 J. Fluid Mech. 69, 453.
52. Ashurst, W. T. 1977 Turb. Shear Flows, Penn State U., p. 11.43.
53. Couët, B., Buneman, O. & Leonard, A. 1977 Turb. Shear Flows, London, p.14.29.
54. Delcourt, B. A. G. & Brown, G. L. 1977 Turb. Shear Flows, London, p. 14.35.
55. Saffman, P. G. & Baker, G. R. 1979 Ann. Rev. Fluid Mech. 11, 95.
56. Acton, E. 1980 J. Fluid Mech. 98, 1.
57. Hussain, A.K.M.F. & Clark, A.R. 1980 A.I.A.A. J. (to appear).
58. Hinze, J.O. 1976 Turbulence: Second Edition, McGraw-Hill.
59. Favre, A., Gaviglio, J. & Dumas, R. 1967 Phys. Fluids 10, S138.
60. Corcos, G.M., & Sherman, F.S. 1976 J. Fluid Mech. 73, 241.
61. Antonia, R.A., Chambers, A.J. & Hussain, A.K.M.F. 1980 Phys. Fluids 23, 695.
62. Hussain, A.K.M.F. & Zaman, K.B.M.Q. 1977 Lecture Notes in Physics 75, Springer, 31.
63. Wygnanski, I., Oster, D. & Fiedler, H. 1979 Turb Shear Flows, London, p.8.12.
64. Townsend, A.A. 1976 Turbulent Shear Flow, Cambridge U. Pr.
65. Davies, P.O.A.L. & Yule, A.J. 1975 J. Fluid Mech. 69, 513.
66. Lin, C.C. 1953 Quart. Appl. Math, 18, 295.
67. Fisher, M.J. & Davies, P.O.A.L. 1964 J. Fluid Mech. 18, 97.
68. Lumley, J.L. 1965 Phys. Fluids 8, 1056.
69. Champagne, F.H. 1978 J. Fluid Mech. 86, 67.
70. Heskestad, G. 1965 J. Appl. Mech. 87, 735.
71. Browand, F.K. & Wiedman, P.D. 1976 J. Fluid Mech. 76, 127.
72. Wygnanski, I. & Champagne, F.H. 1973 J. Fluid Mech. 59, 281.
73. Cantwell, F.J., Coles, D. & Dimotaleis, P. 1978 J. Fluid Mech. 87, 641.
74. Zaman, K.B.M.Q. & Hussain, A.K.M.F. 1980 (submitted to J. Fluid Mech.).
75. Foss, J.F. 1978 Proc. Dyn. Flow Conf., DISA 983.
76. Wills, J.A.B. 1964 J. Fluid Mech. 20, 417.
77. Clark, A.R. & Hussain, A.K.M.F. 1980 (to be submitted).
78. Kleis, S.J., Hussain, A.K.M.F. & Sokolov, M. 1980 (submitted to J. Fluid Mech.).

79. Kleis, S.J. & Hussain, A.K.M.F. 1980 (to be submitted).
80. Lau, J.C. 1978 J. Fluid Mech. (submitted).
81. Lau, J.C., & Fisher, M.J. 1975 J. Fluid Mech. 67, 229.
82. Bruun, H.H. 1977 J. Fluid Mech. 64, 775.
83. Davies, P.O.A.L. & Baxter, D.R.J. 1977 Lecture Notes in Physics 75, p. 125, Springer
84. Saffman, P. G. 1978, J. Fluid Mech. 84, 625.
85. Hernan, M.A. & Jimenez, J. 1979 Turb. Shear Flows, London, p. 7.7.
86. Kibens, V., 1980 A.I.A.A. J. 18, 434.
87. Kolmogoroff, A.N. 1962 J. Fluid Mech. 13, 82.
88. Obukhov, A.M. 1962 J. Fluid. Mech. 13, 77.
89. Monin, A.S. & Yaglom, A.M. Statistical Fluid Mechanics. Vol. 2, 1975, MIT Pr.
90. Klebanoff, P.S., Tidstrom, K.D. & Sargent, L.M. 1962 J. Fluid Mech. 12, 1.

A Comparison of Characteristic Features of Coherent Turbulent Structures Found Using the Variable Interval Time Average (VITA) Technique and Using the Pattern Recognition Technique

Helmut Eckelmann
Max-Planck-Institut für Strömungsforschung
D 34 Göttingen, F.R.of Germany

James M. Wallace
The University of Maryland
College Park, Maryland 20742, USA

Abstract

A quantitative comparison is made of some features of coherent structures called "bursts" found in wall bounded turbulent shear flows. The features have been educted using two entirely different methods: the variable integral time average (VITA) technique of Blackwelder and Kaplan (1976) and the pattern recognition technique of Wallace, Brodkey and Eckelmann (1977). It is found that the instantaneous velocity gradient, $\partial U/\partial y$, normal to the wall is caused, during "bursting", by a phase shift in the velocity perturbations between two neighboring positions separated in the normal direction. This velocity gradient can be determined from the time derivative of the streamwise velocity and a perturbation convection velocity normal to the wall. Finally a qualitative comparison is made between the time-varying ensemble averaged Reynolds stress educted during "bursting" by the two methods and the method of Lu and Willmarth (1973).

Introduction

The existence of coherent structures in turbulent shear flows is today accepted as certain. Although some of their characteristics are reasonably well known now, it is still not possible to conceptualize a complete kinematical picture of these structures in wall bounded flows; even less is held as certain about their dynamics. The methods used to investigate such structures have been quite different, but the existence of particular structure elements (bursts, sweeps, ejections, longitudinal vortices) have been, for the most part, well established. It is striking to us that two completely different methods, namely the VITA technique of Blackwelder and Kaplan (1976) and the pattern recognition technique of Wallace, Brodkey and Eckelmann (1977) obtained both qualitatively and quantitatively good agreement about some of these structural characteristics. Before these are discussed, features of the methods will be briefly described below. Additionally a qualitative comparison with results of Lu and Willmarth (1973) will be made at the end of the paper.

The variable integral time average (VITA) technique uses a detector probe at a distance from the wall of $y^+ = yu_\tau/\nu = 15$. The VITA average of the streamwise velocity $U(x_o,y_o,z_o,t)$ is defined as

$$\hat{U}(x_i,t,T) = \frac{1}{T} \int_{t-\frac{T}{2}}^{t+\frac{T}{2}} U(x_i,t)\, dt \tag{1}$$

where T is the averaging time. If T is large the long-time average is obtained which is independent of t. Blackwelder and Kaplan used the VITA average of the square of the streamwise velocity minus the localized squared mean value which they called the localized variance

$$\widehat{var}\,(x_i,t,T) = \widehat{U^2}(x_i,t,T) - (\hat{U}(x_i,t,T))^2 \tag{2}$$

as a measure of the local turbulent kinetic energy. The velocity signals were always ensembled averaged when the following detection function of the detecting probe had the value of unity.

$$D(t) = \begin{cases} 1 \text{ if } \widehat{var} > k \cdot u^2_{rms} \\ 0 \text{ otherwise} \end{cases}$$

where $$u^2_{rms} = \lim_{T \to \infty} \widehat{var} \;. \tag{3}$$

In their investigation k was always taken as 1.2, but they show how the conditional averages can be made independent of k for a range of $0.9 < k < 2.5$. Since this criterion is fulfilled over short time periods, the velocity signals are also averaged over a period of time before and after it is fulfilled (which is easily done with a digital computer). The ensemble averaged time function of the streamwise velocity component at $y^+ = 15$ that they obtain with this method is shown in the lower part of Figure 1. This time function clearly shows a large slope when the criterion of equation (3) is fulfilled ($\tau = 0$).

The pattern recognition technique of Wallace <u>et al</u>. searches for a pattern in the time function of the streamwise velocity component, U, which is characterized by a relatively weak deceleration immediately followed by a strong acceleration. Such patterns were observed to have a wide range of time scales (1:25). To obtain meaningful ensemble averages, they are therefore normalized to an arbitrary common length by forcing their points of minimum and maximum slope to be located at the same locations along the time axis and expanding or contracting the lengths of the patterns before the averaging was done. In the near wall region at $y^+ = 15$ where turbulence production is a maximum, the average pattern shown in the lower part of Figure 2 was obtained from recognized patterns which accounted for more than 65 % of the total data sample. We believe that this ensembled averaged time function shows the characteristic U-signal during the occurrence of "bursts".

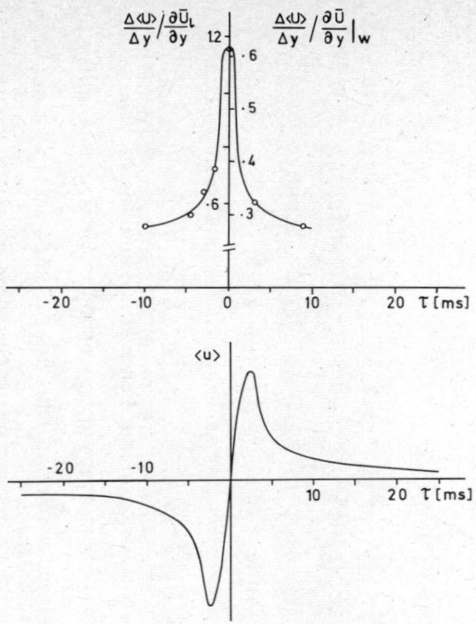

Figure 1: Conditionally averaged streamwise velocity at $y^+ = 15$ from Blackwelder and Kaplan (1976) (bottom). Conditionally averaged velocity gradient obtained from Figure 3 normalized with local mean gradient and gradient at the wall (top).

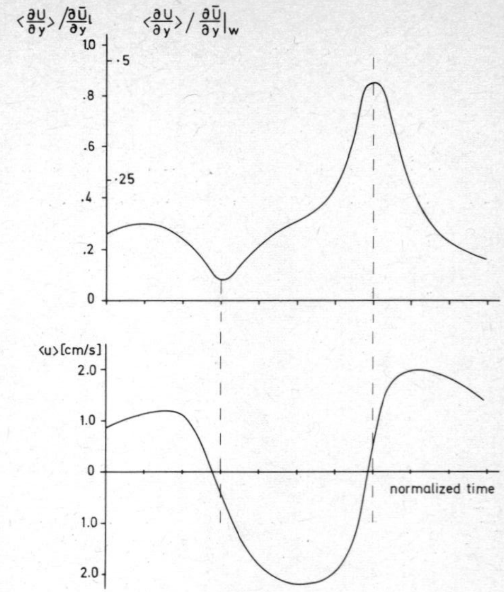

Figure 2: Pattern recognized < u > (bottom). Pattern recognized < ∂u/∂y > normalized with local mean gradient and gradient at the wall (top) from Eckelmann et al. (1977).

Similarly, Blackwelder and Kaplan believe their detection criterion (equation 3) is fulfilled during the occurrence of "bursts". If it is true that both techniques are observing the same process, then similar properties of these processes should certainly be observed by both methods. One such characteristic is the average "bursts" duration. Blackwelder and Kaplan found in a turbulent boundary layer that an ensemble average of over 300 "bursts" gave an average "burst" duration of $T \cdot U_\infty / \delta = 2.8$; Wallace et al. obtained an average pattern duration of $T \cdot U_{CL}/b = 2.4$ in a turbulent channel flow. This striking agreement led us to look for further common characteristics obtained by the two methods.

Results

In Figure 3 conditionally averaged velocity profiles are shown which were measured by Blackwelder and Kaplan during the occurrence of "bursts". At time $\tau = 0$ the detection criterion (equation 3) is fulfilled. The variation of the long-time averaged velocity profiles (the dashed curves in the figure) is the ensembled averaged time function shown in the lower part of Figure 1. By differentiating these profiles in the direction normal to the wall at $y^+ = 15$ and normalizing with the average value of the wall gradient, the local normalized averaged gradient shown in the upper part of Figure 1 can be obtained. One sees that the spatial gradient $\Delta < U >/\Delta y$ has a maximum at exactly

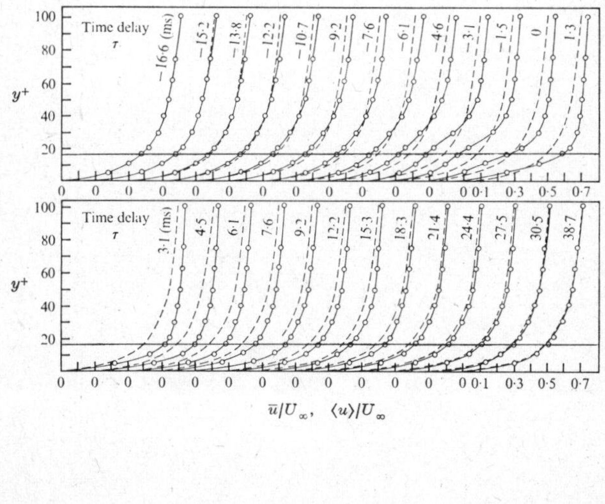

Figure 3: Conditionally averaged and mean velocity profiles with positive and negative time delay τ from Blackwelder and Kaplan (1976).

the same point where the time derivative $\Delta < U >/\Delta t$ shown in the lower part of Figure 1 has its maximum. Eckelmann, Nychas, Brodkey and Wallace (1977) have measured the ensemble averaged spatial gradient $< \partial U/\partial y >$ at various distances from the wall during the occurrence of the "pattern" described above. These are shown in Figure 4. The curve for $y^+ = 15$ is shown again in the upper part of Figure 2 where it is normalized with the averaged wall gradient. Here again it is seen that the maximum of the ensemble averaged gradient occurs at the maximum slope of the time function in the figure below it. Not only is this true; it can also easily be shown that the form of the upper curve can be obtained by time differentiation of the lower curve in Figure 2. This means that during such patterns, Taylor's hypothesis also holds in the direction perpendicular to the wall. This can be expressed as

$$< \frac{\partial u}{\partial y} > = \frac{1}{C_y} \frac{\partial < u >}{\partial t} \tag{4}$$

The convection velocity, C_y, is of the order of u_τ. C_y has been directly measured by Kreplin and Eckelmann (1979) and interpreted as the convection velocity with which a disturbance occurring in a turbulent channel flow is transmitted toward the wall. If one compares the measurements of Blackwelder and Kaplan (Figure 1) with our measurements (Figure 2), it is clear that there is both qualitative and quantitative agreement, i.e. $\frac{\Delta < U >}{\Delta y} / \left. \frac{\partial \overline{U}}{\partial y} \right|_W$ is in both cases of the order of 0.5 where $\left. \partial\overline{U}/\partial y \right|_W$ is the

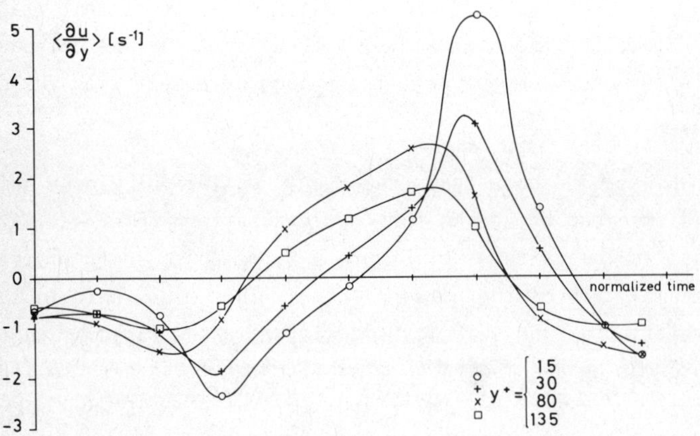

Figure 4: Pattern recognized velocity gradient $< \partial u/\partial y >$ at four different wall distances from Eckelmann *et al*. (1977).

average wall gradient. The gradient measurements of Figure 4 were made by Eckelmann et al. with a five-sensor hot-film probe. The fluctuating component of the gradient normal to the wall, $\partial u/\partial y$, was obtained from the two velocity signals at locations (x_0,y_0,z_0) and $(x_0,y_0 + \Delta y,z_0)$. The spatial separation, Δy, of this probe is 1 mm which in viscous length is $\Delta y^+ = 1.7$. The fluctuating component of the velocity gradient in the spanwise direction, $\partial u/\partial z$, has also been measured with the same probe at the locations (x_0,y_0,z_0) and $(x_0,y_0,z_0, + \Delta z)$. In this case the spatial separation, Δz, is twice as large as Δy, i.e. $\Delta z^+ = 3.4$.

Figure 5 shows the pattern obtained with the pattern recognition technique (solid curve) and the same pattern with the gradient pattern $< \partial u/\partial y > \Delta y$ added to it (dashed curve). The patterns are identical in form but are shifted in phase with the dashed curve leading the solid curve. Figures 6-8 show similar curves for $y^+ = 30$, 80 and 135. In addition to the phase shift, it is seen in these figures that the amplitudes of the patterns obtained from the gradients decrease at increased wall distance. Near the wall the pattern obtained at $y_0 + \Delta y$ is very similar to that at y_0 with only a phase difference. This is in good agreement with the measurements of Kreplin and Eckelmann. They observed that a steep front occurs in the near wall region, out to $y^+ = 50$, which has a convection velocity in the streamwise direction of $C_x \approx 12 - 17\ u_\tau$ depending on wall position. Because of the steep slope of this front, disturbances in the velocity further out from the wall are observed at a later time at locations nearer the wall. This fact gives the effective convective velocity perpendicular to the wall, C_y. This similarity in signals is also seen in the time functions of the streamwise velocity of Blackwelder and Kaplan measured at locations between $y^+ = 5 - 100$ in Figure 9. When "bursts" are observed, as indicated along the time axis of Figure 9, their effect is seen over quite a wide region of the wall layer normal to the wall. The signals at adjacent locations appear very similar. The large spatial gradients, $\partial u/\partial y$, can only then occur because of phase shifts in these signals. The decreasing similarity of adjacent signals further from the wall is in good agreement with the patterns of Figures 7 and 8.

The solid curve in Figure 10 shows again the $< u >$ -pattern of Figure 6 for $y^+ = 30$. The dashed curve is obtained by adding the gradient pattern (Figure 11) to the original pattern. In contrast to the patterns of Figure 5-8, no phase shift is seen here; only a reduction of the amplitude of the pattern occurs. This means that in the spanwise direction the correlation of the velocity signals decreases rapidly and that no "frozen" pattern picture can be assumed. A very similar result is seen in Figure 12 where signals at various z^+ locations parallel to the wall at $y^+ = 15$ were measured by Blackwelder and Kaplan.

Although the results shown here emphasize the similarities obtained with the two techniques, there are significant differences remaining. The detection criterion of Blackwelder and Kaplan detected "bursts" during about 18 % of the total data sample and the

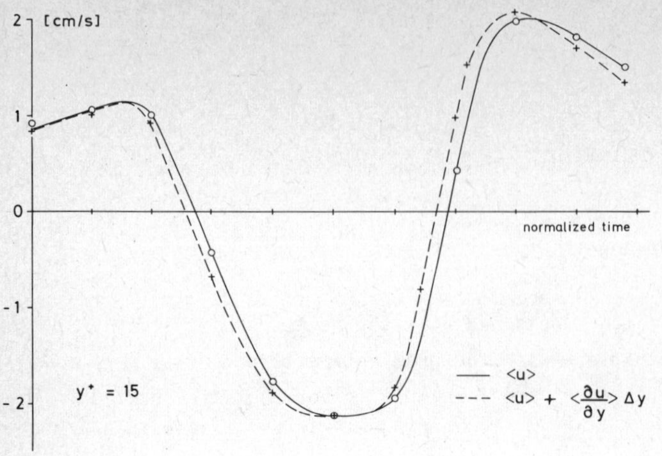

Figure 5: Pattern recognized < u > and added pattern recognized < ∂u/∂y > Δy for $y^+ = 15$.

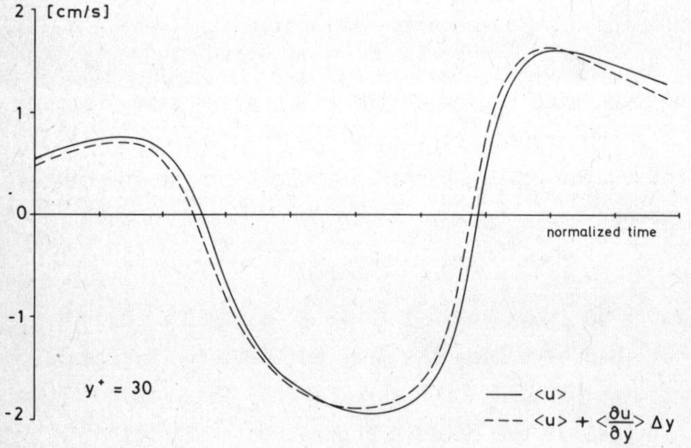

Figure 6: Same as Figure 5 for $y^+ = 30$.

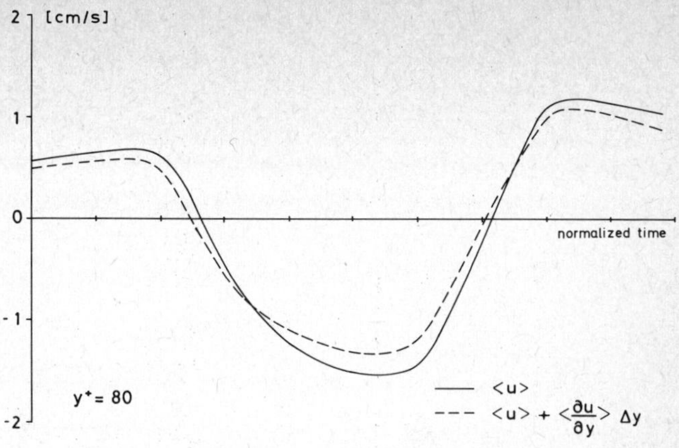

Figure 7: Same as Figure 5 for $y^+ = 80$.

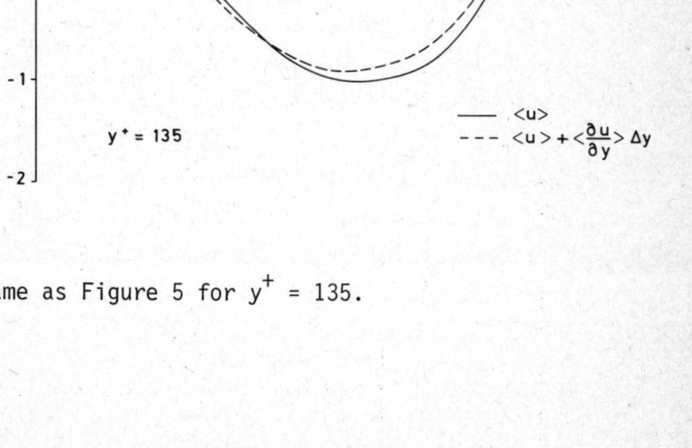

Figure 8: Same as Figure 5 for $y^+ = 135$.

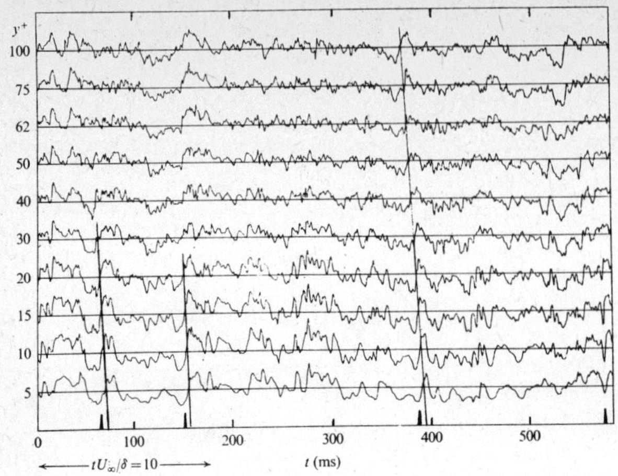

Figure 9: Instantaneous streamwise velocities as a function of the normal wall distance, y^+, from Blackwelder and Kaplan (1976).

maximum of the conditionally averaged Reynolds shear stress is about an order of magnitude greater than the local average at $y^+ = 15$. The pattern recognition technique, on the other hand, found patterns occurring during approximately 65 % of the total data sample but the maximum of the pattern ensemble averaged Reynolds shear stress is only 1.2 times greater than the local average at $y^+ = 15$ and 1.8 times greater at $y^+ = 30$. This would indicate that the VITA technique is a sharper method for detecting Reynolds shear stress producing structures.

The lower level Reynolds stresses obtained by pattern recognition can be explained by the fact that patterns with period lengths ranging over half an order of magnitude were included in the ensemble averages (see Wallace et al.) with the most probable periods being short. The short period, higher frequency fluctuations have lower amplitudes as can be inferred from spectra. Thus the lack of an amplitude criterion in addition to the slope criterion in the pattern recognition technique yields reduced levels of ensembled averaged Reynolds stress. The VITA technique of Blackwelder and Kaplan is basically an amplitude criterion. Either large positive or negative streamwise velocity fluctuations are detected when they occur within the variable interval time averaging window. They, in fact, find that their detection criterion is fulfilled around abrupt accelerations of the streamwise velocity, i.e. regions of large positive slope in the streamwise signal. Thus their amplitude criterion detects periods with

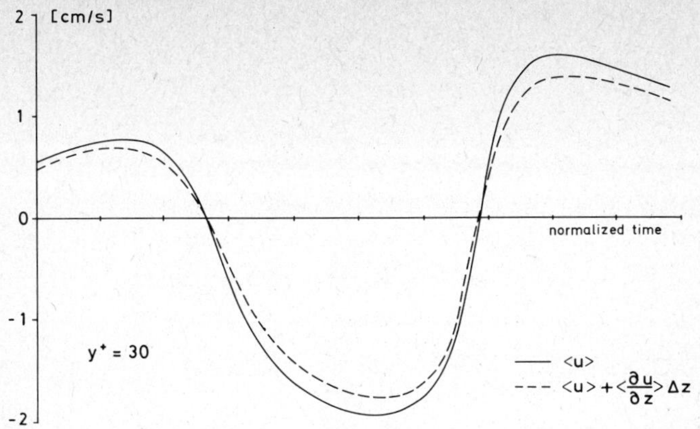

<u>Figure 10</u>: Pattern recognized < u > and added pattern recognized < ∂u/∂z > Δz for $y^+ = 30$.

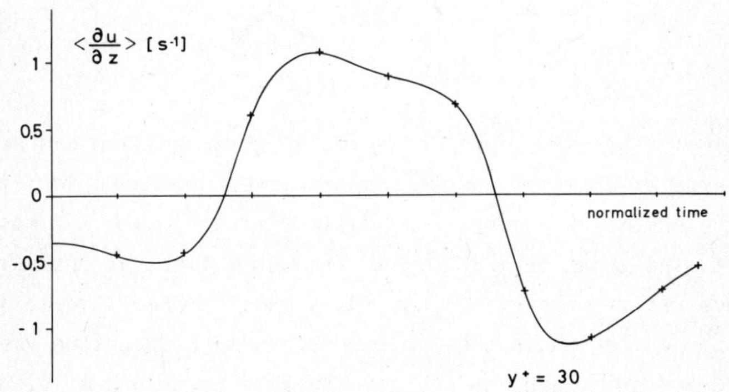

<u>Figure 11</u>: Pattern recognized < ∂u/∂z > at $y^+ = 30$.

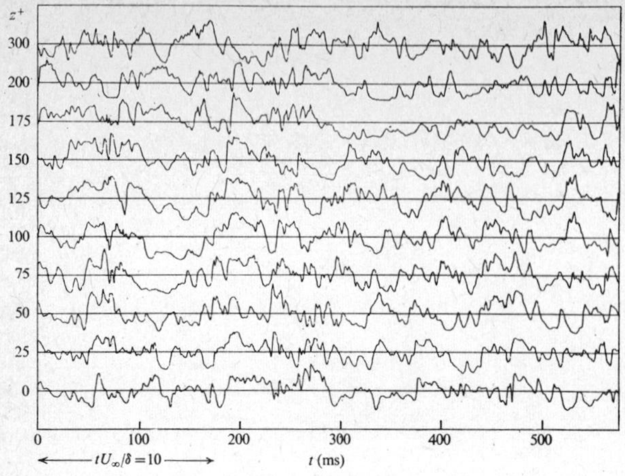

Figure 12: Simultaneous streamwise velocities in the spanwise direction from Blackwelder and Kaplan (1976).

large positive slope, but a slope criterion, as in our case, does not necessarily detect large amplitudes. An improvement in the pattern recognition technique would therefore be to incorporate an amplitude criterion.

Lu and Willmarth (1973) have also measured ensemble averaged Reynolds stresses during "bursting" using a third detection method. "Bursts" are detected at $y^+ = 15$ when the streamwise velocity is decreasing and falls below a preset threshold level. They measured a maximum conditionally averaged Reynolds stress at $y^+ = 30$ of about 2.5 times the local average which is closer to what we find.

Further comparison is planned between these three techniques using a common set of data to see if and when common segments of the signals are detected. Then we will be able to say with certainty whether or not the common characteristics result from the same flow structures being detected.

References

Blackwelder, R.F. and Kaplan, R.E. (1976), J. Fluid Mech., 76, 89.
Eckelmann, H., Nychas, S.G., Brodkey, R.S., and Wallace, J.M., (1977), Phys. Fluids, 20, S 225.
Kreplin, H.-P. and Eckelmann, H. (1979), J. Fluid Mech., 95, 305.
Lu, S.S. and Willmarth, W.W. (1973), J. Fluid Mech., 60, 481.
Wallace, J.M., Brodkey, R.S. and Eckelmann, H. (1977), J. Fluid Mech., 83, 673.

THE EFFECTS OF REYNOLDS NUMBER AND PRESSURE GRADIENT ON THE TRANSITIONAL SPOT IN A LAMINAR BOUNDARY LAYER

I. Wygnanski
School of Engineering,
Tel-Aviv University,
Ramat-Aviv, Israel.

ABSTRACT

The flow associated with the transitional spot is discussed in view of the possible relationship between the spot and the large coherent structure existing in a fully turbulent boundary layer. A brief review of the published results is given, yet the paper stresses new data. The entrainment of non-turbulent fluid into the spot is calculated on the plane of symmetry, taking into account the growth of the spot in the direction normal to the boundary. The effects of Reynolds number on the growth of the spot is discussed, as well as some effect of favourable pressure gradient. It was observed that the spot consists of an orderly array of eddies which give rise to the streaky structure in the spanwise direction.

INTRODUCTION

The role which transitional spots play in the generation of a fully developed turbulent boundary layer stimulates a growing number of researchers.

Visual observations made by Emmons (1951) in water flowing over a smooth table led to the discovery of local spots of turbulence generated randomly in space and time. Schubauer and Klebanoff (1956) charted the general shape of the spot, the propagation velocity of its boundaries, and its lateral rate of spread, while Elder (1960) investigated the interaction of two spots generated simultaneously at two spanwise locations. Since the early seventies the spot has been investigated primarily (though not exclusively) in two laboratories: (1) At the Graduate Aeronautical Laboratories at CALTECH by Professor D. Coles and his colleagues, and (2) at the School of Engineering, Tel-Aviv University.

In the following presentation an attempt is made to review some of the recent results describing the spot and its possible relation to the large coherent structures in the turbulent boundary layer.

RESULTS

(1) The Boundaries of an Ensemble Averaged Spot in Blasius Boundary Layer

Transitional spot may be triggered repeatedly by perturbing the flow at a point for a short period of time. Measurements of the transient phenomenon are usually made at a fixed location downstream of the perturbation while the spot passes a stationary sensor. The shape of the spot is thus determined in coordinates (t,y,z), where t is time, y is distance normal to the solid surface and z is the spanwise distance measured from the plane of symmetry. The boundaries of the spot can be determined with least ambiguity by locating the turbulent-non-turbulent interface. This was done by Wygnanski, Sokolov and Friedman (1976, hereafter referred to as WSF). Coles and Barker (1975) used ensemble averaged velocity records which were conditioned to their disturbance generator and they defined the boundary as the locus of points at which the velocity deviated by ±2% from its free stream value, (i.e. $U-U_{LAM}/U_\infty = \pm 0.02$ where U_{LAM} is the velocity of the unperturbed flow at the particular location and U_∞ is the free stream velocity). The two definitions are in reasonable agreement on the plane of symmetry (Fig. 1) except near the solid surface and near the apex of the spot. The spot has an arrowhead shape in the plan view and a rounded triangular shape in a elevation view.

Although the <u>average boundaries</u> of the spot are smooth, it would be a mistake to conclude that the interface of the spot in a single realization is smooth as well. Pictures of spots using various techniques of flow visualization (Fig. 2) confirm the notion that the spot has a complex substructure which contorts its boundaries in both elevation and plan views, but the simple averaging process is incapable of detecting this substructure. Nevertheless, the average boundaries of the spot appear to be similar in all experiments and independent of the detailed character of the perturbation which triggered them in the first place. WSF showed that the average interface of the spot can collapse onto a universal curve whenever the height of the spot and its length on the solid surface are taken as scaling parameters

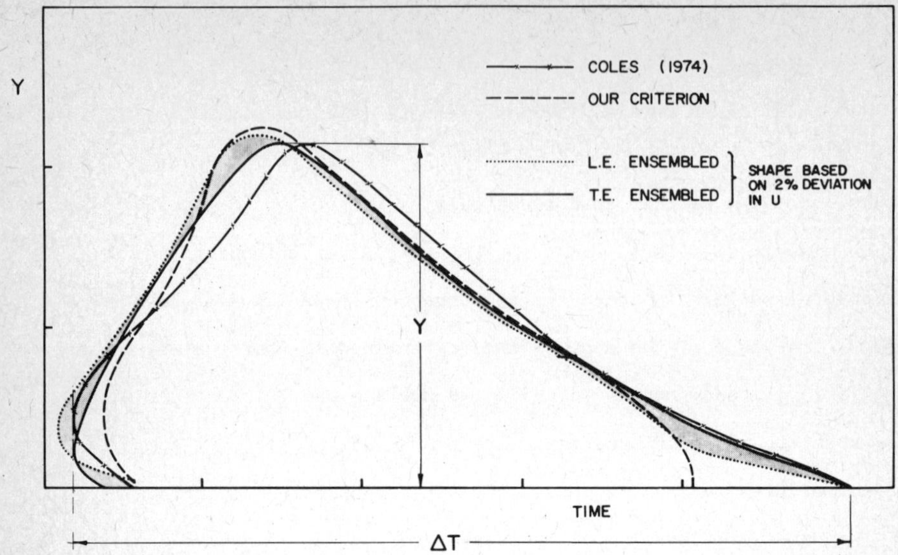

Fig. 1 An elevation view of spot boundaries (WSF)
----intermittency criterion
....±2% perturbation aligned on leading interface
—— ±2% perturbation aligned on trailing interface
x-x Coles and Barker (1975).

(Fig. 3). Since the data used in Fig. 3 was taken at various Reynolds numbers ($R_{e_{\delta^*}}$ is usually used; where $\delta^*$is the displacement thickness of the laminar boundary layer) one may conclude that the gross behaviour of the spot is only weakly dependent on $R_{e_{\delta^*}}$ provided the latter exceeds some threshold level. The effect of $R_{e_{\delta^*}}$ (according to Elder 1960) is merely to serve as an indicator for the amplification of natural disturbances, rather than a criterion for the existence of turbulence. In recent measurements, Wygnanski and Haritonidis 1979 (hereafter referred to a WH) explored the effect of Re_{δ^*} on the rate of growth of the spot. The rate of elongation of the spot with downstream distance may vary by a factor of 2 when the Reynolds number at the perturbation $[R_{\delta^*}]_{PERT}$ is increased from 500 to 1500 (Fig. 4). An increase in the rate of elongation of the spot with downstream dis-

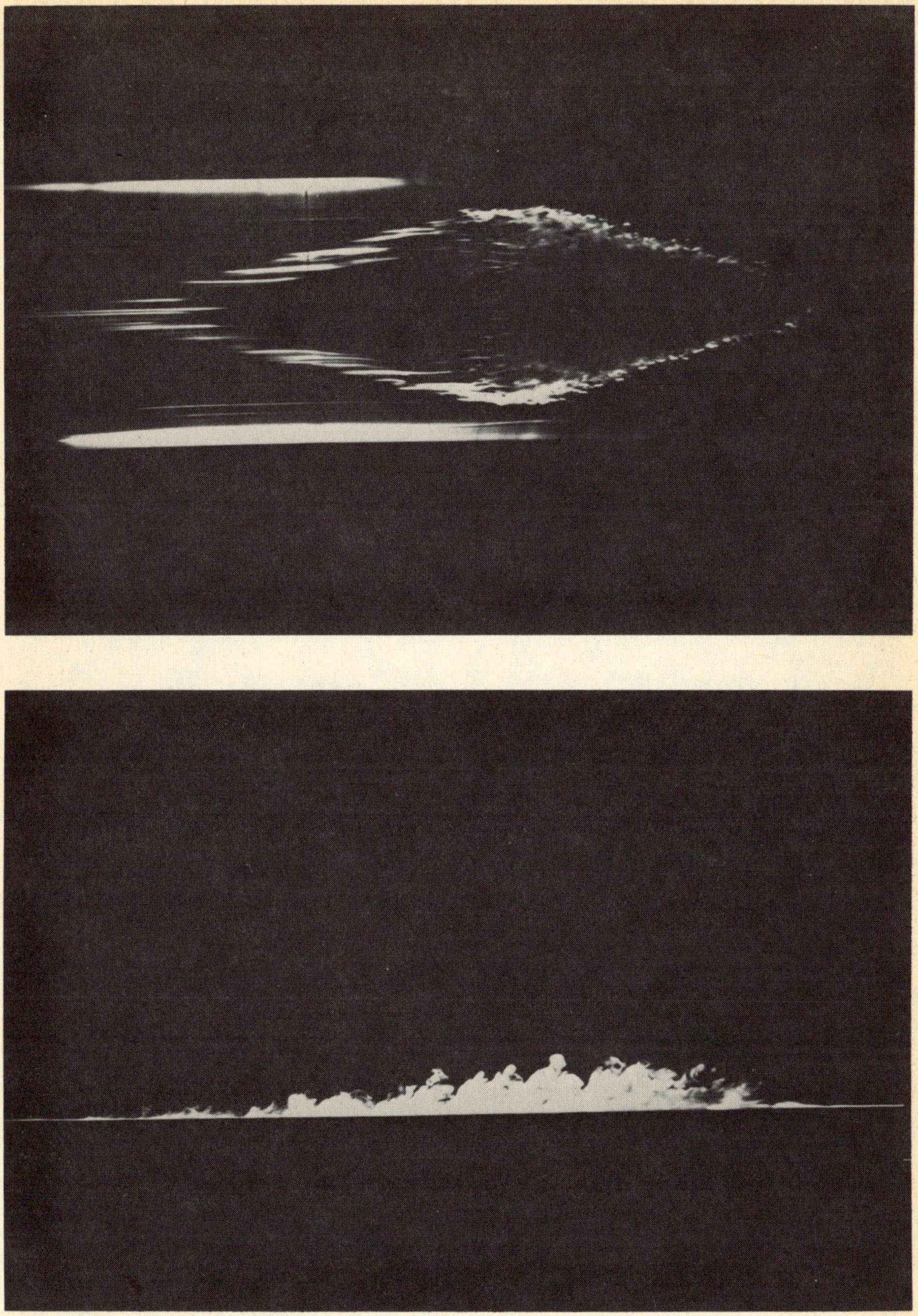

Fig. 2: An elevation and a plan view of a spot triggered by dye (courtesy of M.Gad El Hak - Flow Research Inc.).

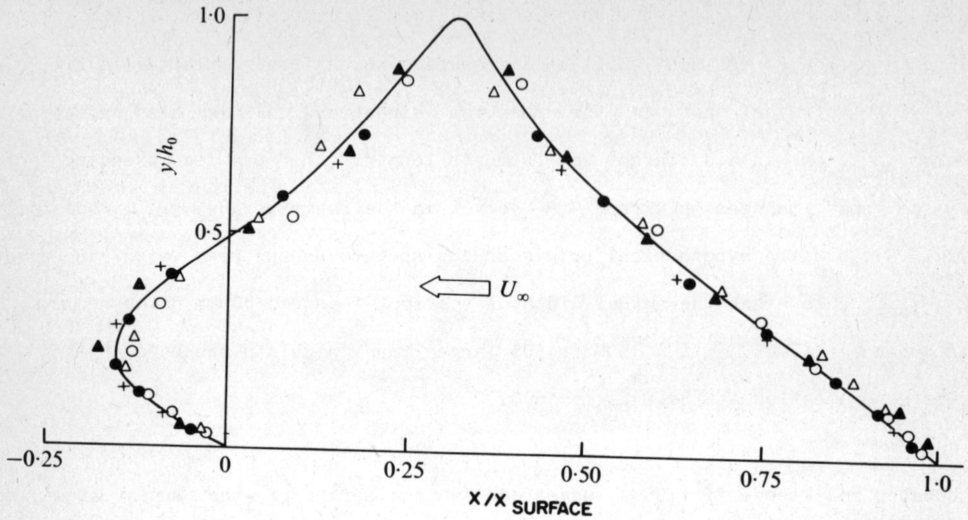

Fig. 3: The similarity of normalized boundaries of a spot (WSF)

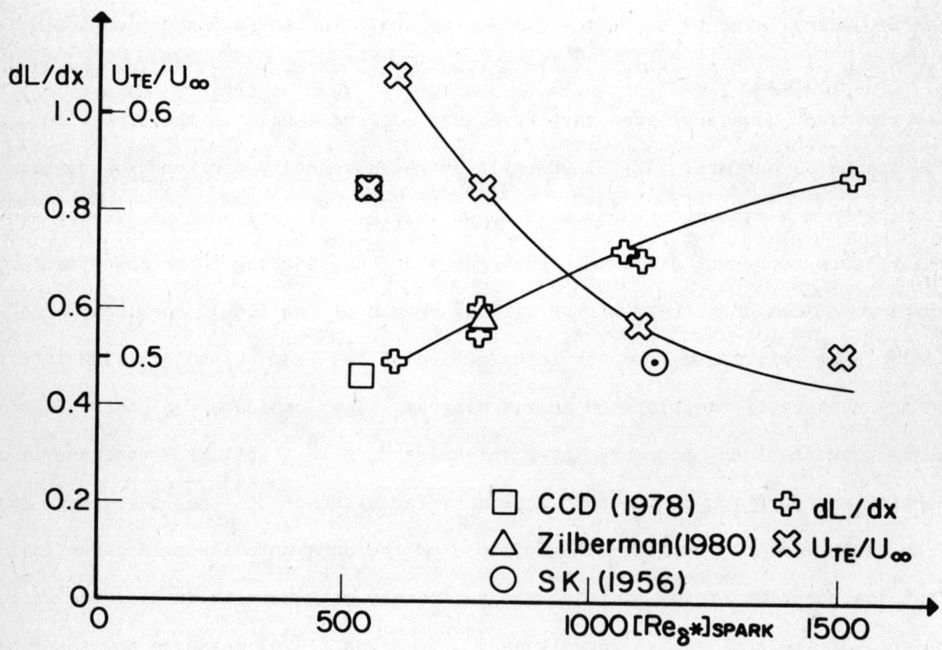

Fig. 4: The longitudinal rate of growth of a spot and the dependence of U_{TE} on $[Re_{\delta}*]$ perturbation (WH).

tance (dL/dx) is associated with a decrease in the convection velocity of the trailing interface of the spot (U_{TE}) at the surface which is also plotted in Fig. 4. The width of the spot in the spanwise direction (z) is also affected by $[Re_{\delta^*}]_{PERT.}$ but in a different way from its length. The spanwise spreading angle of the spot changes only from 9.3^o to 10° in the range of $[Re_{\delta^*}]_{PERT.}$ considered (Fig. 5), but the hypothetical origin of the spot moves upstream with increasing $[Re_{\delta^*}]_{PERT.}$ so that the actual width of the spot measured 800mm downstream of the disturbance at $[Re_{\delta^*}]_{PERT.} = 1500$ is 50% larger than the width measured at the same downstream location at $[Re_{\delta^*}]_{PERT.} = 600$.

Schubauer and Klebanoff (1956) suggested that the height of the spot, above the solid boundary, scales in the same manner as a turbulent boundary layer, developing from the perturbation with an initial height equal to the height of the laminar boundary layer at the location of the perturbation. If one represents the turbulent boundary layer by a 1/7 power law, the above mentioned suggestion implies that the apex of the spot moves away from the surface as $(x-x_{PERT})^{4/5}$ and the height of the turbulent boundary layer may serve as a scaling length in the direction normal to the solid surface. A $\pm$ 2% ensemble averaged velocity perturbation contour, resulting from a passage of a spot is shown in Fig. 6. The various symbols shown in the figure represent different measuring stations starting 800mm downstream of the spark and 730mm downstream of the virtual origin of the flow; the location of the latter was determined from the intersection of the leading and trailing interfaces on the surface of the plate on an x,t diagram. The ordinate in this figure is scaled by the local boundary layer thickness $\delta_T \propto x^{4/5}$ while the abscissa is a dimensionless <u>duration</u> of the spot scaled by the distance of the measuring station from the virtual origin. The collapse of the data onto a single curve indicates that the duration of the spot increases linearly with downstream distance while the height of the spot scales roughly as $x^{4/5}$. Savas (1979) measured the local height of the spot at 6 locations downstream of the disturbance. Replotting his data on a logarithmic scale confirms the $x^{4/5}$ scaling law in the direction normal to the surface, Fig. 7.

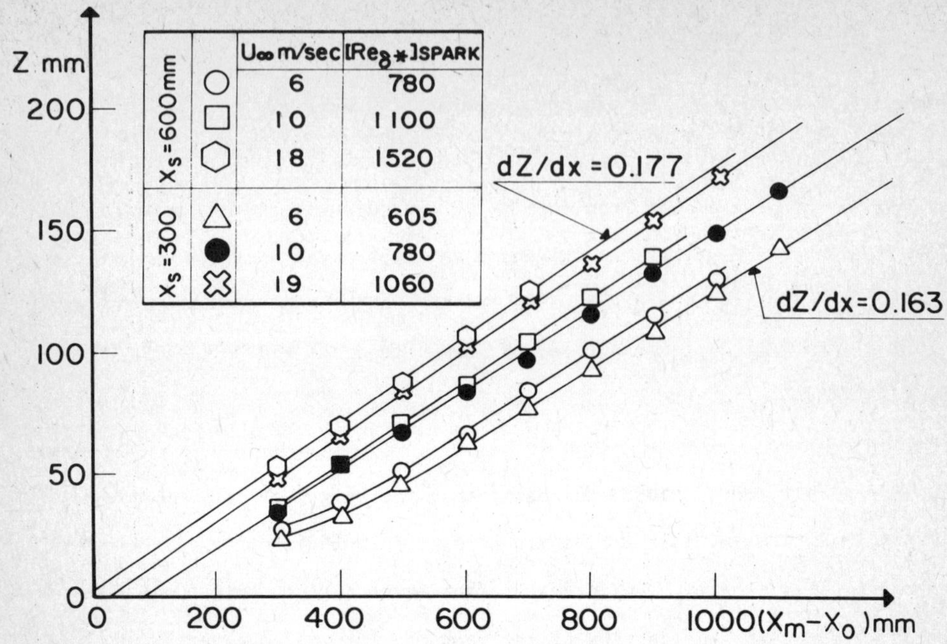

Fig. 5: The spanwise growth of the spot at various $[Re_{\delta}*]_{PERT.}$ (WH).

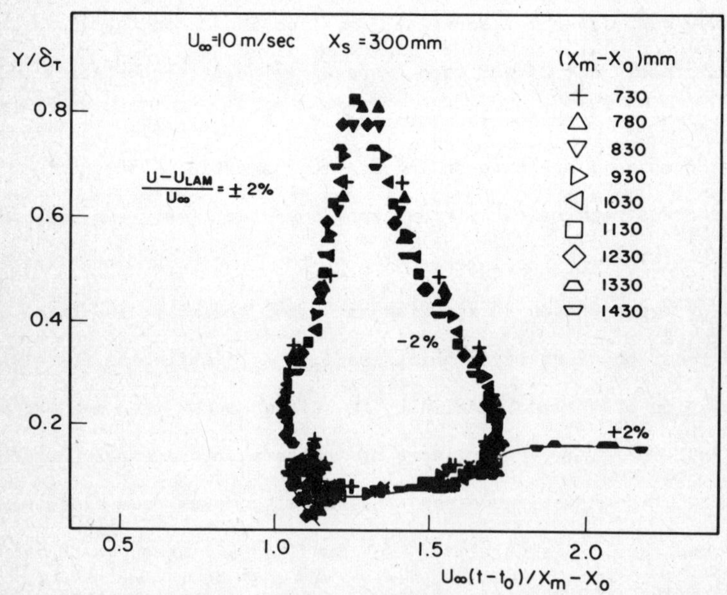

Fig. 6: Perturbation contours of $(U-U_{LAM})/U_\infty = \pm\ 0.02$ plotted in coordinates for which the height of the spot was scaled by $x^{4/5}$ (WH).

Cantwell, Coles and Dimotakis (1978),(hereafter referred to as CCD) suggested that the spot grows conically; namely, all features of the spot can be extrapolated to a single virtual origin and they grow linearly in every direction. The above mentioned experimental results cast some doubt on the validity of conical growth assumption while the dependence of the flow on $[Re_{\delta^*}]_{PERT.}$ limits the usefullness of this assumption.

2. The Entrainment of Non-Turbulent Fluid into an Ensemble Averaged Spot on the Plane of Symmetry

One would like to know the mechanism which is responsible for the growth of the spot in both spanwise and streamwise directions, but in attempting to achieve this goal, one has first to determine the region which most actively entrains non-turbulent fluid. The most obvious first step is to consider the entrainment into an ensemble averaged spot on its plane of symmetry. For this purpose, one should know the shape of the interface and velocity field relative to it. The task is simple in principle but requires tedious and difficult measurements. The first attempt was made by WSF who measured the celerity of the leading and trailing interfaces locally and calculated the function $\int_o^y (U-U_{INTERFACE})dy$ relative to the respective interface. Inferring the entrainment from this procedure implies that the flow relative to the interface is steady and two dimensional [i.e. $(\partial W/\partial z)_{z=o} = 0$]. Furthermore, the growth of the spot in the y direction has been neglected. CCD argued that in laboratory coordinates the stream function depends on x, y and t and thus has a total differential

$$d\psi = - vdx + udy + \frac{\partial\psi}{\partial t} dt$$

which includes a time dependent term. Consequently the integration of $(U-U_{INTERFACE})$ with respect to y is insufficient to calculate entrainment. They estimated the rate of entrainment by assuming conical similarity and calculating

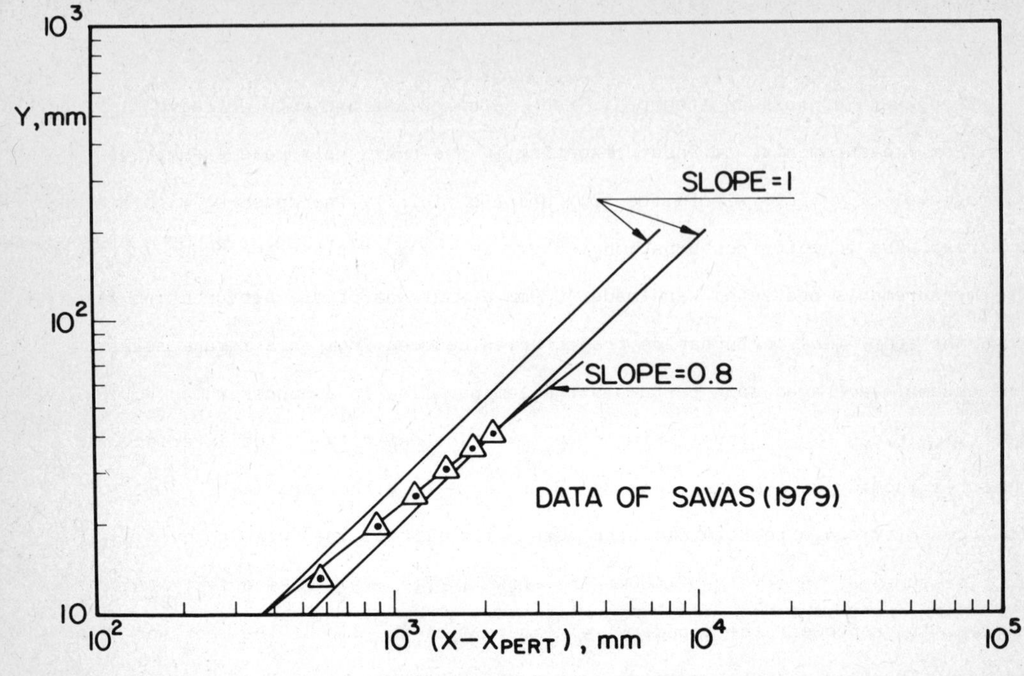

Fig. 7: The growth rate of the apex of the spot with x replotted on logarithmic scale (Savas 1979).

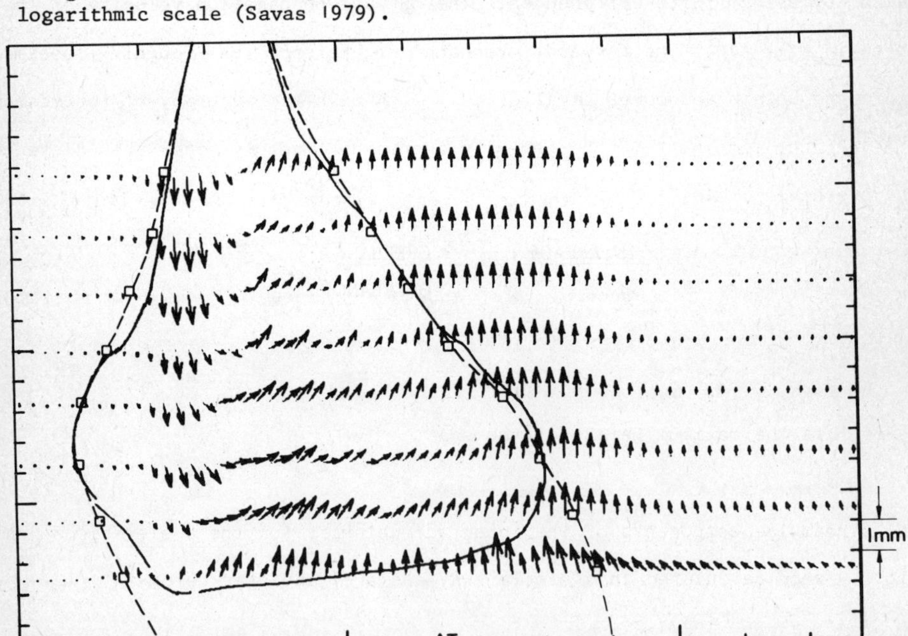

Fig. 8: Velocity perturbation vectors in a transition spot x_{SPARK}=300mm $x-x_{SPARK}$=1030mm U_∞ = 10m/sec (Zilberman 1980).

the particle trajectories relative to an interface.

In the present context an attempt is made to calculate the rate entrainment directly from measurements, without resorting to the previously made assumptions. The average velocity field within the spot and its vicinity was measured with a rake of x wires. The velocity perturbation vectors in the y, t plane are shown in Fig. 8. The measurements presented were made 1030mm downstream of the perturbation generating the spot. The perturbation trajectories deduced from this figure suggest that the ensemble-averaged spot perturbes the laminar flow in a manner which would have been achieved by a large single vortex. The celerity of the interface was not measured locally in a manner done by WSF because only the streamwise component of the celerity could be obtained with some reliability. The outer boundaries of the spot measured at intervals of 100mm are shown in Fig. 9. A 2% defect contour was chosen to represent the boundary. The height of the tip of the spot was measured at every location and the vertical component of the celerity of the tip was calculated by using finite differences. The outward propagation velocity of the tip is given in Fig. 10. The forward 'overhang' of the spot is roughly located y=7mm from the plate and moves parallel to it. Within the accuracy of the measurement, the vertical component of the celerity increases linearly with the distance from the overhang. Thus

$$\frac{V_{INTERFACE}}{V_{APEX}} = \frac{Y-Y_{OVERHANG}}{H-Y_{OVERHANG}}$$

where H is the maximum height of the spot.

The streamwise components of the interface-celerity for the 8 elevations shown in Fig. 9 were calculated in a similar way and are tabulated below

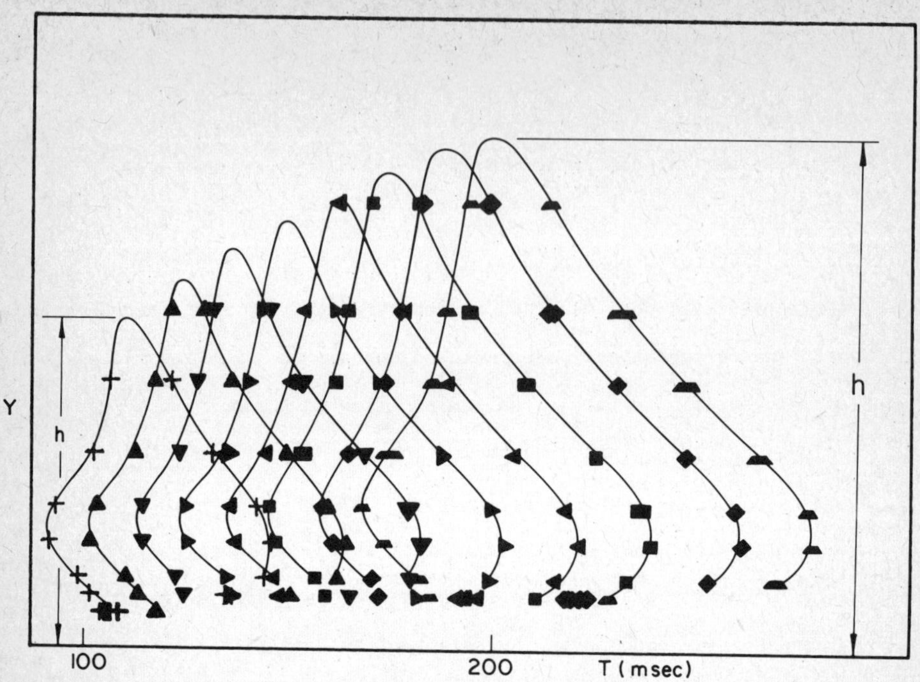

Fig. 9: The boundaries of the spot in (y,t) coordinates , Δx = 100mm between adjacent events x_s=300mm U_∞ =10m/sec

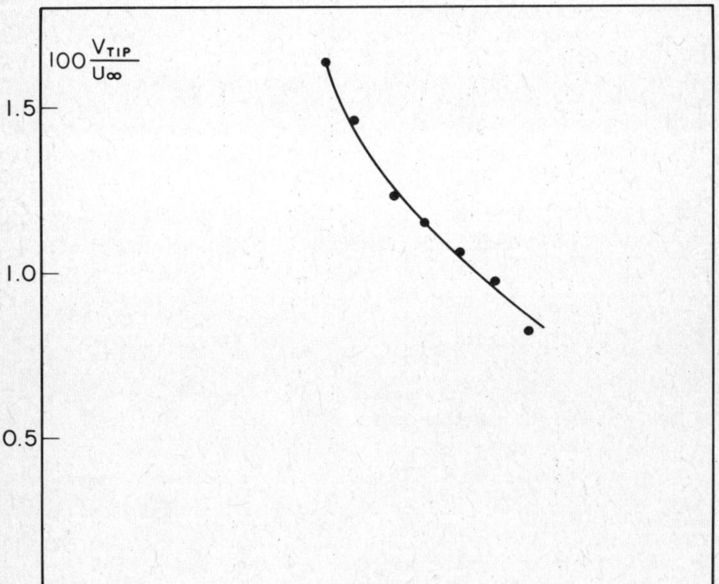

Fig. 10: The variation of the normal velocity of the apex of the spot with x U_∞ = 10m/sec x_{SPARK} = 300mm.

TABLE 1

LEVEL	Y mm	U_{LE}/U_∞	U_{TE}/U_∞	$(V_{INT}/U_\infty)*10^3$
1	2.46	.9	.55	0
2	4.34	.9	.55	0
3	6.21	.9	.55	0
4	8.21	.9	.59	1.18
5	9.96	.85	.63	2.90
6	11.84	.84	.66	4.73
7	13.71	.83	.69	6.56
8	15.71	.82	.72	8.51

In calculating U_{LE} and U_{TE} similar locations on the boundaries of the spot were compared, thus accounting for the growth of the spot in the y direction. The results shown in Table 1 indicate that the celerity of both interfaces is not constant over the entire height of the spot. In fact, only the lower portion of the interface (near the solid surface) is convected at a constant celerity. These results differ from the conclusions drawn by WSF who suggested that the celerity of each interface is a constant independent of y. The difference may stem from the fact that most measurements were made locally, mostly at the lower part ($y/H \leq 0.5$) of the spot (see figure 10 of WSF). Whenever the distances between adjacent measuring stations become large, the growth of the spot in the direction normal to the surface starts to contribute to the estimation of celerity of the interface. The celerity of the leading interface changes by 8% from the surface to the apex of the spot, while the celerity of the trailing interface increases by 30% from its value on the surface. The arrows shown in Fig. 11a represent the velocity vectors relative to the interface. Each tick-mark on the abscissa of this figure represents either a time scale equivalent to 2msec or a velocity scale of 2m/sec. Each tick-mark on the ordinate represents vertical distance equivalent to 1mm or a component of velocity in m/sec equal to $2*1/(U_{LE}*2)$mm. Thus the inclination of the velocity vector is stretched by the same ratio as the slope of the interface. The vertical scale in figure 11a is roughly stretched by a factor of 9:1. The same procedure was used in Fig. 11b relative to the trailing interface. One may integrate the relative velocity component normal to the interface and obtain the average

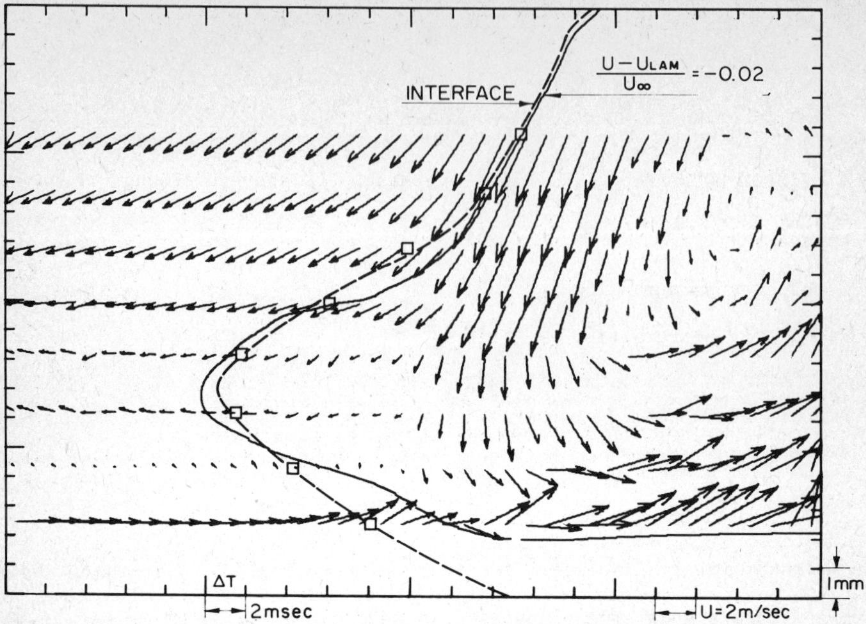

Fig. 11a: Velocity vectors relative to the leading interface

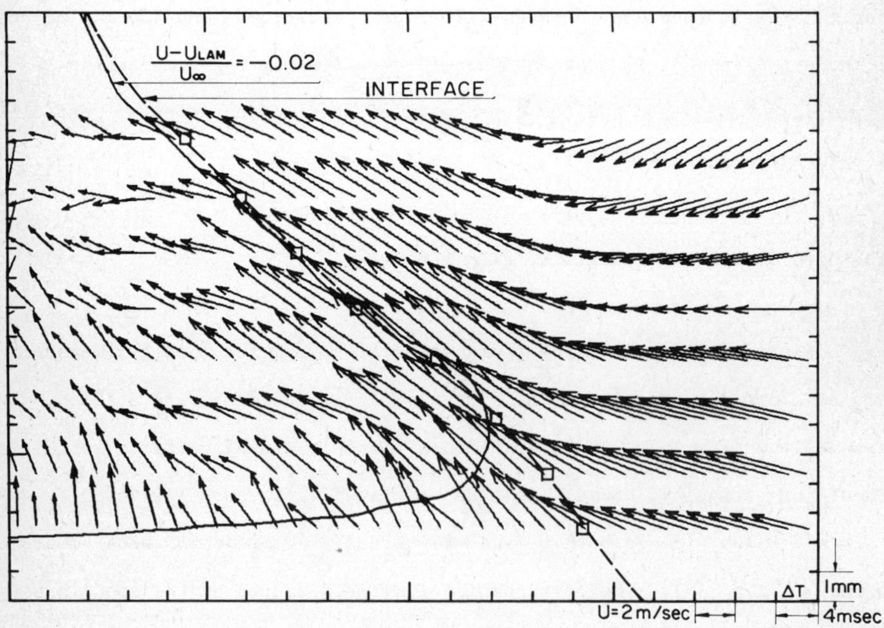

Fig. 11b: Velocity vectors relative to the trailing interface

entrainment distribution into the spot on its plane of symmetry.

TABLE 2

TYPE OF SPOT BOUNDARY	L.E. Below overhang ($y<7mm$)	L.E.above overhang ($y\geq 7mm$)	T.E.
Interface	21%	19%	60%
± 2% velocity perturbation	22%	14%	64%
± 3% velocity perturbation	22%	22%	56%
CCD	16%	5%	79%

The first three rows in Table 2 refer to results obtained by Zilberman (1980), the results of CCD are shown for comparison in the fourth row.

Because the entrainment into the spot depends on the criterion chosen to define its boundaries the comparison between Zilbermans conclusions and CCD may not be precise. Nevertheless, it is apparent that the trailing interface entrains less fluid than previously believed and the outer part of the leading interface entrains more fluid than suggested by CCD.

(3) The Mean Flow Field Associated with the Spot

One can not simply infer the flow field associated with the spot by measuring the streamwise component of velocity at a single streamwise measuring station. Even if one assumes the flow to be two-dimensional, there is a difficulty in estimating the component of velocity normal to the surface (V) because the velocity gradient $\partial U/\partial X$ is not known. Since both U and V velocity components were measured by Zilberman (Fig. 10), the perturbation flowfield caused by the spot in a laminar boundary layer is easily deduced. It appears that the average spot contains a single large vortex which is centered under the apex of the spot. Fluid is drawn towards the surface in the frontal part of the spot and away from the surface near

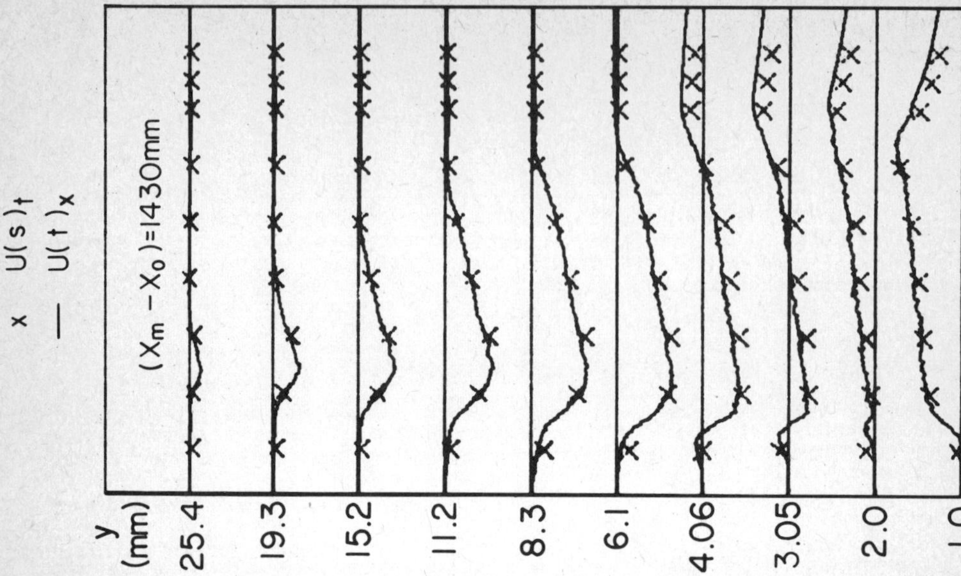

Fig. 12: A comparison between a temporal velocity perturbation record at $x-x_s=1500$ with a spatial velocity perturbation record at a constant time $U_\infty = 10$m/sec x_s = 300m.

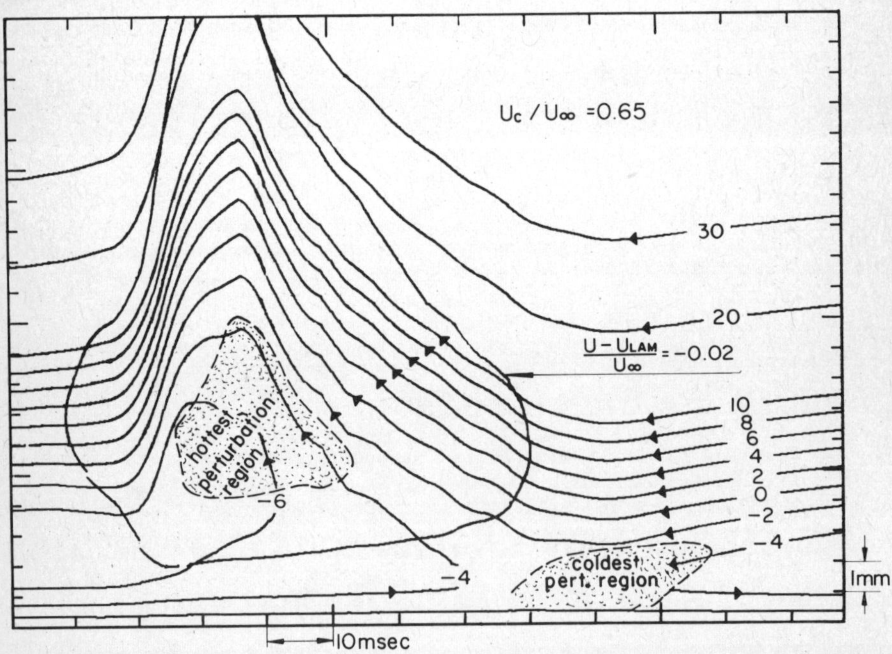

Fig. 13: The streamline pattern relative to an observer moving at a velocity $U_C = 0.65U_\infty$

its rear. One may draw particle trajectories relative to the leading and trailing interface based on data shown in Figures 11a,b. Nevertheless, one can not easily explain the temperature perturbations measured recently by Van Atta and Helland (1979) which are associated with the passage of the spot in a heated laminar boundary layer. It is also difficult to explain these results on the basis of the trajectories calculated by CCD, because the hottest perturbation region inside the spot coincides with a point of accumulation of cold fluid arriving from regions removed from the suface.

WH obtained the average velocity distribution in a spot at a fixed time, (i.e. in spatial coordinates) and compared it to a temporal velocity record measured at the furthest (x) location. The two records are essentially identical implying that the length of the spot measured at a given time is proportional to the duration of the spot at a given measuring station, (Fig. 12), and that the average flow field in the spot is self-similar. One may thus compute the stream function by using a temporal record at a given station assuming it to be equivalent to spatial record at a given time. The resulting streamline pattern relative to an observer moving at $0.65U_\infty$ (this was a convection velocity of the spot determined from two point space-time correlation, see WSF), is plotted in Fig. 13 against the background of temperature perturbation measured by Van Atta and Helland. Fluid is being lifted from the surface at the rear of the spot and moves upwards and towards the spot's leading edge. The lifting process is responsible for the increased temperature under the apex of the spot. The region at which the temperature perturbation is maximum coincides with the highest displacement of the streamlines which were initially adjacent to to surface. Behind the trailing edge of the spot fluid is transported towards the surface reducing the temperature in that region. A different choice of convection velocity $0.5 < U_c/U_\infty < 0.89$ does not lead to a different interpretation of the flow field although the height of the eddy associated with the spot changes significantly.

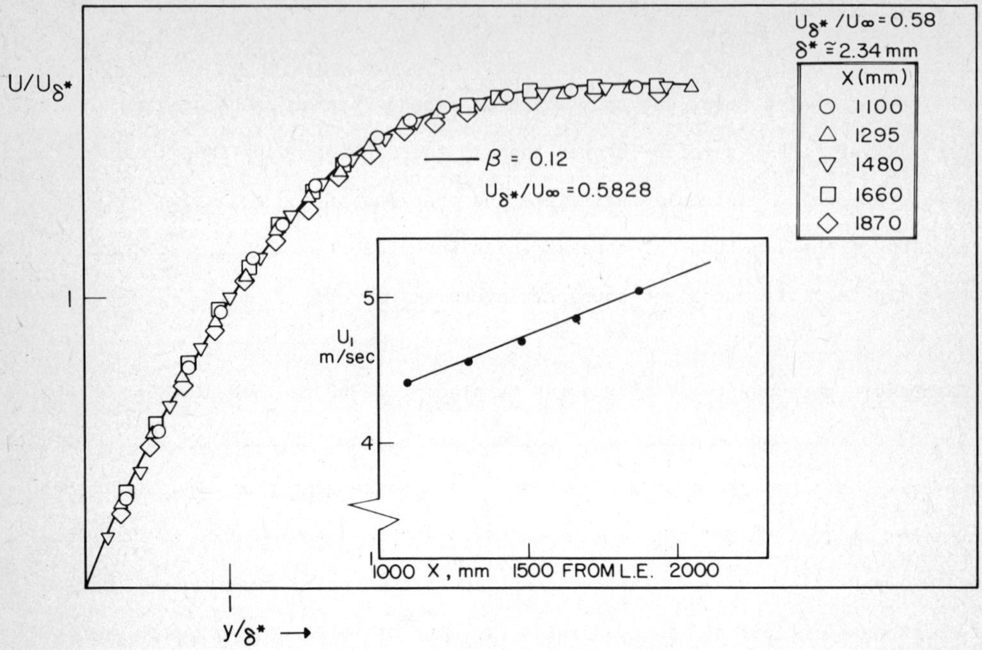

Fig. 14: The laminar velocity distribution on a plate at an angle of incidence (a) The velocity profile (b) The variation of the free stream velocity with x.

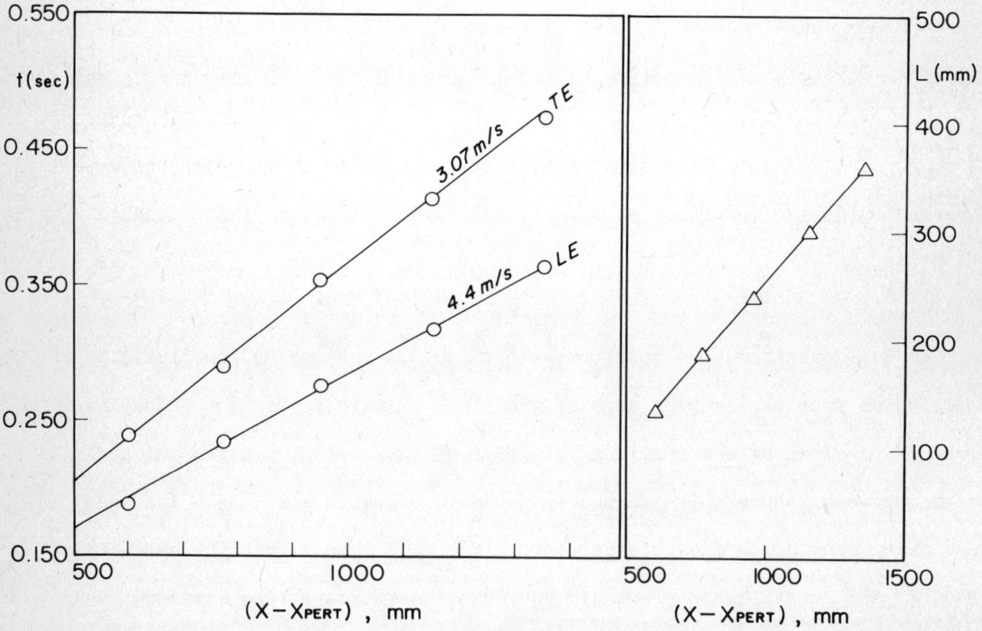

Fig. 15: The longitudinal growth of a spot in favourable pressure gradient

(4) The Structure of the Spot as Favourable Pressure Gradient

It was thought that a turbulent spot which is embedded in an accelerating laminar boundary layer will develop slower than in the corresponding case in absence of pressure gradient. This could facilitate the tracking of the breakdown process and the mapping of the longitudinal streaks which were observed visually in transitional spots and in fully turbulent boundary layers.

The favourable pressure gradient was set by placing a 3600mm long plate at an angle of incidence of approximately 4° to the flow. Since the width of the tunnel is only 612mm, the boundary layer was generated in a convergent channel. Adjustments were made in order to obtain a linear velocity gradient outside of the plate's boundary layer (Fig. 14). This also resulted in a boundary layer whose displacement thickness (δ^*) did not increase with (x), but whose velocity profile resembles a theoretical Falkner-Skan profile with the parameter $\beta = 0.12$. The velocity profiles measured at 5 different stations and normalized by δ and by U_{δ^*} is shown in Fig. 14 and compared with the theoretical profile calculated for $\beta = 0.12$. The velocity profile agrees also with Howarth's calculations for $U(x)=U_0-U_0x^*$ where $x^* = -\frac{x}{U_\infty}\frac{dU}{dx} = -0.1$. However, in this case the shape of the profile does not depend on x. The data presented in this section was taken at reference velocity of 4.5m/sec which was monitored at a station 1000mm downstream of the leading edge of the plate.

A 1.5mm jet generated by a small speaker which was mounted on the back side of the plate 500mm from the leading edge of the plate triggered the spot. The time of arrival of the leading and trailing interface of the spot at 5 measuring stations located between $600\text{mm} \leq (x-x_{PERT.}) \leq 1350\text{mm}$ is shown in Fig. 15. The measurements were made near the surface of the plate. One should note that the celerity of the leading and trailing inerface did not change with downstream distance even though the local free stream velocity increased with x. Thus, the ratio of U_{LE}/U_0 and U_{TE}/U_0 (U_0 is the local free stream velocity) decreased linearly with x. The rate

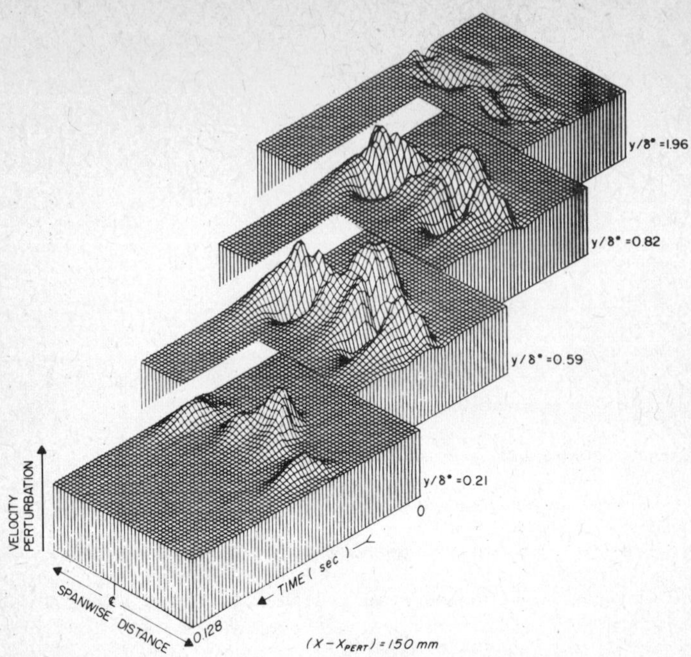

Fig. 16: The velocity perturbation in an incipient spot $(x-x_{PERT}) = 150mm$.

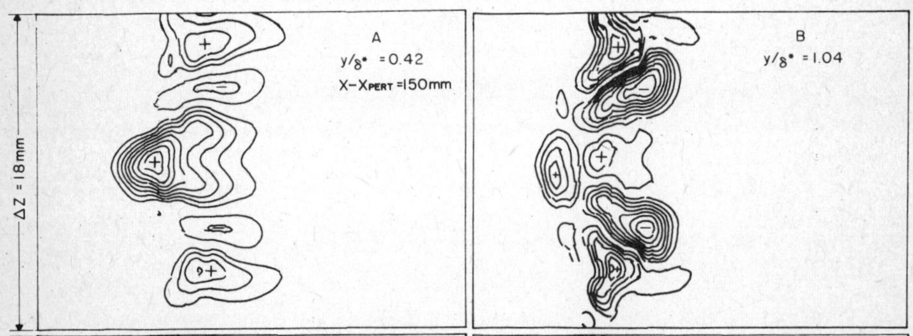

Fig. 17: Velocity perturbation contours in the (t,z) plane at $(x-x_{pert}) = 150mm$. The spacing between adjacent contours is identical throughout.

of elongation of the spot with downstream distance is dL/dx = 0.29. In absence of pressure gradient dL/dx = 0.45 at comparable $[Re_{\delta^*}]_{PERT}$. The spanwise spreading angle of the spot is less than 5° compared to 10° for the Blasius boundary layer. Thus, the spot grows spatially much slower in the presence of favourable pressure gradient than in its absence.

In the second stage of the investigation, the speaker was moved to a position 1000mm from the leading edge for a preliminary investigation of the structure of the flow. The streamwise velocity component was measured with a rake of 10 normal wires located at y=0.5; 0.99; 1.39; 1.91; 2.43; 2.99; 3.66; 4.59; 5.15; and 8.01mm from the bottom of the stem and nominally the surface of the plate. (A photograph of a similar rake is shown in an article by Wygnanski, Haritonidis and Kaplan 1979). The measurements were made at various streamwise and spanwise locations starting 150mm from the speaker in order to explore the breakdown process. The velocity records were ensemble averaged with repsect to the pulse used to trigger to speaker.

In Fig. 16 an isometric view of the perturbation velocities of the incipient (or early) spot are presented, the two horizontal axes are time and spanwise distance (z) with time increasing from right to left in order to accentuate the trailing edge of the spot. Each record represents a different y location. Near the surface, the velocity perturbation is only positive. It is important to note that the <u>velocity perturbation is not homogeneously decaying across the span</u>. One may clearly detect 3 regions of accelerated flow, a large one in the center and two smaller regions at the edges. At a larger distance from the surface, these regions are separated by deep crevases of decelerated flow which become dominant at the outer reach of the boundary layer. At $y/\delta^* = 1.97$, the velocity defect regions join together to give the early spot the characteristic single arrowhead shape. Contours of perturbation velocity in the (z,t) or in the (y,t) planes at various y or z locations respectively, reflect quantitively the structure seen in Fig. 16. At $y/\delta^* = 0.42$ (Fig. 17a) there is a strong positive perturbation region in the

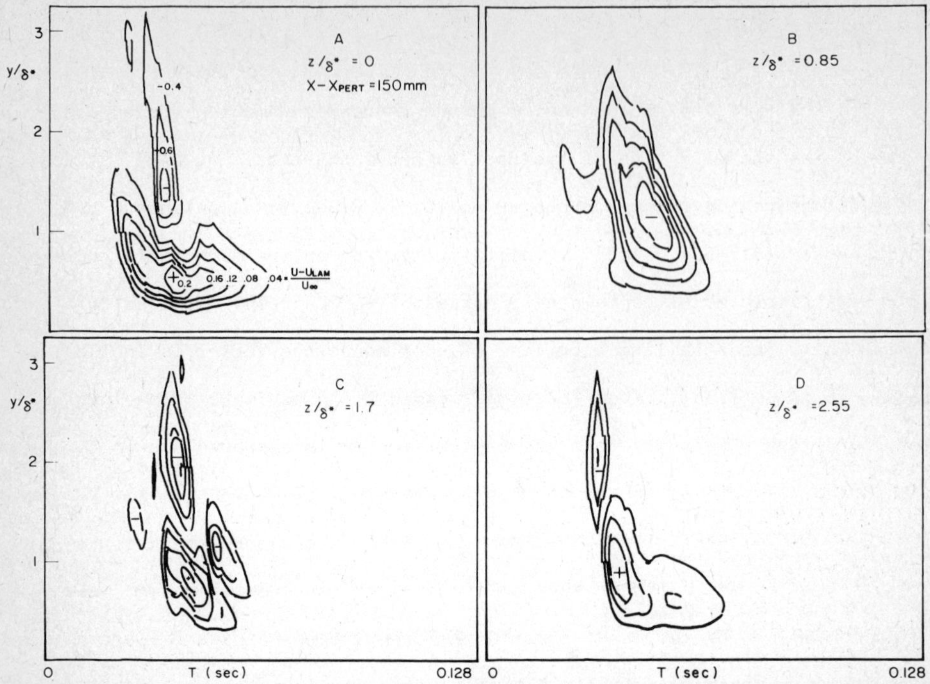

Fig. 18: Velocity perturbation contours in the (t,y) plane at $x\text{-}x_{PERT}$ = 150mm.

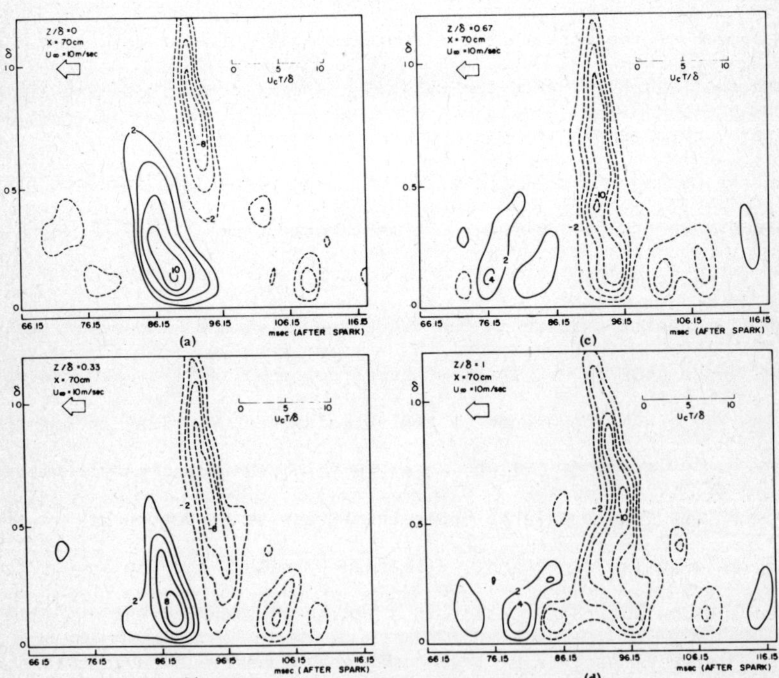

Fig. 19: Velocity perturbation contours in the (t,y) plane in a turbulent boundary layer (Zilberman, Wygnanski and Kaplan 1977).

center which is surrounded by two weak velocity-defect regions which in turn are followed by somewhat stronger velocity-excess regions at larger z. At $y/\delta^* = 1.04$, (Fig. 17b) the velocity excess region in the center is weak and fragmented while the regions at the wing tips of the spot gathered strength. At $y/\delta^* = 1.28$, (Fig. 17c) the defect region becomes stronger and at $y/\delta^* = 1.56$ a continuous defect region is observed, (Fig. 17d). Qualitatively, the entire flow field could be explained by a simple model containing 3 horseshoe (or hairpin) vortices in a typical Λ formation. The trailing portion of each horseshoe vortex is inclined to the surface. It is difficult, from these measurements, to estimate locations of the vortex core but it appears that the central vortex is situated closer to the surface than the two votices at the wingtip. The only explanation for this behaviour stems from the fact that the breakdown to turbulence occurs at the outer part of the boundary layer and not near the wall. The vortices are long and slender, a fact which may not be realized, because the spanwise coordinate is stretched. The time scale may represent a structure whose length is of the order of 100mm while the width is less than 5mm. The aspect ratio of these vortices (i.e. ratio of width to length of the structure) is representative of boundary layer streaks which were observed visually near the surface. There is also a similarity between the flow patterns observed in this experiment and the flow patterns observed by Amini (1978) in the absence of pressure gradient. The comparison suggests that the pressure gradient does not alter the basic structure of the flow.

Representative contours drawn in the y,t plane (Fig. 18) indicate that the same structure tends to repeat itself (though not precisely) at various spanwise stations. The different appearance of the velocity perturbation contours which are separated by $z/\delta^* = 0.85$ (i.e., z = 2mm only) should be noted. It is suspected that in turbulent flow the large coherent streamwise structure may also be quite small in the spanwise direction. Zilberman Wygnanski and Kaplan (1977) initiated, and marked in time a turbulent spot in a laminar boundary layer and allowed it to merge and interact with a natural turbulent boundary layer generated by a row of trips. They were able to follow and recover the ensuing ensemble averaged struc-

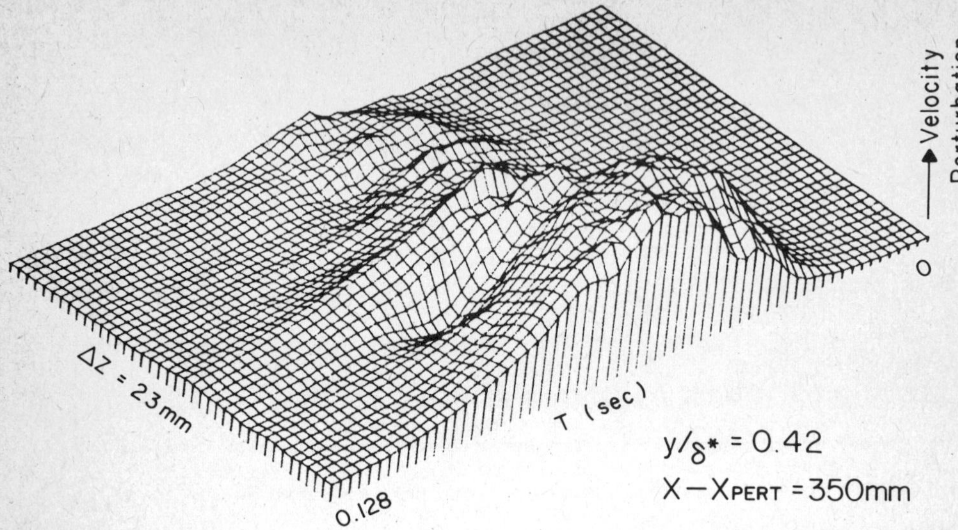

Fig. 20: (a) An isometric view of velocity in a simply ensemble-averaged spot $(x-x_{PERT})$ = 350 mm.

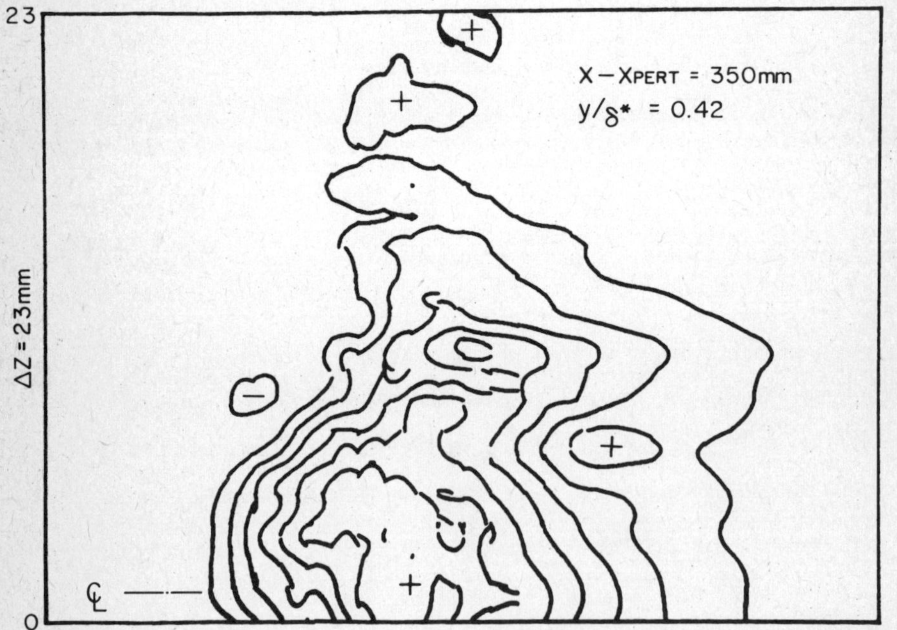

(b) Contours of data shown in (a)

ture although it was immersed in noise (i.e. a turbulent signal) of comparable frequency- spectrum and magnitude. The scale of the structure recovered was much smaller than the scale of a comparable transitional spot in both streamwise or spanwise directions. Since the entire structure had a characteristic convection speed of $0.9U_\infty$ it was postulated that only a narrow region near the leading interface and the plane of symmetry, of a transition spot survived the porlonged interaction with randomly generated simlar structures. In Fig. 19 the perturbation velocity history obtained by Zilberman et al is reproduced. The similarity between the contours shown in figures 18 and 19 suggests that Zilberman et al detected only the leading central structure of their marked spot. It should be pointed out that the large structure referred to in Fig. 19 is a perturbation relative to the turbulent mean velocity profile, whereas the perturbation contours shown in Fig. 18 refer to the laminar velocity profile shown in Fig. 14.

On the basis of the measurements made in an incipient spot one may propose that 3 hair-pin vortices arranged in a Λ formation constitute the spot, one may ask, however, whether a more developed spot contains a larger formation of such vortices or do these three vortices simply grow in the spanwise direction. Repeating the same measurements 350mm from the perturbation, gives an indications how a transition spot growns downstream and what might be the structure of a fully developed spot or a turbulent boundary layer. At $x-x_{PERT}$=350mm the velocity signature observed is fully turbulent and quite similar to the velocity histories recorded in earlier investigations in absence of pressure gradient; the ensemble averaged records are similar as well. The spot was traversed at intervals z=1mm with a ten-wire rake extending from y=0.5mm to y=8mm from the surface. When the data was ensembled, the isometric velocity records and perturbation contours in the (z,t) plane indicated a lack of homogeneity in the spanwise direction but did not prove convincingly the existance of the above mentioned hair-pin vortex structure (Fig. 20a,b). It was realized that at the "wing-tips" of the spot (i.e. at large values of z), the velocity signature was not always turbulent. The the averaging process included events which did not contribute to the velocity perturbation. It became

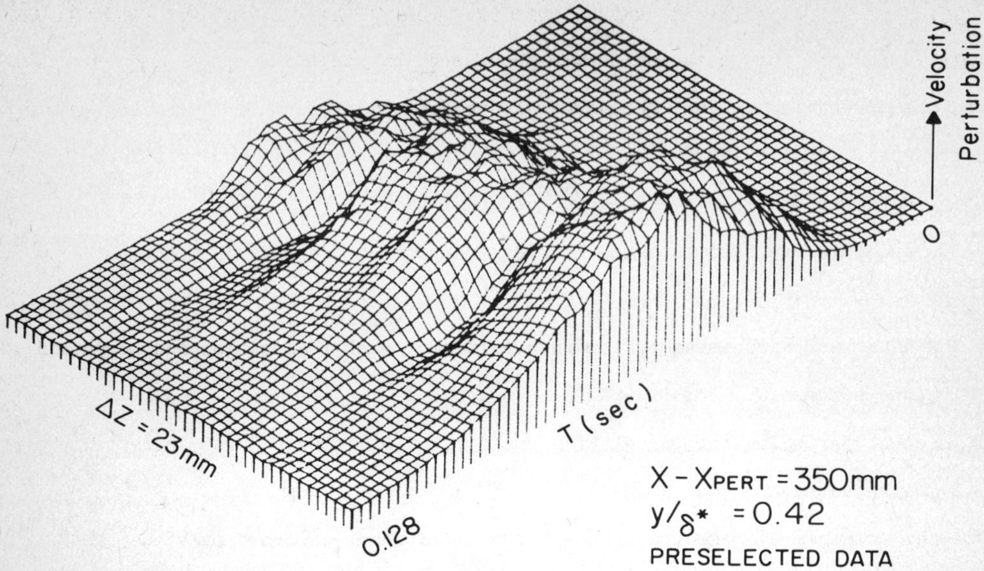

Fig. 21: (a) An isometric view of velocity in an ensemble-averaged large spot at $(x-x_{PERT.}) = 350$ mm.

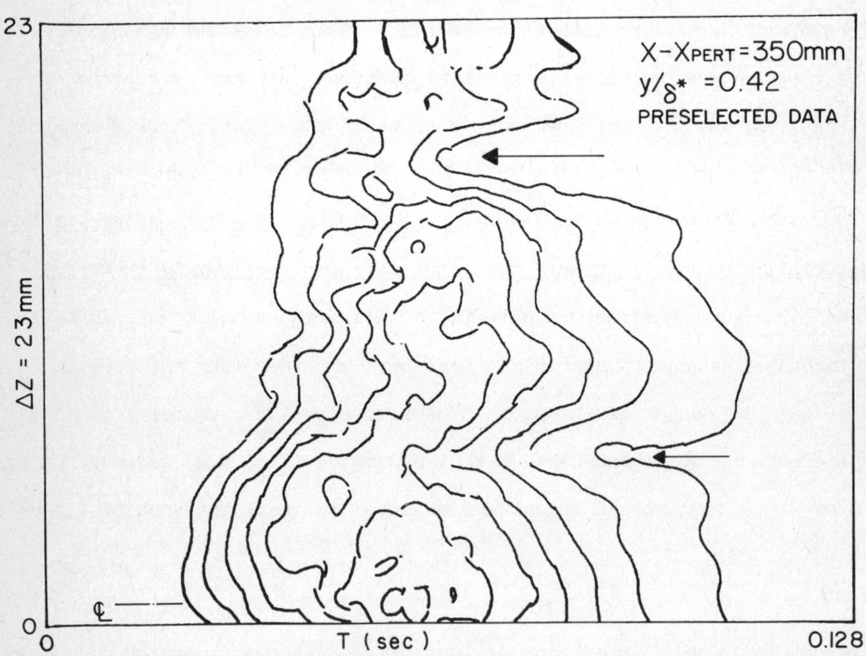

(b) Contours of data shown in (a)

obvious that the spanwise extent of the individual realizations could differ extensively. The same data was then reprocessed, by excluding the smaller spots from the averaging. The procedure adopted is reminiscent of the method used by Zilberman et al. (1977).

(1) The data was ensemble-averaged in the usual way.

(2) The ensemble averaged data was Fourier transformed and kept in memory as a pattern.

(3) The data was reprocessed by low-pass filtering each realization and correlating it with the pattern.

(4) An event for which the correlation coefficient exceeded 0.5 was accepted, the rest were rejected.

(5) A new ensemble averaged pattern was generated which included only the accepted events.

The procedure is biased towards the larger spots but it has the advantage of sifting out irrelevant data. The data shown in Fig. 20 is presented again in Fig. 21 after being reprocessed, showing clearly the spanwise streaky structure of the spot. Since only one half of the spot is shown in Fig. 21a, it is reasonable to suppose that the larger spots at $(x-x_{PERT})$ = 350mm contain at least 5 hairpin vortices while only 3 were detected at $(x-x_{PERT})$ = 150mm. The lateral growth of the spot is associated with an increased number of hair-pin vortices in the formation. It is plausible that the structure detected by Zilberman et al (1977) in the fully developed turbulent boundary layer might have been no other than the leading hairpin vortex in the center of the spot. The almost periodic velocity perturbation streaks are even more distinguishable near the surface (Fig. 22) and may be responsible for the dye streaks which were observed to trail the spot by Elder, CCD and Gad et Hak (1979).

The spot in its fully developed state may not only contain an array of vortices in a Λ formation but may also develop secondary vortices filling the inner space of the Λ , and evolving directly downstream of the leading hairpin vortex. Velocity

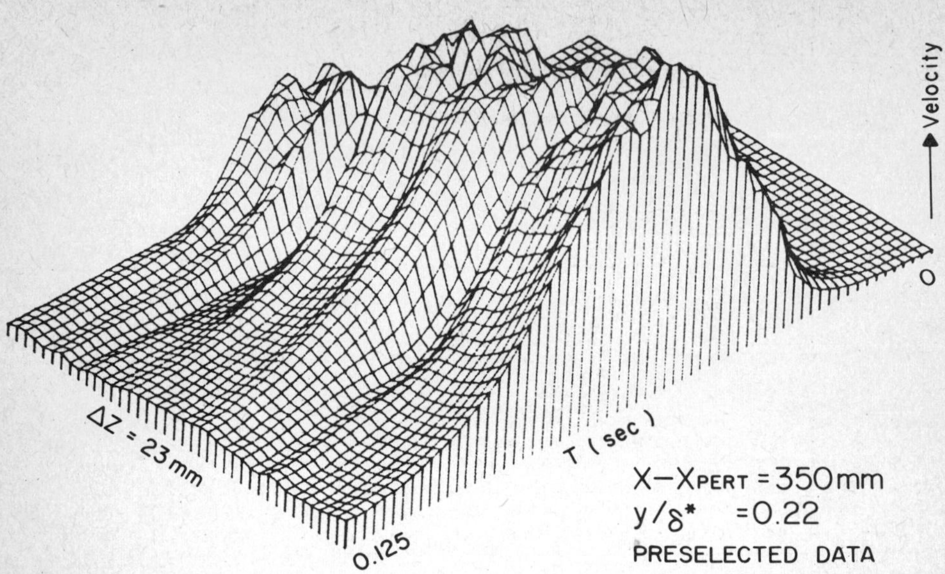

Fig. 22: Velocity perturbation near the surface showing regular longitudinal streaks

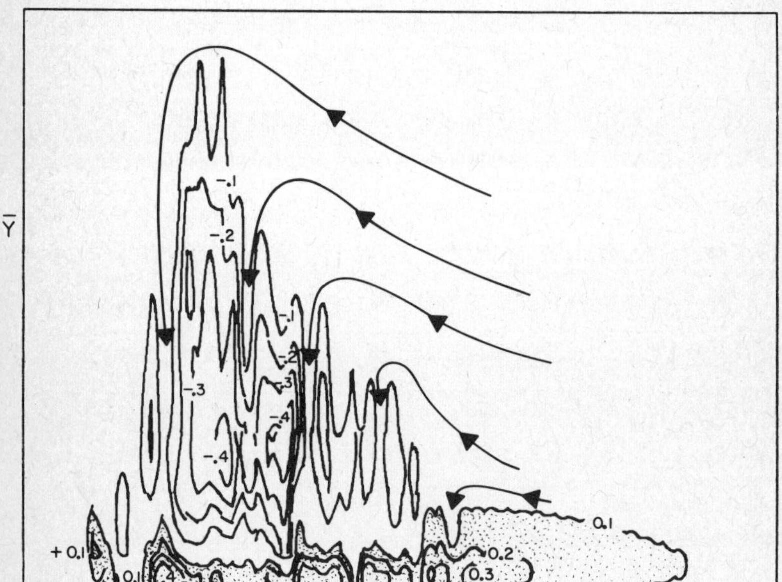

Fig. 23: Velocity perturbation contours for a single event in a plane of symmetry of a spot in absence of pressure gradient

perturbation contours for an individual realization, on the plane of symmetry in absence of pressure gradient, shows that the spot contains more than a single large eddy structure in the streamwise direction as well, (Fig. 23).

CONCLUSIONS

The treatment of the transition spot as a single entity seems to be an over-simplification. The spot appears to contain a hierarchy of large eddies which are repeatable and orderly, giving the spot its peculiar arrowhead shape. The entrainment calculated by considering the average velocity in a spot gives only an overall integral quantity from which very little can be inferred about the kinematics of the process. One may surmise that there is a strong entrainment in the spanwise direction at locations which are removed from the plane of symmetry of the spot. The introduction of favourable pressure gradient, slows down the breakdown process and provides an opportunity for a more detailed analysis of transition than could be available in the Blasius boundary layer in the absence of pressure gradient.

ACKNOWLEDGEMENT

The author is indebted to Dr. Amini and Dr. Einav for the help and cooperation in gathering the data in favourable pressure gradient. The calculations of entrainment were made by Mr. Zilberman. Special acknowledgement is due to Dr. M.Gad-el-Hak for the dye pictures of the spot.

This research was sponsored in part by the U.S. Air Force Office of Scientific Research under Grant No. 77-3275.

REFERENCES

1. Amini, J. (1978) PhD Thesis, Institute de Mecanique de Grenoble.

2. Cantwell, B., Coles, D., Dimotakis, P. (1978) JFM, 87, 641.

3. Coles, D., and Barker, S.J., (1975) in Proceedings of Project SQUID Workshop on Turbulent Mixing in Non-reactive and Reactive Flows, edited by S.N.B. Murthy, Plenum Press, N.Y.

4. Elder, J.W. (1960) JFM, 9, 235.

5. Emmons, H.W. (1951) J. Aero. Sc. 18, 490.

6. Savas, O. (1979) PhD Thesis, CALTECH.

7. Schubauer, G.B. and Klebanoff, P.S. (1956) NACA Rep. 1289.

8. Van Atta, C.W. and Helland, K.N. JFM to be published.

9. Wygnanski, I., Sokolov, M. and Friedman, D. (1976) JFM, 78, 785.

10. Wygnanski, I., Haritonidis, J.H. and Kaplan, R.E. (1979) JFM, 92, 505.

11. Wygnanski, I., Haritonidis, J.H. (1979) to be published.

12. Zilberman, M., Wygnanski, I., Kaplan, R.E. (1977) Phys. of Fluids, 20, S 258.

13. Zilberman, M. (1980) PhD Thesis, Tel-Aviv University.

LEWIS NUMBER EFFECTS ON THE STRUCTURE AND EXTINCTION OF DIFFUSION FLAMES DUE TO STRAIN

A. Liñan

Escuela Técnica Superior de Ingenieros Aeronáuticos
Universidad Politécnica de Madrid
SPAIN

In turbulent diffusion flames chemical reaction and mixing occur simultaneously in thin strained and distorted laminar mixing layers separating fuel from the oxidizer.

If the reaction can be modelled in terms of an irreversible reaction

$$F + n\ O_2 \longrightarrow \text{Products}$$

the relative importance of chemical production and transportation terms in the conservation equations is measured by the Damkohler number, or ratio of the characteristic mixing time t_m and the characteristic chemical time t_c. A typical mixing time is the inverse of the straining rate γ, while t_c^{-1} is proportional to the frequency factor of the reaction times the Arrhenius exponential exp (-E/RT) involving the ratio of the activation energy of the reaction E to the thermal energy TR. Combustion reactions are exothermic and with large values of the ratio E/RT, and, as a result we find in the flow-field regions of low temperature, where the Damkohler number is very small and, therefore, the chemical reaction is frozen. These regions coexist with regions of high temperature where because the reaction is so fast, one of the reactants is depleted, so that either the local concentration of the fuel or of the oxidizer must be zero.

Thus in flows with combustion reactions we find regions in the flow field without fuel separated from regions without oxidizer by thin diffusion flames, where the reactions take place by a process controlled by the rate of diffusion of the reactants towards this surface. These equilibrium regions can be separated from regions of frozen flows, where the reactants coexist, by thin premixed flames that move relative to the unburned mixture leaving behind the reactant that was in excess. See Liñan and Crespo (1976).

For very reactive mixtures the regions of frozen flow cover a small fraction of the flowfields. The diffusion flames are anchored at the injector rim and the chemical reaction is almost everewhere diffusion controlled. However in many cases the flame is blown completely off the flowfield or lifted from the injector rim, so that a region of mixing without significant effects of the chemical reaction appears, bounded from the downstream region of diffusion controlled combustion by premixed flames.

A criterium for the validity of the diffusion controlled assumption, and the conditions leading to local extinction due to flame strain can only be obtained from an analysis of the structure of the thin reaction zones. See Liñan (1974) and Peters (1980) and Peters and Williams (1980). These analyses show that due to finite rate effects the temperature in the reaction zone is lower than the asymptotic flame temperature value associated with the diffusion controlled limit, and that it decreases with decreasing values of the Damkohler number; then due to the large sensitivity of the reaction rate with temperature no diffusion controlled combustion is possible below an extinction value of the Damkohler number.

With the assumption of inifinite reaction rates and if the Lewis number of the reacting species is unity, the passive scalar approach, see for example Bilger (1976), can be used to reduce the problem of describing the reactions in turbulent flows in unpremixed systems to a turbulent mixing problem; this being specially the case if the effects of density changes due to chemical heat release are neglected, although they may be very important in increasing the coherence of the large eddy structures in turbulent combustion.

The purpose of the following is to illustrate the effect of the Lewis number of the fuel in the structures of a strained laminar mixing layer between adjacent eddies of fuel and oxidizer. If a constant positive strain is maintained for some time, the mixing layer will reach a steady state described by the following conservation equations:

$$\gamma Z Y_{oZ} + D Y_{oZZ} + \nu B Y_o Y_F e^{-T_a/T} = 0 \tag{1}$$

$$\gamma Z Y_{FZ} + L^{-1} D Y_{FZZ} + B Y_o Y_F e^{-T_a/T} = 0 \tag{2}$$

$$\gamma Z T_Z + D T_{ZZ} - (Q/c_p) B Y_o Y_F e^{-T_a/T} = 0 \tag{3}$$

to be solved with the boundary conditions

$$Y_o = Y_F - 1 = T - T_o = 0 \quad \text{for} \quad Z \to -\infty \tag{4}$$

$$Y_o = Y_{o\infty} = Y_F = T - T_o = 0 \text{ for } z \to \infty \tag{5}$$

giving the mass fractions Y_O, Y_F of the oxidizer and fuel, and the temperature T, in terms of the distance to the reference stagnation plane. We consider, for simplicity, that the flow field is not modified due to changes in density, so that the velocity in the Z direction is $(-\gamma Z)$, where γ is the straining rate. We also consider the diffusivity of the oxidizer D to be constant and equal to the thermal diffusivity, so that the Lewis number for the oxygen is taken equal to unity, which is not an unreasonable assumption; the Lewis number of the fuel is L different from unity.

The reaction between fuel and oxidizer is modelled by an Arrhenius reaction of frequency factor B, activation temperature $T_a = E/R$, mass stoichiometric coefficient oxidizer/fuel ν, and heat release Q per unit mass of fuel. The temperature of both approaching streams is T.

This problem was analyzed for large values of the nondimensional activation energy T_a/T_o, in the particular case L=1 by this author (1974), and the results were later used by Williams (1975) to describe the structure of diffusion flamelets in turbulent combustion. The analysis showed the existence of a premixed flame regime, in addition to the diffusion controlled regime that appears for large values of the Damkohler number B/γ. Because in most applications the ratio $Y_{O\infty}/\nu$ is small, typically 0.06, it turns out that extinction condition occur in the premixed flame regime that we shall describe, in the following, for $L \neq 1$.

For large values of the nondimensional activation energy T_a/T_o the reaction in the premixed flame regime takes place only at a thin flame region $Z = Z_f$ to be determined as a function of B. The region $Z < Z_f$, toward the fuel side, is a region of near-equilibrium, $Y_O = 0$, while the chemical reaction is frozen in the region $Z > Z_f$, where the fuel and oxidizer coexist. Thus fuel leaks through the reaction zone toward the oxidizer side. The mass fraction of oxidizer on the near-equilibrium side of the flame is small of order T_o/T_a for $Y_{O\infty}/\nu$ of order 1 but becomes negligible if $Y_{O\infty}/\nu$ <</.

The mass fractions Y_O and Y_F are then given in terms of $X = Z\sqrt{\gamma/2D}$ by the relations

$$Y_O = 0, \; Y_F = 1 - A_1 \, (2 - \operatorname{erfc} X\sqrt{L}\,) \tag{6}$$

for $\quad X < X_f = Z_f \sqrt{\gamma/2D}$, and

$$Y_O = Y_{O\infty} \, [1 - \frac{\operatorname{erfc} X}{\operatorname{erfc} X_f}] \, , \quad Y_F = A_2 \operatorname{erfc} X\sqrt{L} \tag{7}$$

for $\quad X > X_f$

The continuity of the fuel concentration at the flame implies that

$$A_2 \text{ erfc } X \sqrt{L} = 1 - A_1(2 - \text{erfc } X \sqrt{L}) = Y_{Ff} \tag{8}$$

The mass consumption rate of oxidizer per unit flame surface is given by:

$$(\rho D\, Y_z)_{z_f} = \rho D\, Y_{o\infty} \sqrt{\gamma/2D}\, \frac{2}{\sqrt{\pi}}\, \frac{\exp(-X_f^2)}{\text{erfc } X_f} \tag{9}$$

Fuel and oxidizer are consumed at stoichiometric proportions at the flame, so that

$$A_1 - A_2 = \sqrt{L}\,(Y_{o\infty}/\nu)\, \frac{\exp\{X_f^2\,(1 - L)\}}{\text{erfc } X_f} \tag{10}$$

and taking into account Eq (8) we obtain

$$2A_1 = 1 + \sqrt{L}(Y_{o\infty}/\nu)\, \frac{\text{erfc}(X_f \sqrt{L})}{\text{erfc } X_f} \exp\{X_f^2\,(L-1)\} \tag{11}$$

thus determining the concentration distributions in terms of the flame position X_f that will be found below in terms of the Damkohler number from an analysis of the reaction zone structure.

To calculate the temperature distribution we notice that $Y_o/\nu + c_pT/Q$ behaves as an inert species, so that

$$c_p(T-T_o)/Q + Y_o/\nu = (Y_{o\infty}/2\nu)(2-\text{erfc } X) \tag{12}$$

In particular, the flame temperature T_f is given by

$$T_f - T_o = (Q\, Y_{o\infty}/\nu c_p)(1 - \tfrac{1}{2}\text{erfc } X_f) \tag{13}$$

Before giving the results of the analysis of the reaction zone structure, let us indicate that the outer structure of the diffusion flame, in the diffusion-contrelled Burke-Schumann limiting case, $B/\gamma \to \infty$, is given by the above equations if we write $A_2 = 0$, to insure that no fuel leaks toward the oxidizer side of the reaction zone. Then, the flame position is $X_f = X_e$, given by the relation

$$\frac{Y_{o\infty}}{\nu} = \frac{\text{erfc } X_e \exp\{X_e^2(1-L)\}}{\sqrt{L}(2-\text{erfc } X_e \sqrt{L})}\ , \tag{14}$$

and the flame temperature $T_f = T_e$ given by

$$c_p(T_e - T_o)\nu/Q\, Y_{o\infty} = 1 - \tfrac{1}{2}\text{erfc } X_e \tag{15}$$

Only in the case L=1 does the flame temperature in the diffusion controlled regime coincide with the adiabatic flame temperature T_{ea}, given by:

$$c_p(T_{ea} - T_o)\nu/Q\, Y_{o\infty} = (1 + Y_{o\infty}/\nu)^{-1}, \tag{16}$$

If $Y_{o\infty}/\nu$ is small the temperature gradient at the flame toward the fuel side is small compared with the gradient toward the oxidizer side. This simplifies the analysis of the reaction zone structure in the premixed flame regime given by Liñán (1974) which can be used without changes in this case to give the following relation

$$\frac{B}{\gamma}\pi\nu \left[\frac{\nu\, c_p\, T_f^2}{Q\, Y_{o\infty}\, T_a}\right]^2 Y_{F_f}\, [\mathrm{erfc}\, X_f]^2 \exp(2X_f^2 - T_a/T_f) = 1 \tag{17}$$

between the Damkohler number and flame position.

In summary the system of algebraic equations (8), (11), (13) and (17) determines in parammetric form, with the flame position X_f as parameter, the relation between the flame temperature and the Damkohler number B/γ involving the straining rate; this relation if, as it has been assumed, T_a/T_o is large is multi-valued for values of $B/_\gamma > \gamma(B/\ )_E$ an extinction value below which neither the premixed flame regime nor the diffusion controlled regime is possible.

The effect of the Lewis number on the relation between the flame temperature and Damkohler number is shown in the figure for the promixed flame regime. The lower branch of each curve corresponds to unstable solutions of the equations.

The multiplicity of solutions would not exist if time derivative terms representing transient effects were retained in Eqs. (1)-(3). The time derivatives must necessarily be retained if the straining rate γ changes with time; however a coordinate transformation as the one used by Corcos (1980) can be used to transform the problem to that of the unsteady unstrained mixing layer between two half-spaces of fuel and oxidizer, with a slight modification of the reaction rate term. Then the analysis by Liñan and Crespo (1976) can be used to describe the mixing layer structure.

Acknowledgement

This research has been partially supported by the U.S. Army European Office under Grant DA-ERO-79 - G 0007.

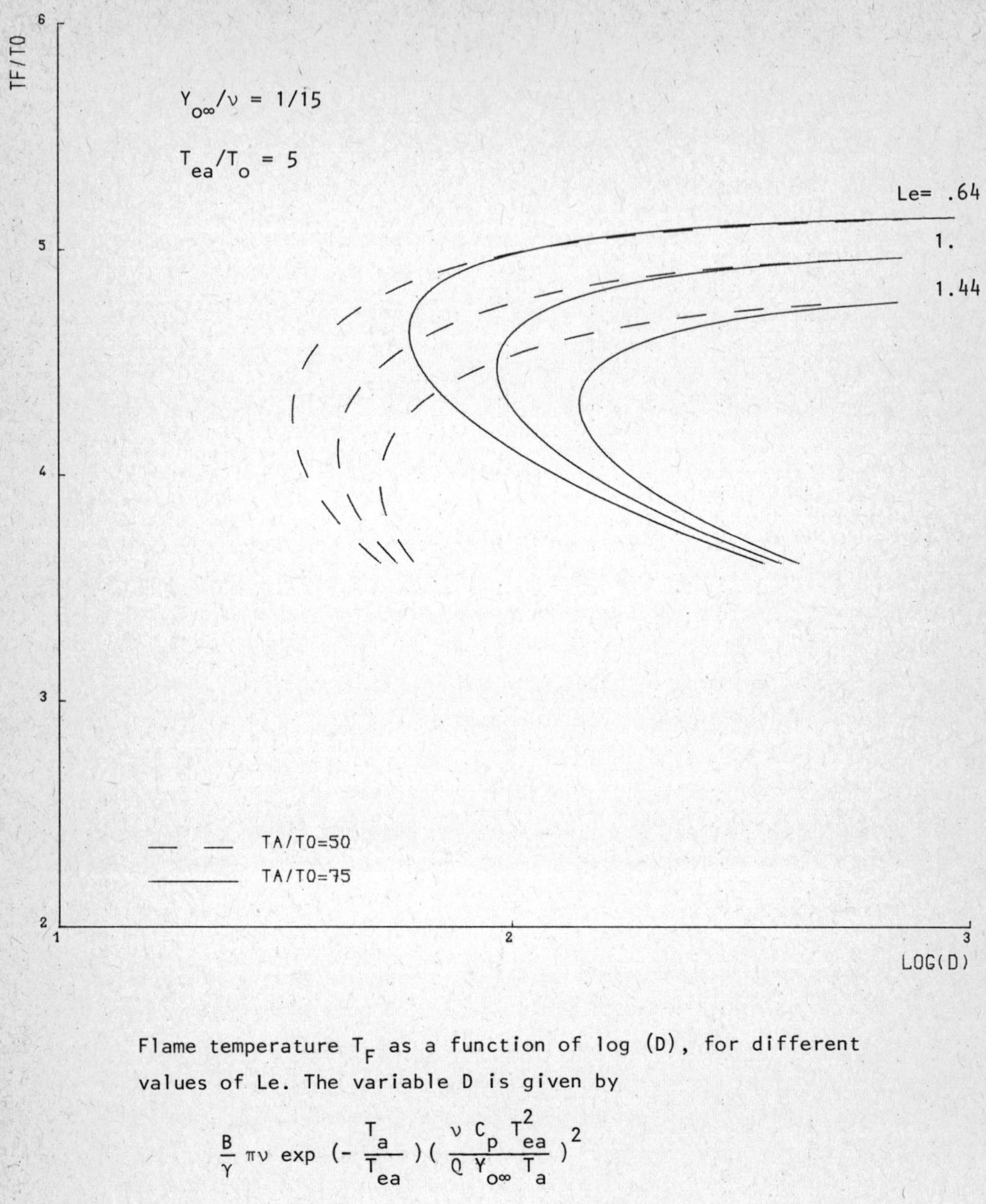

Flame temperature T_F as a function of log (D), for different values of Le. The variable D is given by

$$\frac{B}{\gamma}\,\pi\nu\,\exp\left(-\frac{T_a}{T_{ea}}\right)\left(\frac{\nu\,C_p\,T_{ea}^2}{Q\,Y_{o\infty}\,T_a}\right)^2$$

References

1 Bilger, R.W. (1976), "Turbulent Diffusion Flames". Progress Energy Combustion Sci. 1 pp. 87 - 109.

2 Corcos, G.M. (1980), "The Deterministic Description of the Coherent Structure of Free Shear Layers". In these Proceedings.

3 Liñan, A. (1974), "The Asymptotic Structure of Counter-Flow Diffusion Flames for Large Activation Energies". Acta Astronautica 1, pp. 1007-1039.

4 Liñan, A. and Crespo, A. (1976), "The Asymptotic Analysis of Unsteady Diffusion Flames for Large Activation Energies", Combustion Science and Technology, 14, pp. 95 - 117.

5 Peters, N. (1980), "Local Quenching due to Flame Stretch and Non-Premixed Turbulent Combustion", Western States Section of the Combustion Institute Spring Meeting 21 - 22 April. University of California, Irvine, California.

6 Peters, N. and Williams, F.A. (1980). "Coherent Structures in Turbulent Combustion", In these Proceedings.

7 Williams, F.A. (1975), "Recent Advances in Theoretical Descriptions of Turbulent Diffusion Flames", In: Turbulent Mixing in Non-reactive and Reactive Flow. pp. 189 - 208, Murthy S.N.B. Ed., Plenum Press.

JET NOISE AND THE EFFECTS OF JET FORCING

D. G. Crighton
Department of Applied Mathematical Studies
University of Leeds
Leeds LS2 9JT, England

Abstract

This paper gives a review of the experimental situation regarding the acoustic field of axisymmetric jets subject to controlled excitation. Two very different types of response are found in the literature; in one the response to tonal forcing involves a significant reduction in the broadband noise signal except in frequency bands around the tone and its harmonics and subharmonics, which are strongly amplified, while in the other there is a significant and essentially uniform increase in the entire broadband spectral level at all angles to the jet. In the latter case the forcing tone may or may not protrude above the broadband background in the acoustic far field; if it does, the tone is radiated to the far-field with conservation of acoustic energy at Helmholtz numbers greater than unity, and with a substantial loss of acoustic energy from the incident forcing to vortical instabilities on the jet shear layer at low Helmholtz numbers. It is argued that a turbulent initial shear layer and a Reynolds number $Re_D = U_J D/\nu$ greater than about 10^5 lead to broadband amplification for forcing Strouhal numbers $St_D = fD/U_J$ around 0.5 and to broadband attenuation at much higher Strouhal numbers ($St_D \sim 2$). If the initial shear layer is laminar and $Re_D < 10^5$ the response to tonal forcing at $St_D \sim 0.5$ involves an amplification at this Strouhal number and a suppression of broadband levels, most of the unsteady energy being locked into a periodic oscillation at the forcing frequency. These remarks apply to cold subsonic jets. The experimental evidence on forced jets is also reviewed for hot jets, for hot and cold imperfectly expanded supersonic jets, for cold perfectly expanded supersonic jets and for subsonic hot and cold coaxial jets. Practical implications of all these studies are discussed, and it is emphasised that the broadband amplification associated with a turbulent shear layer and high Re_D is much more likely to be relevant in practice than the highly ordered tonal structure at low Re_D. A brief mention is made of three current theoretical approaches to these phenomena, involving (i) nonlinear stability theory and turbulence closure modelling, (ii) discrete fully nonlinear axisymmetric vortex ring analyses, (iii) acoustic source modelling to simulate nonlinear instability waves and vortex pairing in the presence of turbulent fluctuations.

1. Introduction

It has long been recognised that at low jet exhaust velocities (U_J) internal noise sources within a jet engine must be expected to dominate the acoustically inefficient Lighthill quadrupoles equivalent to the exhaust mixing process. Because of the importance (in the landing approach) of the low velocity condition, a great deal of effort has been devoted to the problem of "internal" or "excess" or "core" noise and many candidate mechanisms proposed, among them combustion noise, the interaction of unsteady flow and combustion inhomogeneities with struts, nozzle guide vanes and the propulsion nozzle, turbine noise, and noise associated with fluctuations of mass flow and thrust. A quantitative connection between these mechanisms and the sound field has been established in only a few cases (high frequency turbine tones, for example) - perhaps not surprisingly in the light of the results of jet forcing studies. Increasingly, however, "excess" noise has been found in surprisingly high velocity conditions, at which internally generated noise with the observed levels in the tailpipe would not be thought capable of dominating the mixing noise. Moreover, the effects of forward flight of the engine have been found to be often difficult to understand, to predict, and to reproduce - even qualitatively - on ground-based external flow rigs such as wind-tunnel simulations.

The "excess" noise position was radically changed by the discovery, by Bechert and Pfizenmaier [4] and Moore [32], of a mechanism altogether more ubiquitous, subtle and insidious than that of internal noise per se. In this mechanism, tailpipe forcing (acoustic or hydrodynamic, tonal or broadband) evokes unstable response of the whole jet column in the form of large scale vortical wavelike motions. Nonlinear response of the instability wave occurs if the forcing level exceeds a very low threshold and causes the broadband levels of jet turbulence and far field radiated noise to rise almost uniformly in frequency and field shape by as much as 8dB. The broadband amplification varies with the forcing frequency, and the forcing level can be small enough to prevent the excitation itself from being discernible in the far field noise spectrum. It is therefore difficult to recognise that a jet is subject to this kind of excitation and harder still to unambiguously diagnose the origin of the forcing and control it in a practical context.

This phenomenon is a manifestation of the presence and dynamic and acoustic significance of so-called "large-scale orderly structure" in the forced jet, and a question which is frequently posed is "how much do these structures contribute to the jet noise field in the absence of forcing?" We regard this question as (a) absolutely irrelevant to the practical (i.e. jet engine) noise problem in the foreseeable future, and (b) perhaps irrelevant in any case, as fluctuations at the nozzle exit are developed as a result of downstream mixing even in the "natural, unforced" jet with a magnitude and coherence which can be estimated to be close to the threshold for broadband amplication. With regard to (a), the position at present is that jet engines have noise fields which are frequently far (up to 10dB in extreme cases) above the scaled-up levels from model jets, the differences being currently ascribed (qualitatively) to a mixture of internal noise mechanisms, broadband amplification and, possibly, imperfectly understood scalings for the mixing noise of hot jets. If we add to this the fact that forward flight often augments the noise levels in the forward arc ahead of the aircraft by several dB in a way which has not yet been satisfactorily simulated on model jets in wind-tunnels, it is clear that the role played by large-scale structures in the noise field of an unforced jet is a matter of no present concern.

This review will, accordingly, consider only the forced jet situation. The principal aim is to classify the numerous and often apparently contradictory experimental results which have been obtained in the past ten years. We start in §2 by describing the two basic instability mechanisms for an axisymmetric jet. Forcing may be applied at the frequency corresponding to either of these instabilities; if the nozzle boundary layer at exit and the initial shear layer are laminar the two instabilities coexist and interact, and the response to harmonic forcing at either characteristic frequency involves the suppression of broadband levels and the locking of most of the turbulent energy into a series of tones at harmonics and subharmonics of the forcing frequency. If the exit boundary layer is turbulent and the jet Reynolds number $Re_D = U_J D/\nu$ is greater than 10^5 (the second of these usually implying the first, but not necessarily vice-versa if the boundary layer is tripped) only the large-scale instability is operative, and forcing at the corresponding frequency leads to large amplification of the instability wave in the jet and also to broadband amplification of the rest of the turbulence spectrum in the jet. The broadband amplification is also found in the far-field acoustic signal, but there is no amplification in the far-field at the forcing tone frequency in the subsonic flow regime, where the instability wave is acoustically inefficient.

Thus we have a criterion which seems (experimentally) to differentiate between the two types of response, though there are several borderline cases which do not altogether conform. We argue that the turbulent initial shear layer case is the one of greatest interest for the jet engine noise problem, and go on to examine the response in more detail for this case, with reference to very recent work on hot jets, supersonic perfectly and imperfectly expanded jets, and coaxial jets. The direct applicability of model results on jet forcing to any real engine has actually yet to be established, but some of the results do indeed explain anomalies observed on static rigs and engines and provide a much more plausible explanation of the mechanisms behind mixing nozzle silencers and tailpipe acoustic liners than has hitherto been given. The results for coaxial jets fail, however, to explain any of the static-to-flight anomalies often observed - and in fact add to the discrepancies, if anything.

Quantitative theoretical understanding of forced jet phenomena is still in its infancy, and we do no more in §4 than catalogue several current, interesting and far from complete approaches. The final section discusses further possible directions for experimental research.

2. Jet Instability and the Two Characteristic Responses to Forcing

A high Reynolds number axisymmetric jet has two significant length scales, the initial momentum thickness θ of the shear layer leaving the nozzle and the nozzle diameter D. In laminar flows at Re_D of order 10^4, θ/D is typically 0.01. Corresponding to these two scales there are two basic instability modes on a laminar jet. The first is a quasi-two-dimensional instability of the initial shear layer, well understood experimentally and theoretically in the linear regime. Maximum spatial amplification rate (for a parallel inviscid hyperbolic tangent shear layer with momentum thickness θ) corresponds to a Strouhal number $St_\theta = f\theta/U_J$ of 0.017 according to the well known theory of Michalke [30], and the corresponding waves are short compared with the overall scale D and may therefore have arbitrary azimuthal variation. These waves can exist only in the first diameter or so downstream from the exit unless they change their frequency and scale to match the local scales of the linearly expanding shear layer, which they can do only in the nonlinear regime, through forming vortex rings which then pair and merge.

A large scale instability - the jet column instability - is possible further downstream (for $D < x < 8D$ say). This involves axisymmetric modes and modes with slow azimuthal variation (exp $i\phi$, exp $2i\phi$ mainly, with ϕ the azimuthal angle about the jet axis) with wavelengths of order D and larger and Strouhal number $St_D = fD/U_J$ for maximum spatial growth rate in the range $0.3 < St_D < 0.6$. This leads to a frequency typically 2-3 octaves lower than that for the shear layer instability. The precise St_D depends very much on the criteria adopted, as the non-parallel nature of the flow is important here and causes significant differences between the growth and growth rates of pressure and velocity [12]. Nonlinearity, and "eddy viscosity" effects of fine-scale turbulence also play a major role in singling out a "preferred mode" in experiments - see [14]. As pointed out by Kibens [23], a third and final instability takes over around and beyond the end of the potential core, and causes the large scale modes with slow azimuthal variation to develop rapid azimuthal change and to break up into fine-scale eddies in the fully developed jet.

If the initial shear layer is turbulent, there is not generally any great change in the mean velocity profile, either immediately downstream of the exit or further down the core. However the initial shear layer instability is no longer operative, as the corresponding waves are well coupled to the shear layer eddies; but the jet column instability persists as for a laminar flow with the mean profile (see [13]) because on this scale the turbulence is fine-grained and behaves as an equivalent "turbular" fluid. It is, of course, possible to use devices in the tailpipe and nozzle to provide a mean profile which is stable to large scale disturbances of a restricted class (such as axisymmetric disturbances); see, for example, Chan and Templin [10] for a specific example, while the action of multielement suppressor nozzles offers other examples, as shown by Moore [34].

In order to study these instabilities, and the underlying order which they contribute to the turbulent flow, numerous investigators have forced the jet with controlled excitation around the preferred instability frequencies. Most commonly, acoustic excitation has been used, with loudspeakers in the plenum chamber, or feeding a slot in the tailpipe or around the nozzle exit. However the acoustic wavelength is long compared with all scales of interest, and the forcing provides an incompressible surging - normally uniform across the nozzle exit but also sometimes with controlled azimuthal variation. This spatial uniformity of excitation is not possessed in practical cases, and this has led some workers (e.g.[40]) to claim that the forced jet studies are irrelevant to jet engines. That may be true of the high forcing levels used by, for example, Crow and Champagne [14], involving velocity fluctuations u_e uniform across the nozzle exit and equal to 2% of the mean exhaust velocity U_J. On the other hand, local turbulence intensities in engine jet pipes have been found to be of order 10%, and if these intensities were allied to a correlation area $10^{-2}D^2$ (as can be estimated for low speed flows, see [32]) they would produce a correlated velocity fluctuation of 0.1% U_J - which is above the level required to produce a significant broadband noise increase (see later). This is supported by the increasing amount of evidence for the presence of broadband amplification in jet rigs and engines, even for hot flows and high velocities.

On the other hand, significant changes are produced in the jet structure itself by far lower forcing levels than 0.1% U_J. These changes involve the <u>linear</u> growth and decay of the instability wave at the forcing frequency, a process which is well understood and on which practically all investigators agree. Forcing at linear amplitudes of the jet column instability has been frequently studied ([9], [14], [28], [32] among others). Pressure and velocity fluctuations at the forcing tone are greatly amplified between the nozzle and x = 3D where, typically, the peak levels are attained ; the pressure may be amplified by as much as 20 dB, the velocity by 30dB [32], the difference indicating that although this is a linear phenomenon, non-parallel flow effects lead to substantial differences between the evolution of different physical quantities. Reasonable calculation can be carried out with simple non-parallel inviscid flow theory to predict growth rates and phase speeds at different locations, for different Strouhal numbers and for different azimuthal variation of each of the various physical quantities, [12], [41]; the decay of the waves is a more difficult matter and may need an eddy viscosity (here appropriate) for effects on the wave of fine-grained turbulence.

Although big changes take place in the large-scale structure in the linear forcing regime, these have no effect on <u>either</u> the broadband turbulence and sound fields <u>or</u> on the radiated sound at the forcing frequency if the instability phase speed is less than the ambient sound speed a_o. Indeed, it has been shown experimentally, [6], [32], that at low Helmholtz numbers fD/a_o there is a substantial (up to 15 dB) conversion of the incident forcing acoustic energy to vortical energy of the instability wave at the nozzle, and this is not re-radiated by the growing and dying instability. This mechanism has now been thoroughly and entirely satisfactorily studied in recent theoretical work, [7], [8], [19], [39], and excellent agreement is obtained between theory and experiment. The agreement tends to confirm the assumption crucial in the theory, that a Kutta condition prevails on the unsteady flow at the nozzle lip, and this provides a local coupling between acoustic forcing and vortical response. There is no coupling elsewhere, the sound radiates from the nozzle, and the vortical waves grow and die downstream as free instability waves. (See also the experiments by Bechert and Pfizenmaier on the Kütta condition, [3].)

The fact that at subsonic phase speeds the jet column instability wave is acoustically weak sheds much doubt on the relevance of large eddy structures to the sound field unless the eddies interact nonlinearly. An experiment was, however, conducted by Crow [15](see also [11]and [18]) on the tonal forcing of a large hot model jet, and a "gain" of 34 dB was found at the peak noise angle and at the tone frequency, together with increases all over the field shape which (in contrast to the assertion of several authors, e.g. [6], [32]) rules out the possibility that the gain is merely associated with mean flow convection and refraction. Crow also gave a simple calculation for the sound field to be expected from a growing and decaying instability; his expression (although highly sensitive to the precise form of the amplitude variation) in fact confirms the view that the instability with subsonic phase velocity radiates only weakly - though it is difficult to reconcile that with Crow's application of his expression to his own experiment, where good agreement was obtained even though in the experiment the phase speed was not clearly in the supersonic range. A convincing explanation of Crow's experiment has yet to be given.

The large scale instability wave may be more efficient acoustically if the broadband fluctuation level is unusually low. This happens in the low Reynolds number (up to 8700), high Mach number experiments conducted by McLaughlin's group [38] , where the low Reynolds number arises from expansion of the jet into a vacuum tank at low pressure. Then the internal and far-field spectra consist of narrowband components around a dominant mode with $St_D \sim 0.16$. Exponential growth of this mode (always with azimuthal variation $\exp i\phi$ or $\sin\phi$) was noted over many diameters, and the presence of this component in the near and far field is evidence of direct radiation from the instability, although only at the highest Mach number ($M_J = 2.5$) was the phase speed greater than a_o, leading in that case to a distinctively directional Mach wave field. The low Re_D situation is, however, not representative of the high Re_D case in one important respect which will be mentioned in a moment.

With regard to linear forcing of the high frequency shear layer instability, there has been no systematic investigation of the differences between linear and nonlinear response such as has been done for the jet column instability, and the acoustic power balance has not been checked either. Thus while it is common in subsonic flows with laminar shear layers and forcing at St_θ to observe substantial acoustic energy at the forcing frequency (cf [22]), it is not known whether this energy is radiated directly by the excitation (with acoustic convection, refraction and diffraction effects) or whether it is augmented by direct radiation from the instability. At supersonic speeds it is clear that the instability does radiate directly (provided phase speed exceeds a_o); this can be seen from the fact that the Mach waves associated with the high frequency shear layer oscillations are confined sharply to a wedge-shaped sector on either side of the jet. It can be shown theoretically that an instability wave generated at the nozzle would excite a field of this kind, and detailed calculations by Tam [43] for helium and nitrogen jets exhausting into air are in good agreement with experiment on the value of the wedge angle (which is determined by the Mach number and the thermodynamic properties of the gas). The growth rates of these shear layer instabilities are much lower at high speed than at low subsonic Mach numbers, and the wave is able to retain its identity without nonlinear break-up for much longer at high speeds.

That ends the discussion of linear instability response to forcing; the response is confined to the excitation frequency and leads to large internal fluctuations but no "acoustic gain" due to the response,except when the phase speed exceeds a_o.

Nonlinear aspects of the response seem to be noticeable at a forcing threshold around 0.01% U_J for the axial exit velocity fluctuation and the optimum value of St_D. The form taken by the response has one of two characteristics: either the broadband levels decrease and the fluctuation energy (which may actually decrease under application of forcing) is bound up into a highly ordered structure consisting of the forcing tone and its harmonics and subharmonics, or the broadband level, hydrodynamic and acoustic, may be amplified, leading to a large increase in the overall levels. With these two possibilities in mind, consider Table I, which attempts to summarise the conclusions of a number of jet forcing investigations. We note the range of values of Re_D for each, whether the initial shear layer was laminar (L) or turbulent (T), and whether the broadband levels increased (+) or decreased (-) under forcing at a frequency characteristic of the initial shear layer (St_θ) or of the jet column (St_D). There is no implication that the values of either of these Strouhal numbers were the same in any two experiments, or that the values were optimised to maximise the response in any way. Also it should be remarked that in most of the experiments at $Re_D > 5 \times 10^5$, say, the nozzle boundary layer was already naturally turbulent, while in others such as [14] at rather lower Re_D a boundary layer trip was used.

Investigator	Re_D	Shear Layer	St	Broadband Response	Remarks
Bechert/Pfizenmaier [4]	6×10^5	T	St_D	+	0.4% - 1% forcing
Chan [9]	2.6×10^5	T	St_D	Not measured	Instability wave only measured
Crow/Champagne [14]	$1.1 \times 10^4 - 1.2 \times 10^5$	T	St_D	−	1% - 2% forcing
Favre-Marinet/ Binder [17]	$1.0 \times 10^4 - 4 \times 10^4$	L	St_D	+(small)	Forcing up to 40%
Jubelin [20]	$6 \times 10^5 - 1.3 \times 10^6$	T	St_D	+	Similar results for hot jets & supersonic jets
			St_θ	−	Larger reduction for hot jets
Kibens [22]	5×10^4	L	St_θ	−	
	1.0×10^5	L	St_θ	−	Reduction at high frequencies only
Kibens [24]	$1.0 \times 10^5 - 8 \times 10^5$	T	No forcing	Not measured	Development of coherent energy at St_D only, not at St_θ
Morrison/ McLaughlin [38]	$3.7 \times 10^3 - 8.7 \times 10^3$	L	St_D	−	Supersonic jets, $M_J = 1.4, 2.1, 2.5$
Moore [32]	$1.3 \times 10^5 - 1.0 \times 10^6$	T	St_D	+	0.2% maximum forcing
			St_θ	−	
Zaman/Hussain [44]	$1.0 \times 10^4 - 4 \times 10^4$	L	St_θ	−	
		T	$St_D - St_\theta$	No change	$1 \le St_D \le 3$

TABLE I: Broadband jet response to single frequency forcing at amplitudes for which the instability wave behaviour is nonlinear

In all the experiments in Table I except those by Morrison and McLaughlin on supersonic jets, the forcing levels were large enough that the instability wave response was nonlinear. With forcing at St_D the nonlinear behaviour takes the form of the development and amplification of several higher harmonics of the basic signal in all cases, together with a subharmonic at ½ the fundamental frequency in some cases. The subharmonic was observed by Crow and Champagne [14] for forcing at St = 0.6, when the resulting subharmonic at St = 0.3 caused a very rapid increase in the jet spreading rate and also, for example, by Jubelin [20] for hot jets (T_J=900°K) forced at St = 0.5 - 0.6 but not on cold jets forced at similar levels at similar Strouhal numbers. The subharmonic is also not detectable in the far-field spectra of Bechert and Pfizenmaier [4] or in the near and far-field spectra of Moore [32], where again the jet is cold and St = 0.48 but where nonlinear instability behaviour and a broadband noise increase were both observed. Thus the circumstances in which the subharmonic arises are not clear when the fundamental forcing is around $St_D \sim 0.5$, and in particular these remarks throw some doubt on the connection sometimes alleged between the "vortex-pairing" mechanism represented by the subharmonic and the broadband turbulence and noise increases found in [4, 20,32]. As for the higher harmonics always observed, these seem to correspond to a weakly nonlinear process, as the spectral levels usually decrease by at least 10 dB from one harmonic to the next. However, the presence of these harmonics, and the deformation of the mean flow by the wave, are sufficient to greatly reduce the "gain" experienced by the fundamental component as compared with the gain in the linear regime. For example, the axial velocity gain in the linear regime can amount to a factor of as much as 10^3 [32], whereas at forcing levels of 4%, as in [14], that factor has dropped to about 5.

When the nonlinear forcing is applied at St_θ and the shear layer is laminar, an instability wave develops rapidly at the fundamental frequency, forms a first subharmonic, then a second and even a third (see[14 p. 555] and [22, 23]). After three such events - which have, from flow visualisation, been positively identified as "vortex pairings" - the frequency has dropped to a value in the jet column instability range of frequencies, the energy has been transferred to the large scale instability mode, and the sequence of pairings stops. The presence of tones, in the near and far-field, at frequencies f, f/2, f/4, f/8 is clearly shown in [23](f being the forcing frequency), along with other tones generated by quadratic nonlinearity such as (f/8 + f/4), 2(f/8 + f/4), [(f/8 + f/4) + f/4], etc. If the initial shear layer is turbulent there is naturally no multi-tone structure of this kind, though forcing still produces significant effects.

A point of interest here is that in Kibens' experiments [22] the tone frequencies were exactly the same for both the jet near-field and the acoustic far-field, with no Doppler frequency shift. This indicates, as stressed by Laufer and Monkewitz [27], that although the instability waves are convected structures (for which a Doppler frequency shift would be expected in the acoustic field) the radiation takes place from fixed locations. These locations must, of course, be the locations (fixed, when forcing is applied) at which the vortex pairings take place, though no source location schemes have yet been used to confirm that. Inasmuch as the preferential fore-and-aft directivity of jet noise is usually ascribed to Doppler frequency shifts on quadrupole sources this absence of shifts is a very significant feature.

Return now to Table I and consider the results for $Re_D > 5 \times 10^5$ ([4], [20], [32]). Here tonal forcing at the jet-column range of instability frequencies ($0.3 < St < 0.7$) leads (in addition to large amplification of the tone component) to a large increase in the broadband turbulence spectrum and in the far-field acoustic spectrum. The increase is almost uniform in frequency over many octaves, and the directivity pattern of the overall sound pressure is very similar to the "unexcited" jet noise spectrum - but increased in level by as much as 8 dB. If the excitation is a plane acoustic wave, then the tone protrudes from the spectrum in the far-field, at least for cold jets but not necessarily for hot jets [20]. However, the far field tone amplitude is linearly related to the amplitude of the incident wave in the jet pipe, although the instability wave responds nonlinearly [32]; and further, the sum of the far-field tone power plus the power reflected back down the jet pipe is very close to the incident acoustic power at Helmholtz numbers above unity (roughly) and less than the incident power by as much as 15 dB at lower Helmholtz numbers - just as in the case of linear instability response (see [6], [32]). This implies that the propagation of the acoustic tone to the far-field is linear (at any rate at forcing levels up to about 1%), but that interaction of the tone with the nozzle launches a nonlinear instability wave whose interaction with the background turbulence in the jet leads to the broadband amplification.

If, in the same high Reynolds number range, the forcing is applied at high Strouhal numbers (say $St_D > 2$) then there is no tonal response at all (as the shear layer is invariably turbulent at these Re_D), and the effect of forcing is to reduce the broadband turbulence and noise spectra. The reduction is small (not more than 1-2 dB for cold jets [20 , 32], a little more for hot jets with exit temperature $T_J = 600^oK$ [20]), and confined to frequencies below the forcing frequency.

Contrast this high Reynolds number situation with the results of experiments at, say, $Re_D < 5 \times 10^4$. In these experiments, whether the forcing is applied at the column frequency or the shear layer frequency, whether the shear layer is laminar or tripped into turbulence, and whether the jet is subsonic or supersonic, the effect of excitation is always qualitatively the same. It is to suppress the broadband spectrum and to lock much of the turbulent energy into a series of spectral spikes at the forcing tone and its harmonics and subharmonics. The most highly ordered structure arises when the shear layer is laminar and high frequency forcing is applied; then the repeated vortex pairings lock the whole first 6 or 8 diameters of the flow into a coherent, periodic flow, as in [22 , 44]. This suppression of broadband levels due to high frequency forcing is not, we believe, to be regarded as similar to that occurring at much higher Re_D where there can be no natural structure (like the closely-packed vortex rings) at high frequencies. A less organised structure is formed when the jet is forced in the column instability mode, but nonetheless, at the lower Re_D the forcing tone phase-locks much of the random fluctuations at lower frequencies into the coherent large scale structure. Such control is clearly not possible at the higher frequencies, and accordingly the broadband suppression observed in [14 , Figure 30] does not extend to those higher frequencies. Further, it must be emphasised that the broadband suppressions are always small compared with the broadband gains at high Re_D; those gains amount to several dB for forcing levels of 0.1% in [32], whereas the suppressions amount usually to less than 3 dB for forcing levels of 2% in [14].

Our claim is thus that the difference between the two types of response - namely broadband suppression and broadband amplification - is associated with Re_D less than or greater than 10^5, roughly speaking. It might be thought that the difference was principally associated with a laminar or turbulent initial shear layer, respectively, but that alone does not seem to be the deciding factor. For example, in the Crow and Champagne experiments the nozzle boundary layer was tripped, but broadband suppression was observed, while Moore, at a much higher Re_D, found very similar broadband amplification for both a long parallel nozzle with thick turbulent boundary layer and for a conical nozzle with a thin boundary layer "of indeterminate type". Not all cases fit into the classification according to Reynolds number, however; Zaman and Hussain [44] found no broadband suppression with a turbulent boundary layer and Reynolds number close to those of Crow and Champagne, while Favre-Marinet and Binder [17] measured a small broadband turbulence increase (though they showed no spectra) even with a laminar shear layer - though in this case enormous forcing levels of up to 40% at the nozzle exit were used. The second entry for Kibens [24] is included in Table I because, although no forcing was used and no turbulence spectra are presented, it includes measurements at much higher Re_D than those in [22] and reports a type of coherent structure different from that in [22]; no subharmonic cascade from high frequencies is possible with high Re_D and a turbulent boundary layer, and coherence is developed only in the jet column instability range. The entry [38] is included in Table I because it tends to show that the different types of broadband response are reflections of different Reynolds numbers rather than Mach numbers; in these experiments, broadband suppression was observed, with very weak forcing from a glow discharge, up to exit Mach numbers of 2.5, but in all cases the value of Re_D did not exceed 8700.

Consider now a typical aeroengine exhaust condition, with, say, D = 1m, U_J = 500m/s and $T_J = 900^oK$. The corresponding Re_D is about 2×10^6, and the initial shear layer is undoubtedly turbulent, as also is the flow in the jet pipe and potential core. It seems certain then that conditions will in practice favour the broadband amplification process, and as it has recently been shown that this process operates in many high Re_D situations (beyond the cold subsonic single stream jets which mostly figure in Table 1) it is appropriate next to outline the properties of it.

3. Broadband amplification at high Reynolds numbers

In the original papers of Bechert and Pfizenmaier [4] and Moore [32], broadband amplification of a plane wave pure tone excitation was demonstrated for cold subsonic jets with $Re_D > 10^5$, providing the forcing Strouhal number lies in the range $0.2 < St < 0.8$, say. Coherent velocity fluctuations of 0.1% U_J are sufficient to provoke nonlinear instability wave response and broadband gain, and the gain is found in the internal turbulence spectra (on the centreline and in the shear layer) and throughout the distant acoustic field. The amplification is almost uniform in frequency and in angle (in the acoustic field), and amplification up to 8 dB has been measured at moderate subsonic Mach numbers. The amplified jet noise, is apart from the tone and its harmonics in the spectrum,"morphologically similar to ordinary jet noise, but the levels are higher" (Deneuville and Jacques, [16]). As the broadband lift is a nonlinear function of the forcing velocity, and as the latter does not change linearly with U_J, the amplified jet noise does not have the U_J^8 intensity scaling for pure jet noise which is realised down to very low velocities on model rigs with very carefully controlled upstream conditions; rather, it has something like a U_J^6 scaling, which is often found in "excess" aeroengine noise (excess noise being simply noise above the pure jet mixing noise level at the given engine exhaust conditions).

Flow visualisation studies by Moore [32] have shown that for cold jets there is a more or less fixed location, around x = 3 - 4D, at which vortex pairing takes place between axisymmetric vortex rings which develop out of the nonlinearly growing instability wave, the position of vortex pairing being localised by the excitation. The presence of the subharmonic cannot be clearly detected in Moore's spectra, nor in those of Jubelin [20] for similar conditions; for hot jets, however, Jubelin found that a spectral peak was present at the subharmonic frequency, and that it was entirely stable and repeatable (the exit temperatures were 600°K and 900°K). Moore [33] has also used an acoustic telescope technique to show that the apparent source position for all frequencies is, in the case of amplified noise, around the vortex pairing location, whereas, when the forcing is linear and the jet noise not amplified, the sources are distributed over many diameters of the exhaust, with high frequencies coming from near the nozzle, low frequencies from sources well down in the fully developed jet. Thus there is considerable evidence for the claim that the broadband turbulence and noise increase is produced by the vortex pairing mechanism first postulated by Laufer [26], and that little sound is radiated by the instability wave except in the pairing phase. Just how a pairing between almost identical vortex rings at Strouhal numbers around 0.5 can produce so uniform an increase in the broadband level is hard to understand (in the unforced jet there are presumably vortex interactions taking place all over the jet on a variety of scales) and it is also not known why the subharmonic is not seen in cold jets but so clearly seen in hot jets.

Moore [32] makes brief reference to excitation of a broadband nature itself, and it has been found that broadband amplification still results in this case (although in a given experimental facility the amplification will usually be less because of weaker response of the loudspeaker system to broadband drive). Deneuville and Jacques [16] quote several examples which show that rather uniform amplification of several dB is found on either side of the frequencies occupied by broadband forcing. They also emphasise that although in the subsonic cold jet experiments the forcing is easily discerned in the far-field, where generally the tone level exceeds the jet noise OASPL, there are many cases in which it is impossible to detect the forcing. As an example they quote results for a jet with $M_J = 1.0$, $T_J = 300^{\circ}K$ in which the only irregularities in the spectrum are to be seen at 120° to the jet exhaust, and there they are sufficiently slight as to deserve no comment - but the jet actually suffers from broadband amplification by a pure tone. Bechert and Pfizenmaier [5] used an array of loudspeakers feeding into the jet pipe to force a cold subsonic jet with a first order helical mode. The frequency was such that the mode was cut-off in the jet pipe and therefore not detectable in the far-field, although it generated a nonlinear instability wave in the jet and broadband noise amplification very much as in the plane-wave case. The upshot of these results is that the presence of irregularities in the far-field spectra is not a satisfactory criterion by which to judge whether jet noise amplification has taken place.

Jubelin [20] has studied forcing effects on hot jets, with temperature up to $900^{\circ}K$ and Mach number 0.5. Broadband amplification results, much as for cold jets, but with the following differences:- (i) the amplification is not as uniform in angle for the hot jet, being small close to the exhaust direction and much larger in the forward arc, (ii) the amplification is less uniform in frequency for the hot jet, being smaller at low frequencies than for the cold jet and generally more concentrated around the forcing frequency, (iii) the hot jet is less sensitive to excitation than the cold in the sense that the slope of the jet noise power vs excitation curve is less for the hot jet, (iv) the threshold for amplification may, for the hot jet, be as much as 25 dB less in power than the jet noise power, whereas for the cold jet the excitation power must, for broadband amplification, be much closer to the jet noise power, (v) as noted before, a subharmonic at $\frac{1}{2}$ the forcing frequency is a pronounced and stable feature of the spectra of excited hot jets, but not of cold jets, (vi) the broadband decrease which accompanies high frequency forcing is more noticeable on hot jets than cold (though in neither case does it exceed 2-3 dB). These results are of particular importance, as model jets at these high temperatures and with tail pipe excitation are the best simulation of a jet engine that can be achieved.

In the same paper, Jubelin also showed the presence of broadband amplification for warm (i.e. exit temperature 300°K) supersonic jets ($M_J = 1.2$). Jets were excited at this Mach number both in underexpanded flow through a convergent nozzle and in perfectly expanded flow through a con-di nozzle. In the first case the spectra at all angles from 60° to the exhaust round into the forward arc are dominated by high levels of broadband shock-associated noise, generated by the interaction of shear layer turbulence with the cellular wave-pattern in the exhaust jet; in the latter case there is some slight evidence of shock-associated noise but the main spectral energy is associated with supersonic mixing noise (though not Mach wave radiation from supersonically convected eddies as the value of M_J is not high enough for this at ambient temperatures). In both cases there is strong broadband amplification under forcing at $St_D = 0.46$, with no evidence of the forcing in the spectra and rather large amplifications at high frequencies and at high angles to the exhaust. That shock-associated noise and mixing noise should both be amplified is not surprising, as both are generated by the broadband turbulence in the jet; no account has yet been taken, however, in prediction schemes, of the fact that each may suffer broadband amplification of 5 dB or more in spectral ranges which at full-scale are extremely important in the determination of Perceived Noise levels (PNdB). A further point which should be observed is that in none of these studies has any "saturation" effect been found for the broadband amplification; the maximum amplification obtained has been limited by the available drive power, and there might well be greater amplification if forcing levels higher than 0.2%-1% could be applied. On the other hand, as remarked by Moore [32], jet engine noise levels correlate reasonably from one engine to another, and far better than they correlate with scaled-up model data, which might be taken to imply that all jet engines are in fact excited to the point of saturation.

The broadband amplification effect is thus ubiquitous in the noise and turbulence of model jets; it occurs when forcing levels exceed a low threshold, over a wide range of Strouhal numbers, over the temperature range $300\text{-}900^{o}$K, and at subsonic or supersonic speeds, provided only that $Re_D > 10^5$. Under these conditions it is eliminated, or greatly alleviated, by two mechanisms - coaxial flow, and use of a convoluted suppressor nozzle. Moore and Brierley [34] and Jubelin [20] have studied the effects of a secondary coannular flow on the noise of an excited primary jet, with possible application not only to turbofan engines but also to the noise of a single stream turbojet in flight. There are several mean flow parameters here additional to the Mach number M_J and temperature T_J of the primary, namely the temperature, area and velocity ratios T_S/T_J, A_S/A_J and U_S/U_J of secondary to primary flow, and only a very limited number of configurations has been examined so far; the experiments of [20] at SNECMA have $T_S = 330^{o}$K, $T_J = 600$ or 800°K, $A_S/A_J = 3.5$, $M_J = 0.5$ or 0.75 and $U_S/U_J = 0.4$ or 0.6, while the Rolls-Royce experiments [34] have

U_S/U_J = 0 to 1.2, A_S/A_J = 0.9 or 6.0, M_J = 0.7, and cold flow in both streams. In each case, the behaviour of the primary alone under excitation was as expected, but in [20]no significant amplification of the dual-flow jet was found in any case, regardless of whether the primary stream forcing was tuned to the preferred frequency of the primary or secondary stream. In the Rolls Royce experiments the amplification decreased to a minimum (effectively zero) at U_S/U_J = 0.5, and for A_S/A_J = 6 remains zero at high velocity ratios, while for the small area ratio (0.9) the amplification increased again beyond U_S/U_J = 0.5, probably because of the small secondary area and the fact that the frequency then tends to be tuned to the preferred frequency of the secondary flow. High speed films showed ([34] and Strange [42]) that at U_S/U_J = 0.5 there was no evidence of any vortex pairing, and in fact the vortices in the primary shear layer were convected downstream with little development for more than 10D without loss of identity; Strange also modified the analysis of [12] to account for an infinite co-flowing external stream and showed that at the condition U_S/U_J = 0.5 the amplification rates and total growth of linear stability theory are indeed very small for St_D = 0.5. Thus the fairly definite conclusion must be that external flow inhibits instability wave growth and prevents the vortex pairing which seems to be the source of broadband amplification. The position may be different if the forcing is injected into the secondary flow, rather than the primary, but that situation is irrelevant if the secondary flow is the infinite ambient medium in the engine-in-flight case, and less interesting anyhow in the turbofan engine case where one would expect secondary excitation to be small. Note that these results apparently do not help to explain observed static-to-flight differences; in fact, the static-to-flight difference will be greater for an excited jet than for an unexcited, whereas the need is for mechanisms which reduce this difference (and even change its sign in the forward arc).

Several authors have considered tailoring the mean flow profile in a single stream jet so as to inhibit instability wave growth. Chan and Templin [10] used a nonuniform gauze to produce a bell-shaped profile at exit (this according to well-known results in stability theory being stable to axisymmetric disturbances), and showed that all such disturbances decayed rapidly downstream; nonaxisymmetric modes can, however, still grow on the modified profile, but were not considered. Bechert and Pfizenmaier [4] showed that a simple annular ring inside the nozzle reduced the broadband amplification, though how that was achieved is not clear as the ring does not change plane wave forcing nor does it significantly change the mean profile. Moore and Brierley [34] examined the forcing of a cold jet from a 6-chuted and a 7-lobed suppressor nozzle and compared the noise fields with those from their parallel circular nozzle (at the same mass flow); only very small levels of amplification were found for the suppressor nozzles as against the 8 dB or so for the plain nozzle. The explanation given is that there is no thin shear layer on which coherent structures can grow, but probably most important is the fact that the suppressor nozzles destroy all azimuthal coherence. In any event, the lack of amplification may well be the prime mechanism behind the silencing action of these nozzles (for which many different claims have been made) and may explain the great variability of effectiveness of them in different situations. And, finally, Kibens [21] has shown that an "inverted profile" jet (not exhausting from a coaxial arrangement but formed by flow through a sharp edged orifice), leading in the first few diameters to high velocities in the shear layer and lower velocities on the centreline, cannot sustain any large-scale azimuthally coherent structures of the kind formed at the same Re_D on a jet from a conical nozzle. In all cases, however, the use of a tailpipe device to modify the mean jet profile advantageously with regard to large-scale structures incurs a thrust loss which, when restored, may lead to little noise benefit at constant thrust.

There is increasing indirect evidence for amplification phenomena in practically important cases, whether model jets, or full-scale combustion and turbine rigs, or complete engines. Many such pièces of evidence are quoted in a delightful paper by Deneuville and Jacques [16], who point out that the model and rig tests fully represent practical conditions. Among those discussed in [16] are the following:-

(1) Frequent occurrence of noise fields with characteristics of pure jet noise, but with higher levels, while a model rig was being set up and adjusted.

(2) A 2 dB broadband noise increase from a hot jet rig when the mass flows in the fuel sprays were imperfectly balanced, leading to flame instability and low frequency parasitic tones. In the example chosen the jet was supercritical, with mixing and shock-associated noise in different angular and spectral ranges - and both fields were amplified.

(3) In this hot jet rig, a large nozzle was fitted and the combustion turned off. For a given nozzle pressure ratio the large nozzle leads to a much greater flow velocity through the combustion chamber, and to higher, though similar, noise spectra at all angles than in the case of a small nozzle scaled to the same area. In this case there is no evidence of any tones in the spectra, but it is thought that there is an amplification effect in response to large fluctuations developed in the combustion chamber.

(4) A conical nozzle flow emerging into a cylindrical ejector shroud (with only weak secondary flow) can, in some flow regimes, excite open-tube axial resonances in the shroud. These react upon the jet, causing broadband amplification distinguishable (from spectral and angular properties) from broadband noise generated by hydrodynamic interaction of the jet with the ejector.

(5) The Concorde Olympus 593 engine has a field shape generally 3-5 dB above jet noise prediction, especially in the foward arc, and though the spectra contain a lot of mid and high frequency internal noise components it is possible that much of the low frequency excess noise is amplified jet noise.

(6) A turbojet engine was run at the <u>same running point</u> with a small and a large nozzle, with the aid of upstream throttling. With the large nozzle, jet noise levels

are low and what is measured is essentially internal noise (with a flat spectrum, rather different from that of jet noise). Then the internal noise levels are corrected for nozzle area and compared with the measured noise for the small nozzle - for which it is found that the spectra are like those of jet noise, but much higher in level than according to prediction. The internal noise levels are rather accurately known from the large nozzle measurements, and cannot explain the discrepancy which, it is concluded, is caused by jet noise amplification by the internal noise.

(7) A perforated honeycomb tailpipe liner, tuned to attenuate high frequencies, was fitted to a Larzac engine on which jet noise levels were very low and internal noise levels high; the liner attenuated only the expected high frequencies. When the same liner was fitted to the Olympus 593 at a condition where the predicted jet noise and measured noise were comparable, the liner gave significant attenuations (2-6 dB) over a very wide frequency range, and in particular a uniform 2 dB attenuation below 400 Hz. Acoustic absorption by the liner is not possible at such low frequencies and the effect is attributed to a reduction of upstream forcing levels, and hence a reduction of jet noise amplification, by the liner. An "egg-box" flow straightener fitted to the Olympus 593 also leads to significant broadband reductions, particularly in the forward arc, which might be expected if the action of the device is to reduce tailpipe fluctuations in a hot jet.

Finally, in this catalogue of practical examples, Moore [33] shows noise levels measured on a large (D = 0.3 m) hot jet rig with internal combustion noise; the spectra are similar to predicted pure mixing noise spectra, with some irregularities at frequencies in the 200-400 Hz range characteristic of combustion noise, and up to 10 dB higher. Moore estimated the internal excitation levels by extrapolating the energy in the combustion noise spectral bump back to the nozzle exit, and then used his (cold jet) results to predict that such excitation would produce a broadband increase of 8.5 dB - rather close to the observed difference.

Enough has now been said to convince the reader of the frequent occurrence of broadband amplification in many practical cases with naturally occurring types and levels of excitation. Most cases in which its occurrence is suspected do, however, involve static single stream jets, and the work of [20, 34] on dual flow jets makes it appear unlikely that broadband amplification can play such a significant role in engines in flight, or on turbofan engines.

4. Theoretical Studies

The primary issue is the development of large scale structures and their interaction with the mean flow and background turbulence; subsequently we need to calculate the sound field, in itself a difficult and sensitive task. Two main lines of attack have been used so far with the hydrodynamic problem, reflecting emphasis on two aspects of the experiments which are hard to reconcile with each other.

On the one hand, spectra of forced jets show a sequence of tones corresponding to the fundamental and several of its integral harmonics, in addition to the broad-band amplification (see, e.g., Figure 2 of [4]). Only rarely is any subharmonic evident (see [20]) while the harmonics decrease in level by 10 dB or so from one to another. This suggests a model based on weak nonlinearity of the coherent forcing signal, involving coupling between the fundamental, a second harmonic, the mean flow and the fine-scale background turbulent field. Approaches of this kind have been developed by Chan [9] , Morris [36] and Liu [29] among others. They all essentially reduce the problem to one of integration of three ordinary differential equations for the mean flow momentum thickness $\theta(x)$, the fundamental amplitude $A(x)$ and the radially integrated turbulence energy as functions of the axial coordinate, and they differ primarily in the closure assumptions used to calculate the "wave-induced-stresses" and to relate the fine-grained Reynolds stresses to the mean flow. To reduce the partial differential equations to ordinary ones the equations are integrated radially and a "shape assumption" employed, to the effect that the transverse structure of the fundamental wave is given locally by linear parallel stability theory and only the axial development involves nonlinearities and nonlocal coupling with the mean flow. Strange [42] has recently shown that the "shape assumption" is well satisfied even at forcing levels well into the nonlinear instability response range and even at spatial locations around $x = 3D$ where vortex pairing occurs; moreover, he dealt with a mode with azimuthal mode number 1 for which all three components of velocity and the pressure were found to be locally given by the shape assumption when the level of one quantity was specified at one radial location.

These calculations, particularly those of Mankbadi and Liu [29] , yield rather good predictions of the instability wave response, but no calculations yet published predict a sufficiently large increase in turbulence levels to agree simultaneously with experimental results for the turbulence and the instability wave. Also, of course, they make no prediction about the spectral uniformity of the turbulence increase at any single point, but only about the overall level of the radially integrated fine-scale turbulence at each axial location. However they do have the advantage of allowing a discussion of jet response to higher order (e.g. $n = 1, 2$) azimuthal forcing, as in [5, 42], which is simply not possible in the discrete vortex modellings. Indeed, a principal conclusion of [29] is that because of the azimuthal variations with $n = 1$ the wave-induced Reynolds stresses are increased over their values for $n = 0$, and the energy exchange between large-scale and fine-scale components is correspondingly greater.

Calculations of the sound field have yet to be satisfactorily carried out, largely because a knowledge of turbulence intensities alone is far from sufficient information. The sound from the fundamental large-scale component could, perhaps, be calculated, but so far this has only been done in the linear instability regime, by Morris and Tam [37] . Calculations are presented for a cold jet with $M_J = 1.5$, so that phase speeds are subsonic and the radiated field is caused by the growth and decay of the instability wave on a diverging jet flow. Since the theory is essentially inviscid the decay involves a transfer of energy back to the mean flow - a feature also found in [29]. For an appropriate choice of initial instability wave amplitude the SPL contours for a single wave with $St_D = 0.25$ agree well with measured contours for the SPL in a 1/3 octave band around this Strouhal number, suggesting that at high speeds the linearly growing and decaying instability wave may contribute significantly to the overall SPL, though the OASPL does not seem to be dominated by the instability wave unless the phase velocity, rather than the mean velocity, is supersonic.

The second approach focusses on the concentrated vortex picture revealed by flow visualisation, and represents the jet entirely by sequences of axisymmetric ring vortices, released at a fixed origin and allowed to convect and interact under their mutual coupling fields. Models of this kind have been studied by many authors; see [1, 2, 25, 35] and references therein. These studies are restricted at the outset to incompressible flow which is axisymmetric at every instant, but the vortical interactions are fully nonlinear.

The computed flows resemble those seen in flow visualisation studies of jets in the way in which vortices group together to form structures of larger scale, and the random ingredient implicit in the numerics soon scrambles the flow into a fully turbulent state (though one which is still completely axisymmetric). Acton [1, 2] calculated the distributions of mean velocity and Reynolds stress in an azimuthal plane and found good agreement with measured distributions. Calculation of the sound field is not satisfactory. Morfey [35] carried out the calculation for a flow field modulated at a Strouhal number St_D of 0.45 and 0.66 and with an "exit-plane" velocity fluctuation equal to 5% of the mean velocity and found only small effects of this rather large forcing when comparing the computed noise spectra with those for an unmodulated jet. Moreover each of the computed spectra lay some 30 dB above the spectra measured for model jets at the same mean velocity. This large discrepancy may, as suggested by Morfey, be due to azimuthal coherence of the fluctuations which is not possessed in reality, even though the axisymmetric model is capable of predicting one point mean values rather well.

The final approach may be termed "source-modelling". It does not attempt to calculate the acoustic source terms from hydrodynamic theory, but rather, postulates forms for those terms which incorporate effects seen in experiments and tries to correlate the resulting calculated acoustic field with measurements. An example is Crow's [15] controversial calculation of the sound field of a source with an amplifying and then decaying pure tone wave structure. This model was extended by Ffowcs Williams and Kempton [18] , who included a small random component in the phase speed of the wave and showed that the ensuing radiation would be narrowband or broadband depending on whether the accumulated phase distortion at the maximum amplitude position was a negligible or substantial fraction of 2π. For this wave model they also showed that for supersonic phase speeds the field radiated at the Mach angle would be narrowband regardless of the degree of phase distortion, whereas away from the Mach angle the field could again be broadband. Ffowcs Williams and Kempton also simulated vortex pairing by taking a source strength equal to $K \cos k_o(x - U_o t)$ for $x < g(t)$ and equal to $2K \cos \frac{k_0}{2}(x - U_o t)$ for $x > g(t)$, K being a constant and $g(t)$ a normally distributed random function with variance equal to the square of an eddy scale. It was shown that details of the pairing should not be important provided the pairing takes place in a distance small compared with k_o^{-1}, which seems reasonable from observations. In this model the radiated sound is in the form of discrete tones, at the radian frequencies $k_o U_o$ and $\frac{1}{2}k_o U_o$, if the variation in pairing positions is small in the sense $g_o k_o << 1$, where g_o is the r.m.s. value of $g(t)$. If $g_o k_o >> 1$ then the radiation is broadband.

This type of approach clearly reveals a number of features in qualitative agreement with observation, though it is difficult to recommend it as a reliable way of getting quantitative information as the values of numerical coefficients for the acoustic field are very sensitive to details of wave shapes, etc. It should, however, be very useful in suggesting quantities that should be measured in experiments or calculated from hydrodynamic theory.

5. Conclusions

More measurements are still needed in many areas of jet forcing. A study in rigs which have shown the high Re_D amplification effect is needed at lower Re_D to confirm the change in response character around $Re_D = 10^5$, and to examine the role played by turbulent or nonturbulent nozzle boundary layers. The effect of external flow on forced jets needs to be examined with a very large area external flow to properly simulate the effect of flight on a single stream jet (though there is still some concern as to whether a large wind-tunnel really can simulate flight effects on jet noise properly). A conclusive proof of the existence of forcing and broadband response on a jet engine needs to be obtained - perhaps by the use of large and small nozzles to get a good estimate of internal noise levels, as in the SNECMA work [16]. External aerodynamic excitations are possible in the practical case, due to landing gear and flap wakes, a thick boundary layer on the outside of the engine nacelle, etc. It is not known whether these effects can provoke broadband amplification, but this again is hinted at by Deneuville and Jacques in a reference to external excitation of a jet mounted on the Aérotrain ground vehicle.

Great progress has, however, been made in the past few years, and a new insight gained into many previously disparate and puzzling results in aeroacoustics and even, perhaps, into the basic mechanisms of noise generation in subsonic flows.

References

[1] E. Acton (1976) "The modelling of large eddies in a two-dimensional shear layer" J. Fluid Mech. 76, 561-592

[2] E. Acton (1976) "A modelling of large jet eddies". Ph D dissertation, Cambridge University.

[3] D. Bechert & E. Pfizenmaier (1975) "Optical compensation measurements on the unsteady exit condition at a nozzle discharge edge" J. Fluid Mech. 71, 123-144

[4] D. Bechert & E. Pfizenmaier (1975) "On the amplification of broadband jet noise by a pure tone excitation". J. Sound Vib. 43, 581-587

[5] D. Bechert & E. Pfizenmaier (1976) "On the amplification of broadband jet noise by a pure tone excitation" AIAA Paper 76-489

[6] D. Bechert, U. Michel & E. Pfizenmaier (1977) "Experiments on the transmission of sound through jets" AIAA Paper 77-1278

[7] D. Bechert (1980) "Sound absorption caused by vorticity shedding, demonstrated with a jet flow". J. Sound Vib. 70, 389-405

[8] A. Cargill (1979) "Low frequency sound radiation due to the interaction of unsteady flows with a jet pipe". "Mechanics of Sound Generation in Flows", Ed. E.-A. Müller, Springer, Berlin, pp 19-25

[9] Y.Y. Chan (1974) "Spatial waves in turbulent jets . Parts I and II" Phys. Fluids 17, 46-53 and 1667-1670

[10] Y.Y. Chan & J.T. Templin (1974) "Suppression of spatial waves by distortion of jet velocity profile". Phys. Fluids 17, 2124-2125

[11] D.G. Crighton (1975) "Basic principles of aerodynamic noise generation." Prog. Aerosp. Sci. 16, 31-96

[12] D.G. Crighton & M. Gaster (1976) "Stability of slowly diverging jet flow" J. Fluid Mech. 77, 397-413

[13] D.G. Crighton (1979) "Why do the acoustics and the dynamics of a hypothetical mean flow bear on the issue of sound generation by turbulence?". "Mechanics of Sound Generation in Flows", Ed. E.-A. Müller, Springer, Berlin, pp 1-11

[14] S.C. Crow & F.H. Champagne (1971) "Orderly structure in jet turbulence" J. Fluid Mech. 48, 547-591

[15] S.C. Crow (1972) "Acoustic gain of a turbulent jet". Paper IE.6 Am. Phys. Soc. Fluid Dynamics Meeting, Boulder, Colorado, Nov. 1972

[16] P. Deneuville & J. Jacques (1977) "Jet noise amplification - a practically important problem" AIAA Paper 77-1368

[17] M. Favre-Marinet & G. Binder (1979) "Structure des jets pulsants". J. de Mécanique, 18, 356-394

[18] J.E. Ffowcs Williams & A.J. Kempton (1978) "The noise from the large-scale structure of a jet" J. Fluid Mech. 84, 673-694

[19] M.S. Howe (1979) "Attenuation of sound in a low Mach number nozzle flow" J. Fluid Mech. 91, 209-229

[20] B. Jubelin (1980) "New experimental studies on jet noise amplification" AIAA Paper 80-0961

[21] V. Kibens (1976) "Large-scale turbulent structure of an inverted profile jet" Bull. Am. Phys. Soc. Nov. 1976 Paper DB4

[22] V. Kibens (1979) "Discrete noise spectrum generated by an acoustically excited jet" AIAA Paper 79-0592

[23] V. Kibens (1979) "On the role of vortex pairing in jet noise generation" "Mechanics of Sound Generation in Flows", ed. E.-A. Müller, Springer, Berlin, pp 174-180

[24] V. Kibens (1980) "Interaction of jet flow field instabilities with flow system resonances" AIAA Paper 80-0963

[25] C. Kitaplioglu & V. Kibens (1980) "Discrete vortex modelling of an axi-symmetric jet flowfield" AIAA Paper 80-1003

[26] J. Laufer (1973) Discussion contribution at AGARD Noise Mechanisms Conference. Brussels, 1973: see AGARD CP-131

[27] J. Laufer & P. Monkewitz (1980) "On turbulent jet flows: a new perspective" AIAA Paper 80-0962

[28] B.H.K. Lee (1976) "Some measurements of spatial instability waves in a round jet" AIAA Journal, 14, 348-351

[29] R. Mankbadi & J.T.C. Liu (1980) "A study of the interactions between large-scale coherent structures and fine-grained turbulence in a round jet" Phil. Trans. R. Soc. (in press)

[30] A. Michalke (1965) "On spatially growing disturbances in an inviscid shear layer". J. Fluid Mech. 23, 521-544

[31] A. Michalke (1971) "Instabilität eines runden kompressiblen Freistrahls unter Berücksichtigung der Freistrahlgrenzschichtdicke" Z. Flugwiss. 19, 319-328

[32] C.J. Moore (1977) "The role of shear-layer instability waves in jet exhaust noise". J. Fluid Mech. 80, 321-367

[33] C.J. Moore (1977) "The effect of shear layer instability on jet exhaust noise" Structure and Mechanisms of Turbulence Vol. II pp. 254-264, Springer Lecture Notes in Physics, 76, Ed. H. Fiedler

[34] C.J. Moore & D.H. Brierley (1979) "Shear layer instability noise produced by various jet nozzle configurations" "Mechanics of Sound Generation in Flows" ed. E.-A. Müller, Springer, Berlin, pp 48-54

[35] C.L. Morfey (1979) "Sound radiation from a simulated jet flow", "Mechanics of Sound Generation in Flows", ed. E.-A. Müller, Springer, Berlin, pp 12-18

[36] P.J. Morris (1980) "A model for broadband jet noise amplification" AIAA Paper 80-1004

[37] P.J. Morris & C.K.W. Tam (1979) "On the radiation of sound by the instability waves of a compressible axisymmetric jet". "Mechanics of Sound Generation in Flows" ed. E.-A. Müller, Springer, Berlin, pp 55-61

[38] G.L. Morrison & D.K. McLaughlin (1979) "Noise generation by instabilities in low Reynolds number supersonic jets". J. Sound Vib. 65, 177-191

[39] R.M. Munt (1977) "The interaction of sound with a subsonic jet issuing from a semi-infinite cylindrical pipe" J. Fluid Mech. 83, 595-604

[40] R.A. Pinker & W.D. Bryce (1976) "The radiation of plane-wave duct noise from a jet exhaust, statically and in flight" AIAA Paper 76-581

[41] P. Plaschko (1979) "Helical instabilities of slowly divergent jets" J. Fluid Mech. 92, 209-216

[42] P.J.R. Strange (1980) Ph D Thesis, University of Leeds

[43] C.K.W. Tam (1971) "Directional acoustic radiation from a supersonic jet generated by shear-layer instability" J. Fluid Mech. 46, 757-768

[44] K.B.M.Q. Zaman & A.K.M.F. Hussain (1980) "Turbulence suppression in free shear flows by controlled excitation" J. Fluid Mech. (in press)

CHEMICALLY REACTING MIXING LAYER

G.L. Brown

California Institute of Technology

Pasadena

U.S.A.

Note: Not available at the time of publication.

COHERENT STRUCTURES IN TURBULENT COMBUSTION

by

N. Peters
Institut für Allgemeine Mechanik
RWTH Aachen, West-Germany

F.A. Williams
Department of Applied Mechanics and
Engineering Sciences, University of California
San Diego, La Jolla, 92037, Ca., U.S.A.

Abstract

Existing data on the interaction of coherent structures with combustion are reviewed. The combustion characteristics are quantified using the concept of stretched laminar flamelets in a turbulent flame. The difference between premixed and diffusion flamelets is established using results that have been obtained by activation-energy asymptotics.
The effect that coherent structures may have on flame stability is discussed in terms of the quenching conditions of laminar flamelets and their statistics. Numerical methods that incorporate the essential features of coherent structures such that they could predict the effect of combustion on the structures are presented. Combustion-driven instability phenomena are described. Experimental data indicate that the probability density functions of conserved scalars are strongly changed by combustion. It is concluded the coherent structures may be more important in combustion systems than in other practical flow systems.

1. Introduction

Ever since the existence of large coherent structures in turbulent shear flows became evident to a large community, the effect that they may possibly have on combustion processes has been subject to speculation. It was realized that those large eddies that are formed in a mixing layer in a quasi-orderly fashion would change the picture of the turbulent mixing process dramatically. Concentration measurements with a small, fast-response probe by Brown and Roshko /1/ and Brown and Rebollo /2/ had shown that irrotational fluid from both sides was carried all the way across the mixing layer were it was digested by small-scale turbulence. This "new look" (Roshko /3/) at the mixing process was thought to be able to resolve an old question in turbulent combustion about the role that the so-called "unmixedness" had on modeling turbulent flames. It stimulated the formulation of new models, the "coherent flame model" by Marble and Broadwell /4/ as well as, less explicitly, the ESCIMO (Engulfment, Stretching, Coherence, Interdiffusion and Moving-Observer) model by Spalding /5/. Both models accentuate a genuine property of turbulent flows that is evidently enhanced by large coherent structures, the stretching of local flame elements due to the straining motion of the flow. They also regard the flame elements as quasilaminar structures, subject to the strain rate - in Marble and Broadwell's analysis as diffusion-flame structures, in Spalding's model as premixed flames.

In this paper we shall consider two aspects of the interaction of large coherent structures with combustion, namely: the effect that coherent structures have on the combustion and the effect that combustion has on coherent structures. The first aspect is, of course, associated with the stretch of local flame elements up to the point where they are extinguished. To quantify this effect we use the asymptotic method of large activation energy that has been developed in the recent years by a number of authors (Bush and Fendell /6/, /7/, Liñán /8/, Sivashinsky /9/, /10/, Clarke /11/, Joulin and Clavin /12/, Clavin and Williams /13/, Williams /14/, Law /15/, Peters /16/, Mitani /17/, Buckmaster /18/, Kapila and Ludford /19/, just to name a few of the papers). The asymptotic method leads to the definition of an asymptotically thin flamelet and allows determination of the inner structure of the flamelets and the way they respond to "flame stretch". The chemical reaction is confined to the thin inner layer (in the sense of matched asymptotic expansions), and this inner layer is embedded between outer layers that provide the reactants and receive the combustion products by convection and diffusion.

We must distinguish two different structures of the flamelets, depending on nature of the mixing process: There are premixed flamelets and diffusion flamelets. This classification is common for laminar flames and turns out to be very useful in determining their structure. We shall see the different responses that premixed and diffusion flamelets have with respect to flame stretch.

While the laminar flamelet model and the tool of activation-energy asymptotics will give us access to the influence of the fluid dynamics on the combustion, there is no equivalent analytical tool that helps us to understand the influence that combustion has on the large coherent structures. However, there is a numerical method based on vortex dynamics, developed by Chorin /20/. Ashurst /21/, /22/ has shown that the method simulates the large-scale motion in mixing layers and may thus contribute to their understanding. The model recently has been extended to non-constant-density flows by Ghoniem, Chorin and Oppenheim /23/ with the aim of modeling combustion in coherent structures. This method and its potentials together with recent calculations of probability density functions (p.d.f.'s) by Thies and Peters /24/ will also be discussed.

The existing experimental data will be reviewed separately for premixed and diffusion flames. Motion pictures show large coherent structures in premixed flames (Ganji and Sawyer /25/) and in diffusion flames (Altgeld /26/). A computer movie by Thies demonstrates numerical simulation of mixing in a shear layer.

2. Premixed Flames

2.1 Experimental Results

A recent and well-documented study of coherent structures in a premixed combustor has been made by Ganji and Sawyer /25/. They observed large structures dominating the mixing layer that develops behind a step under nonreacting and reacting conditions (fig.1). They concluded that "the dynamics of vortex pairing" ... "is definitely one of the mechanisms for introduction of fresh reactants into the shear layer of the reacting flow." As in previous investigations they found that the upstream boundary layer affects the formation and persistence of vortices. Tripping the boundary layer by a wire ahead of the step could trigger turbulent transition in the boundary layer such that the large coherent structures were not observed. In comparing reacting and nonreacting flows, they found that "the reacting eddies have a lower growth rate, are more closely distributed in space and have a slightly smaller rate of coalescence than nonreacting eddies". However, the differences are much less important than one might expect. The spreading rate of the mixing layer is about 30% less in the reacting case (cf. also Pitz and Daily /27/). This is the same reduction that Brown and Roshko /1/ obtained in their He-Ne mixing layer; the density ratio between He and Ne of 1:7 is close to that induced by combustion. The eddy spacing, which is a measure of the coalescence rate, is about 30% less in the reacting flow than in the nonreacting flow.

The influences that large coherent structures have on combustion are particularly evident through the stability investigations of Ganji and Sawyer /25/. If the velocity is increased or the equivalence ratio is reduced (the mixture was always fuel lean) the reaction ceased first in the downstream part of the combustor. Then blow-out gradually propagates upstream until the reaction in the recirculation zone behind the step is finally extinguished. The extinction time of the whole flame was about an order of magnitude larger than the main flow passage time through the reactor. Tripping of the upstream boundary layer, which tends to destroy the large coherent structures, shifts the blowout limit to richer mixtures (equivalence ratios as much as 8% higher) or to velocities lower by about a factor of two. There are also data on flashback, but these will not be considered in the present context. Since no comparable experiments have been performed in other configurations, the degree of generality of these results is unclear.

2.2 Theoretical Concepts

In a premixed flame combustion takes place at the interface between the unburnt and the burnt mixture. The interface, called a laminar premixed flamelet, changes its position due to the motion imposed by the flow-field and due to the flame propagation itself, as it moves into the unburnt mixture. The fluid motion may not only convect the interface but also distort it, stretch it and thereby increase its area. Thus the supply of reactants to the interface and the removal of heat increase. It is this rate of increase in area that causes what it has been introduced by Karlovitz /28/ as flame stretch. If there is a definite influence of large coherent structures on combustion in a premixed turbulent flame - and Ganji and Sawyer's experiments show that there is at least under certain conditions - then its cause is most likely related to the difference in the way that the flame is distorted and stretched. We therefore discuss these effects in the light of recent results obtained with activation-energy asymptotics.

Sivashinsky /10/ has analysed a distorted flame front for one-step Arrhenius reaction kinetics under steady-state conditions. His model involves temperature-dependent density and conductivity and a constant Lewis number Le (ratio of diffusivities of heat and matter) of the deficient reactant. The flamelet consists of an infinitely thin inner reactive-diffusive layer (reaction zone) separating a convective-diffusive layer (preheat zone) on the unburnt side and a chemical equilibrium layer on the burnt side. A cut normal to the flame front is schematicly shown in fig.2 for a Lewis number larger than, equal to and smaller than unity. If the diffusivity of heat is larger than the diffusivity of the deficient reactant, the temperature in the reaction zone drops below the adiabatic value. Since the reaction rate and the flame propagation velocity depend exponentially on the flame temperature, they are considerably decreased. By integrating over the thin reaction zone, Sivashinsky obtained a relation between the shape of the interface, the velocity through it and its temperature:

$$(1) \qquad \vec{v}\cdot\vec{n} = \exp\left(\frac{1}{2}\,\frac{(T - T_b)}{RT_b^2}\,E\right)$$

Here v denotes the velocity vector on the burnt side of the interface, and n the unit vector normal to the interface. The velocity and the space coordinates are normalized by the flame velocity and the flame thickness of a plane flame without stretch. On the r.h.s. T is the actual temperature at the flame front, T_b is the adiabatic flame temperature, E the activation energy and R the universal gas constant. The flame velocity itself is given by

$$(2) \qquad (\vec{v}\cdot\vec{n})\ \ln(\vec{v}\cdot\vec{n}) = \frac{1}{2}\,\vec{n}\ I\ \mathrm{curl}\ \frac{\vec{v}\times\vec{n}}{\vec{v}\cdot\vec{n}}$$

where in the case of constant conductivity the parameter I is given by

$$(3) \qquad I = \frac{E(T_b - T_u)}{RT_b}\ \frac{(Le - 1)}{Le}$$

If this relation is analysed for a plane stretched flamelet, lying in the y-z plane characterized by a single stretch parameter γ, one obtains the form

$$(4) \qquad I\gamma = -\,u^2 \ln u^2 , \qquad \gamma = \frac{\partial v}{\partial y} + \frac{\partial w}{\partial z}$$

Here u is the normal velocity in x-direction and v and w the velocities in y and z direction, respectively. This relation is plotted in fig.3. It is seen that for $Le > 1$, $I > 0$ there is a maximum flame stretch for which a solution exists. This value corresponds to extinction. A similar analysis has been made recently for the combined influence of curvature effects and heat loss by Mitani /17/. His result, derived under the assumption of constant density, reduces for a single-reactant flame to

$$(5) \qquad I\ \frac{u}{R_c} + Q = -\,u^2 \ln u^2$$

Here $1/R_c$ is the flamelet curvature (positive when convex towards the fresh mixture) and Q is a volumetric heat-loss parameter. The comparison with eq.(4) shows that flame extinction occurs for a sufficiently large curvature as well as for a sufficiently large heat loss. Generally speaking, since flame stretch and curvature enhance the diffusion of reactants into the thin reaction zone and heat loss from the thin reaction zone into the preheat zone, they correspondingly enhance the reduction of the flame temperature in the reaction zone for a Lewis number larger than unity. Thus they have the same effect as heat loss and lead finally to extinction of the premixed flamelet.

2.3 Interpretation of the Experimental Results by Ganji and Sawyer

Since the experiments were carried out with lean propane-air mixtures, the Lewis number is larger than unity. Also, since the walls of the combustion chamber were non-adiabatic, the burnt gas was partially cooled to temperatures much lower than the adiabatic temperature. Both effects tend to quench the premixed flamelets locally since flame compression is less probable than flame stretch in a highly strained flow field. Curvature effects, although positive and negative, will add to the spreading of the distribution of the temperature, conditioned at the interface, in the turbulent flame. Thus the temperature of a larger number of flamelets drops below the extinction temperature. At conditions where blow off occurs downstream, flame stretch, curvature effects and heat loss (which is more important in the downstream part of the combustion chamber than immediately behind the step) quench the flamelets locally at a larger rate than burning flamelets are convected downstream. Therefore blow off starts downstream and moves progressively upstream until the whole flame is extinguished.

The observation that the tripping of the upstream boundary layer destabilizes the flame may be explained by the higher turbulence level and consequently higher local flame stretch and curvature effects. For a better understanding of this effect data about the distributions of stretch, curvature, eddy size and heat loss, conditioned at the interface, would be needed. Alternatively, the temperature distribution at the interface would be of great interest. How such data could be used will be discussed for the case diffusion flames, where probability density functions are available.

3. Non-Premixed Combustion

3.1 Experimental Results

There are many visual observations of large turbulent structures in jet diffusion flames, but very few quantitative results. Yule and Chigier /29/ reported coherent structures in jet flames impinging on a flat plate. Photographs were taken through the plate showing characteristic structures that probably originate in the upstream jet. The transition region and the turbulent flow region of non-reacting jets have been investigated by Yule /30/. He finds that coherent structures, which develop in the transition region by Kelvin-Helmholtz instabilities as vortex rings, are distorted and produce large eddies in the downstream turbulent region. Chigier and Yule /31/ have analysed the transition region of unconfined jet diffusion flames. They find the length of this region, which is about 3 diameters for nonreacting jets, to extend up to 20 diameters for reacting jets. They attribute this behaviour to the large increase of the kinematic viscosity as a result of heat release. In a recent paper /32/ Yule et al. compare the transition region of non-burning and burning jets for different Reynolds numbers and equivalence ratios. In flames they find at least two combustion-driven instabilities, an inner high-frequency instability and an outer low-frequency phenomenon. They detect a number of distinct peaks in the frequency spectra that are independent of the Reynolds number.

In turbulent flames the low frequency phenomenon develops as a quasi-laminar structure in the upper part of the flame ($x > 20d$). This is clearly seen in a picture taken by Donnerhack (fig.4) for a propane flame at Re = 12200, d = 5.5 mm. Regular tongue-like structures emerge from the bulk of the turbulent flame brush. The frequency range of these structures is 10 - 20 Hz. Grant and Jones /33/ point out that frequencies of the order of 10 Hz were reported in the early work of Chamberlin and Rose /34/ and that this frequency is independent of the dimensions of the burner, the fuel flow rate and the fuel. They show that existing theories based on a Tollmien-Schlichting instability mechanism would predict frequencies that depend on the flow rate. Ballantyne and Bray /35/ used Schlieren deflection techniques to derive the frequency spectra of diffusion flames. They found a fundamental frequency at 13 Hz for different Reynolds numbers between 1380 and 7320 in an 8 mm diameter methane/air diffusion flame. The mechanism of these combustion driven instabilities is not well understood at the present time. Since they occur in vertical as well as in horizontal jet flames, bouyancy is not an essential mechanism. However, it is interesting to note that in horizontal flames they develop mainly in the lower part of the jet, less than in the upper (fig.5). This seems to indicate that the radial velocity component is important for the instability mechanism. It points towards a Landau-type flame-surface instability /36/ where the velocity component normal to the flame is accelerated by density changes. The constant-frequency phenomena are genuine properties of flames; they do not occur in nonreacting jets. They may be regarded as "combustion-driven coherent structures".

Large turbulent coherent structures of the type observed in two-dimensional shear layers were reported by Altgeld /26/ in a hydrogen-air flame (Re = 2200, d = 6 mm) using high-speed Schlieren photography and laser doppler anemonetry. He also observed a laminarisation effect due to higher viscosity in the flame (as compared to a hydrogen jet at the same initial conditions) in the first 20 diameters of the jet. Figures 6 and 7 are Schlieren photographs for the hydrogen flame and the hydrogen jet, respectively. In the flame at $x \sim 20d$ a sudden burst-like transition into a turbulent large structure with nearly circumferential symmetry occurs. These structures seem to be very regular. The corresponding nonreacting hydrogen jet at the same Reynolds number is fully turbulent exhibiting no visible coherent structures. A quantitative support of this observation is given by the measurement of the turbulent kinetic energy on the axis (fig.8). The maximum occurs at about twice the distance from the nozzle for the flame as compared to the inert H_2 - jet. It may be concluded that combustion in nonpremixed flames seems to conserve coherent structures in the flow by delaying transition to turbulence.

3.2 Theoretical Concepts

A very convenient description for non-premixed combustion is known as the conserved-scalar approach /37/. It requires the assumption of equal diffusion coefficients for all n species. Since elements are conserved during the chemical reaction, the differential equations for the element mass fraction Z_i contain no source term; in cartesian tensor notation

$$(6) \qquad L(Z_i) = \rho \frac{\partial Z_i}{\partial t} + \rho v_k \frac{\partial Z_i}{\partial x_k} - \frac{\partial}{\partial x_k} (\rho D \frac{\partial Z_i}{\partial x_k}) = 0$$

Similarly, if radiative heat transfer, external work and viscous dissipation are neglected and the diffusivities of heat and matter are equal, i.e. Le = 1, the enthalpy equation becomes

(7) $$L(h) = 0$$

If there are but two uniform reactant feeds and the zero-flux condition is assumed at all other boundaries, the element mass fraction and the enthalpy can be normalized to yield the mixture fraction

(8) $$\xi = \frac{Z_i - Z_{i2}}{Z_{i1} - Z_{i2}} = \frac{h - h_2}{h_1 - h_2}$$

The subscripts 1 and 2 refer to the composition in the two feeds. For the following analysis we need to assume that one feed is fuel rich (index 1) and the other is fuel lean, such that the stoichiometric value of the mixture fraction lies between 0 and 1. We picture the turbulent flame, schematically shown in fig.9, as an ensemble of stretched laminar diffusion flamelets. The concept of diffusion flamelets in a turbulent flow has been introduced in /38/. Here we will follow an extension of this idea that has been recently developed /39/. A diffusion flamelet is defined as a structure in which a thin reaction zone is embedded between two convective-diffusive zones at the surface of stoichiometric mixture. Let us assume that the mixture fraction ξ is given in the flow field as a function of space and time by solution of eq.(6). Then the surface of stoichiometric mixture can be determined from

(9) $$\xi(x_k, t) = \xi_s$$

This is shown schematically for a lifted jet diffusion flame in fig.9. The structure of the diffusion flamelet may be described in terms of its temperature profile:

For the case of a one-step undirectional reaction

(10) $$\nu_F' F + \nu_{O_2}' O_2 \longrightarrow \text{products}$$

with the reaction rate $\dot{w}$ (moles per volume and time) the temperature equation is written as

(11) $$L(T) = \frac{(-\Delta h)}{c_p} \dot{w}$$

Here $(-\Delta h)$ is the (positive) heat of reaction and c_p the heat capacity, both assumed constant for simplicity. To determine the temperature profile across the flamelet, let us introduce a coordinate system such that one coordinate, x_1, is replaced by the mixture fraction ξ, while the others, x_2, x_3 and t, remain unchanged. Applying formally the transformation rules, one obtains with $L(\xi) = 0$ /39/

$$\rho \left(\frac{\partial T}{\partial t} + v_2 \frac{\partial T}{\partial x_2} + v_3 \frac{\partial T}{\partial x_3}\right) - \rho D \left[\left(\frac{\partial \xi}{\partial x_k}\right)^2 \frac{\partial^2 T}{\partial \xi^2} + 2 \frac{\partial \xi}{\partial x_2} \frac{\partial^2 T}{\partial \xi \partial x_2} + 2 \frac{\partial \xi}{\partial x_3} \frac{\partial^2 T}{\partial \xi \partial x_3} + \frac{\partial^2 T}{\partial x_2^2} + \frac{\partial^2 T}{\partial x_3^2} \right] = \frac{(-\Delta h)}{c_p} \dot{w} \tag{12}$$

If the reaction zone is asymptotically thin (a limit of either large activation energy or large Damköhler number) all terms involving derivatives with respect to t, x_2 and x_3 are of higher order in the reaction zone and can be neglected to lowest order. One obtains

$$- \frac{\chi_{st}}{2} \frac{\partial^2 T}{\partial \xi^2} = \frac{(-\Delta h)}{c_p} \left(\frac{\dot{w}}{\rho}\right)_{st} \tag{13}$$

where

$$\chi_{st} = 2D \left(\frac{\partial \xi}{\partial x_k}\right)^2_{st} \tag{14}$$

is the instantaneous scalar dissipation rate at the flamelet position, conditioned at stoichiometry. The solution of this equation yields the temperature profile in the reaction zone. The temperature profile is schematicly plotted in fig. 10 as a function of the normal coordinate x and the conserved scalar ξ.

We may look at χ_{st} as the inverse of a characteristic diffusion time. If χ_{st} is large, heat will be conducted to both sides of the flamelets at a rate that is not balanced by the heat production due to chemical reaction. Thus the maximum temperature will decrease until the flamelet is quenched at a value of $\chi_{st} = \chi_q$. If the activation energy is large, a plot of the maximum temperature with respect to χ_{st}^{-1} will show the well-known S-shaped curve of fig.11. Burning of the flamelet corresponds to the upper branch of the curve. If χ_{st} is increased, the curve is traversed the left until χ_q is reached, beyond which value only the lower, nonreacting branch exists. Thus at

$$\chi_{st} = \chi_q \tag{15}$$

quenching of the diffusion flamelets occurs. The value of χ_q can be determined for known kinetics by an asymptotic analysis in the limit of large activation energies. This has been done in /39/ using the results of Liñán's diffusion-flame regime /8/. If the kinetics are not known, χ_q must be determined from extinction conditions of laminar diffusion flames. The turbulent mean reaction rate can be calculated by integrating eq.(13). To do this we need to know the probability distribution of the instantaneous dissipation rate under stoichiometric conditions. By methods of Papoulis /40/, for example, it may be shown that

$$\tilde{P}(\chi, \xi \mid \xi = \xi_{st}) = \tilde{P}_\xi(\xi_{st})\, \tilde{P}_\chi(\chi \mid \xi = \xi_{st}) \tag{16}$$

Here the tilde denotes density-weighted Favre averaging /37/. Integration to χ_q yields /39/ the mean turbulent reaction rate

$$\begin{aligned} \dot{w} &= \frac{\bar{\rho}}{2} \int_0^{\chi_q} \chi_{st}\, \tilde{P}_\xi(\xi_{st})\, \tilde{P}_\chi(\chi \mid \xi = \xi_{st}) \frac{\partial^2 T}{\partial \xi^2}\, d\chi_{st}\, d\xi \\ &= \frac{\bar{\rho}}{2} \left(\frac{Y_{F1}}{\nu'_F M_F} + \frac{Y_{O2}}{\nu'_O M_O} \right) \tilde{P}(\xi_{st})\, \tilde{\chi}_q \end{aligned} \tag{17}$$

where Y_{F1} and Y_{O2} are the fuel and oxygen concentrations in the feeds and the M_i are the molecular weights. We also need to know

$$\tilde{\chi}_q = \int_0^{\chi_q} \chi_{st}\, \tilde{P}_\chi(\chi \mid \xi = \xi_{st})\, d\chi_{st} \tag{18}$$

The result is a generalisation of Bilger's /37/ expression obtained under the fast-chemistry assumption, where $\tilde{\chi}_q$ is replaced by the mean value $\tilde{\chi}_{st}$.

3.3 Interpretation of Experimental Results

From the preceding discussion it is seen that the influence that large coherent structures have on combustion can be quantified by determining the influence that they have on the joint p.d.f. of ξ and χ, conditioned at stoichiometry. The only measured p.d.f.'s in an inert mixing layer with coherent structures are those by Konrad /41/, reported in /3/. These p.d.f.'s have strong intermittency spikes at both edges. Also it is seen that on one side of the mixing layer a large amount of material from the other side is present, leading to a strong bimodal character of the p.d.f. This is due to the large-scale motion. It is interesting to compare these shapes with the p.d.f.'s of Batt's /42/ mixing layer, which do not have any visible large-scale structures. His temperature p.d.f.'s are much more Gaussian for the middle 50% of the shear layer, and in the other parts the intermittency spikes are narrower.

It would be important to know how combustion changes the p.d.f. of the conserved scalar. Kennedy and Kent /43/ reported both conventional and density-weighted p.d.f.'s for hydrogen flames at Re = 6400 and Re = 3960. Close to Gaussian on the axis, these p.d.f.'s show a bimodal character at the edges of the jet. This structure does not appear in the recently reported p.d.f. data of Lockwood and Moneib /44/ for a heated round jet, which show only single peaks.

There is further evidence that the p.d.f. of the conserved scalar changes due to combustion. Bilger /37/ finds a correlation between higher moments and intermittency for a number of chemically inert free-shear flows (mixing layers, jets, wakes). If Kennedy and Kent's density-weighted data for skewness, kurtosis and intermittency are inserted into his plot, fig.12, a definite deviation from the correlation for inert flows is observed. There is a question as to whether a strong bimodal character of the p.d.f. is itself an indicator of large coherent structures. In this case Kennedy and Kent's data show definite, combustion-dependent structures that are not easy to detect by visual observation. This supports the hypothesis that due to the increase of viscosity the transition characteristics are stabilized and persist as large coherent structures in the flame.

The theoretical considerations show that the entire p.d.f. of the conserved scalar is not needed; its value at stoichiometric conditions is sufficient. For a beta-function p.d.f. Janicka and Peters /45/ have calculated this value, which also determines the thermal NO-production in turbulent flames. Figure 13 shows for a hydrogen flame a plot of $\tilde{P}(\xi_{st})$ as a function of the mean of $\tilde{\xi}$ with the variance $\widetilde{\xi''^2}$ as parameter. It is seen that the maximum shifts to the right as the turbulence intensity increases. This corresponds to the "rich shift" observed by Kent and Bilger /46/ for the NO production in diffusion flames. The important question is: How does $\tilde{P}(\xi_{st})$ behave for bimodal p.d.f.'s? This question has been raised recently in a paper on NO production in turbulent diffusion flames /47/. There it was found that the total NO production depended on a turbulent length scale as well as the residence time. The hypothesis was made that large coherent structures are responsible for this dependence.

To determine the effect of local quenching on the reaction rate, eq.(18), the conditioned p.d.f. of χ is needed. Since χ_{st} is increased by strain, considerable changes are to be expected for flows exhibiting large coherent structures. No such p.d.f.'s have been reported. From Kolmogoroff's third hypothesis /48/, the general shape of $P(\chi)$ may be assumed to be a lognormal distribution. This would have to be verified for flows with large coherent structures. Lognormality tests for other flows have been carried out by Gibson and Masiello /49/. If the lognormal shape is sufficiently accurate, the Favre mean and the variance of χ_{st} are sufficient to determine $\tilde{P}(\chi_{st})$. Such data would be useful for determining lift-off and blow-off conditions in turbulent non-premixed flames /39/ and the effect that coherent structures have on turbulent flame stability.

4. Numerical Methods Predicting Large Coherent Structures

Numerical methods based on semi-empirial turbulence models describe turbulent transport by some sort of a gradient flux approximation. Such a formulation involves the turbulent integral length scale as a local quantity and ignores the effect of much larger structures that may exist in a shear layer. Parker and Sirignano /50/ have recently compared a mixing-length model and two two-equation turbulence models (Launder and Saffman) concerning their capability to calculate Brown and Roshko's data for the spreading rate and the mean velocity and density profiles. They also compared density-weighted and conventional averaging. The results show that the transport of matter across the mixing layer is considerably underpredicted by all methods and formulations, while the prediction of the mean velocity profile is not as bad. This supports the point of view that the effect of large coherent structures on scalar mixing cannot be predicted by methods that use localized turbulent transport models such as the scalar flux approximation.

A numerical method that explicitly avoids turbulence modeling is the random vortex method by Chorin /20/. The inviscid flow is broken into discrete vortices (no finite-difference mesh is required) while the viscous effects are incorporated via a random walk for each vortex. The method is formulated for two-dimensional problems, but an extension to three-dimensional calculations is possible. Each vortex induces a rotating velocity field which is derived from the solution of a single vortex in an infinite field. The contributions of all discrete vortices are superimposed to calculate the local velocity components at each single time step. A random motion is added such that the total displacement $\vec{l}$ at each time step is

$$\vec{l} = \vec{u}\,\Delta t + \vec{l}_{random} \tag{19}$$

The direction and step length of the random motion is calculated by analogy between diffusion and random walk using pseudo-random numbers.

Ashurst /21/ has applied the method to the non-reacting free shear layers of Brown and Roshko /1/ and Winnant and Browand /51/. The interaction of the discrete vortices leads to a rearrangement into large vortex clusters subject to large-scale vortex pairing. The mixing layer grows due to the roll-up and pairing process. The computer plots visually appear quite similar to those of the experiments. The numerical simulation of aperiodic flow entrainment and vortex coalescence is remarkable. Following this work, Thies /24/ has added a large number of closely spaced marker particles of different material on both sides of the flow in order to simulate the mixing process. A shear layer with a velocity ratio 1:2 (upper to lower side), constant density and a Reynolds number 1000 was taken as an example. In fig.14 the marker particles from both sides are shown, deliberately taken apart to facilitate the visual observation. The engulfment of material from both sides and its transport through the mixing layer is clearly recognized. If both parts of the picture are fitted together again, fig.15, the distribution of marker particles is not uniform, and holes are apparent at the highly convoluted interface of the two materials. This is due to the high strain rates that stretch those regions and compress others nearby. In highly compressed regions marker particles are plotted upon each other such that they do not show up in the figure.

A simple model was devised to account for the transport on the molecular level /24/. Essentially, at each time step mass exchange between two particles with different concentrations is calculated using the error-function solution for the interdiffusion at an interface. Between time steps, rapid internal homogenization is assumed such that each interdiffusion starts with a step function as initial condition. The model is used to calculate the p.d.f. of the conserved scalar at different positions in the mixing layer. In fig.16 the p.d.f.'s at $x = 35$, $y = 2, 1, 0, -1$ are plotted. They show a similar behaviour as Konrad's p.d.f. measurements, particularly the strong bimodal character. For quantitative comparisons a refinement of the molecular transport model would be necessary.

A calculation of the flow field in the combustion chamber of Ganji and Sawyer /25/ has been performed by Ashurst /22/. His computer-generated movies show a close visual similarity to the high-speed Schlieren cinematography of the experiment. Compared to the free shear layer calculations, the presence of walls introduces important three-dimensional effects, which are beyond the scope of the two-dimensional method. In an independent study Ghoniem, Chorin and Oppenheim /23/ have added expansion and flame propagation effects to the random vortex flow-field calculation of the same combustor. They use a flame advection and propagation algorithm developed by Chorin /52/. This algorithm approximates the advection of a thin flame front by the fluid motion and the propagation in the direction of its own normal using a constant flame velocity. It does not take into account the accelaration, deceleration or extinction due to stretch, curvature or heat loss, discussed in section 2. The expansion due to heat release is simulated by introducing discrete source blobs in analogy to the discrete vortex blobs. The results show a qualitative agreement with the experimentally observed behaviour of the flow in the combustion chamber.

It appears that the random-vortex method provides an idealized turbulence structure that simulates the essential features of large coherent structures and their effects on turbulent mixing. In this sense it is a promising method that could be employed to provide some insight into the general shape of the p.d.f.'s that were identified in the preceding sections.

5. Discussion

It has sometimes been argued that large coherent structures are only of academic interest, since they occur only in flows at moderate Reynolds number which have extremely well-controlled inlet conditions. In practical high Reynolds-number systems with high turbulence levels the onset of three-dimensionality would rapidly distroy these structures. One may object that, if this were correct, then flows in combustion chambers would be those for which large coherent structures are the most important. Review of the existing data shows that combustion stabilizes existing coherent structures by an increase of molecular viscosity. This is particularly evident for jet diffusion flames. On the other hand coherent structures tend to stabilize the combustion in a premixed combustor.

The influence that large structures have on flame stability makes their study in combustion systems very important. It is known from high-speed gas-turbine combustion chambers as well as from low-speed utility boilers that the stability characteristics even of equally designed burners may be extremely different. Awareness of coherent structures and their dependence on inlet flow conditions and turbulence levels may provide an essential insight into the problem. The influence of the type of flow field on flame stability becomes understandable from the way that laminar flamelets are quenched locally. It was seen that Lewis number effects and heat loss may quench a premixed flamelet. Diffusion flamelets are quenched at large instantaneous local dissipation rates. The two mechanisms are different and lead to different descriptions of turbulent flames. To predict the stability conditions of the entire flame, statistical information about the conditions that quench the individual flamelets is necessary. For premixed flames probability density distributions of stretch, curvature and heat loss would be needed, while for diffusion flames data about the joint p.d.f. of the conserved scalar and the instantaneous scalar dissipation rate are required. As long as experimental data are not available, calculations using the random vortex method could provide some knowledge about the general shapes of these p.d.f.'s.

6. Conclusions

Existing data show that coherent structures are potentially more important in combustion systems than in other technical flow systems. This is due to the strong influence that they have on the turbulent mixing of reactants and to the stabilization of existing structures by combustion.

Existing semiempirial turbulence models seem not to be able to incorporate the important features. More physical understanding and more data on p.d.f.'s under various flow conditions would be needed to design future predictive methods.

Acknowledgements

This work has been supported partly by the Deutsche Forschungsgemeinschaft under Grant Number Pe 241/3 and partly by the United States Air Force Office of Scientific Research under Grant Number AFOSR 77-3362.

7. References

1 Brown, G.L., Roshko, A., J. Fluid Mech. 64, 775-816 (1974)

2 Brown, G.L., Rebollo, M.R., AIAA J. 10, 649-652 (1972)

3 Roshko, A., AIAA J. 14, 1349-1357 (1976)

4 Marble, F.E., Broadwell, J.E., The Coherent Flame Model for Turbulent Chemical Reactions, Project SQUID Technical Report, TRW-9-PU (1977)

5 Spalding, D.B., 17th Symposium (Int.) on Combustion, p.431-440, The Combustion Institute, Pittsburgh 1979

6 Bush, W.B., Fendell, F.E., Combust. Sci. Techn. 1, 421-428 (1970)

7 Fendell, F.E., J. Fluid Mech. 56, 81-95 (1972)

8 Liñán, A., Acta Astronautica 1, 1007-1039 (1974)

9 Sivashinsky, G.I., Combust. Sci. Techn. 15, 137-145 (1977)

10 Sivashinsky, G.I., Acta Astronautica 3, 889-918 (1976)

11 Clarke, J.F., J. Fluid Mech. 94, 195-208 (1979)

12 Joulin, G., Clavin, P., Combust. Flame 35, 139-153 (1979)

13 Clavin, P., Williams, F.A., J. Fluid Mech. 90, 589-604 (1979)

14 Williams, F.A., A Review of Some Theoretical Considerations of Turbulent Flame Structure, AGARD CP 164, III-1-25 (1975)

15 Law, C.K., Law, H.K., J. Fluid Mech. 92, 97-108 (1979)

16 Peters, N., Int. J. Heat Mass Transf. 22, 691-703 (1979)

17 Mitani, T., Asymptotic Theory for Extinction of Curved Flames with Heat Loss, Combustion, Science and Technology, 1980 to appear

18 Buckmaster, J., Combustion Flame 26, 151-162 (1976)

19 Kapila, A.K., Ludford, G.S.S., Combustion Flame 29, 167-176 (1977)

20 Chorin, A.J., J. Fluid Mech. 57, 785-796 (1973)

21 Ashurst, W.T., Proc. of the 1st Symposium on Turbulent Shear Flows, p.402-413, Springer-Verlag, Berlin 1979

22 Ashurst, W.T., Vortex Simulation of a Model Turbulent Combustor, Proc. of the 7th (Int.) Coll. on Gasdynamics of Explosions and Reactive Systems, Göttingen 1979

23 Ghoniem, A.F., Chorin, A.J., Oppenheim, A.K., Numerical Modeling of Turbulent Combustion in Premixed Gases (1980) unpublished

24 Thies, H.-J., Peters, N., Die Berechnung turbulenter Diffusionsflammen mit einem netzfreien numerischen Verfahren, DFG-Schlußbericht, Pe 241/1 (1980), unpublished

25 Ganji, A.R., Sawyer, R.E., An Experimental Study of the Flow Field and Poolutant Formation in a Two-Dimensional Premixed Turbulent Flame, AIAA Paper No. 79-0017 (1979), also NASA Contractor Report 3230 (1980)

26 Altgeld, H., Laser-Doppler Messungen in einer turbulenten Wasserstoff-Luft-Diffusionsflamme, Dissertation am Institut für Technische Thermodynamik, Aachen 1979

27 Pitz, R.W., Daily, J.W., Second Symposium on Turbulent Shear Flows, Imperial College, London 1979

28 Karlovitz, A.G., Denninston, D.W., Knapschaefer, D.H., Wells, F.E., 4th Symposium on Combustion, p.613-620, Williams and Wilkins, Baltimore 1963

29 Yule, A.J., Chigier, N.A., Thompson, D., Coherent Structures in Combustion, Symposium on Turbulent Shear Flows, Penn State University, 1977

30 Yule, A.J., J. Fluid Mech. 89, 413-432 (1978)

31 Chigier, N.A., Yule, A.J., The Physical Structure of Turbulent Flames, AIAA Paper 79-o217, 17th Aerospace Science Meeting, New Orleans 1979

32 Yule, A.J., Chigier, N.A., Ralph, S., Boulderstone, R., Ventura, J., Combustion-Transition Interaction in a Jet Flame, AIAA Paper 80-0077, 18th Aerospace Science Meeting, Pasadena, Ca. 1980

33 Grant, A.J., Jones, J.M., Combustion and Flame 25, 153-160 (1975)

34 Chamberlin, D.S., Rose, A., 1st Symposium on Combustion, Ind. Eng. Chem. 20, 1013-1016 (1928)

35 Ballantyne, A., Bray, K.N.C., 16th Symposium (Int.) on Combustion, p.777-787, The Combustion Institute, Pittsburgh 1977

36 Markstein, G.H., Non-Steady Flame Propagation, Perganon Press, Oxford 1964

37 Bilger, R.W., Prog. Energy Combust. Sci. 1, 87-109 (1976)

38 Williams, F.A., Recent Advances in Theoretical Descriptions of Turbulent Diffusion Flames. In: Turbulent mixing in non-reactive and reactive flow, p.189-208, Murthy, S.N.B. ed., Plenum Press (1975)

39 Peters, N., Local Quenching due to Flame Stretch and Non-Premixed Turbulent Combustion, Paper WSS 80-4, Western States Section of the Combustion Institute, Irvine, Ca. 1980

40 Papoulis, A., Probability, Random Variables and Stochastic Processes, p.176, Mc Graw-Hill, New York 1965

41 Konrad, J.H., An Experimental Investigation of Mixing in Two-Dimensional Turbulent Shear Flows with Application to Diffusion Limited Chemical Reactions, Ph. D. Thesis, California Institute of Technology, Pasadena, Ca. 1976, also Project SQUID Technical Report CIT-8-PU

42 Batt, R.G., J. Fluid Mech. 82, 53-95 (1977)

43 Kennedy, M., Kent, J.H., 17th Symposium (Int.) on Combustion, p.279-287, The Combustion Institute, Pittsburgh 1979

44 Lockwood, F.C., Moneib, H.A., Combust. Sci. Techn. 22, 63-81 (1980)

45 Janicka, J., Peters, N., Combust. Sci. Techn. 22, 93-96 (1980)

46 Kent, J.H., Bilger, R.W., 14th Symposium (Int.) on Combustion, p.615-625, The Combustion Institute 1973

47 Peters, N., Donnerhack, S., Structure and Similarity of Nitric Oxide Production in Turbulent Diffusion Flames, 18th Symposium (Int.) on Combustion, to appear

48 Kolmogoroff, A.N., J. Fluid Mech. 13, 82-85 (1962)

49 Gibson, C.H., Masiello, P.J., Statistical Models and Turbulence, Lecture Notes in Physics 12, 427, Springer Verlag 1972

50 Parker, S.F., Sirignano, N.A., Comparisons amoungst various theories for turbulent, reacting planar mixing layers, Proc. of the 7th Int. Coll. on Gasdynamics of Explosions and Reactive Systems, Göttingen 1979

51 Winnant, C.D., Browand, F.K., J. Fluid Mech. 63, 237-255 (1974)

52 Chorin, A.J., Flame Advection and Propagation Algorithms, J. Computational Phys. 34, 1980 to appear

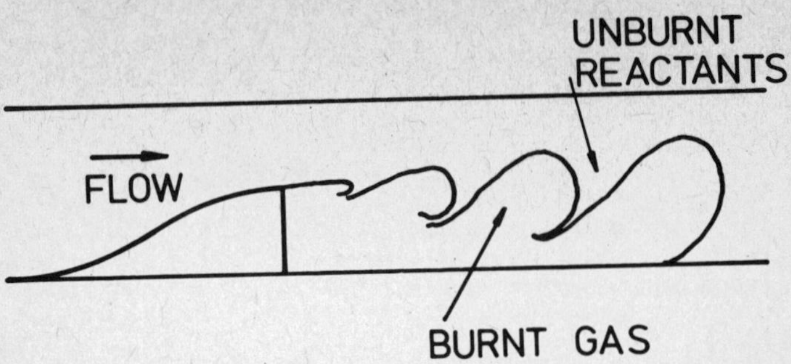

Fig. 1: Schematic of Ganji and Sawyer's premixed combustor

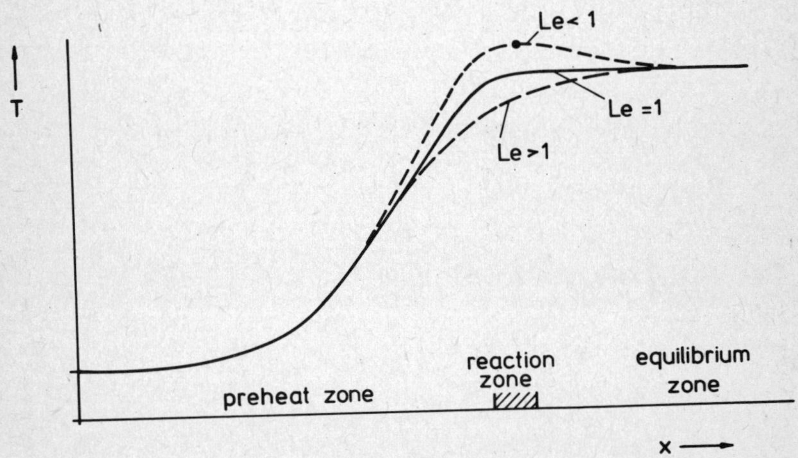

Fig. 2: The structure of a premixed laminar flamelet

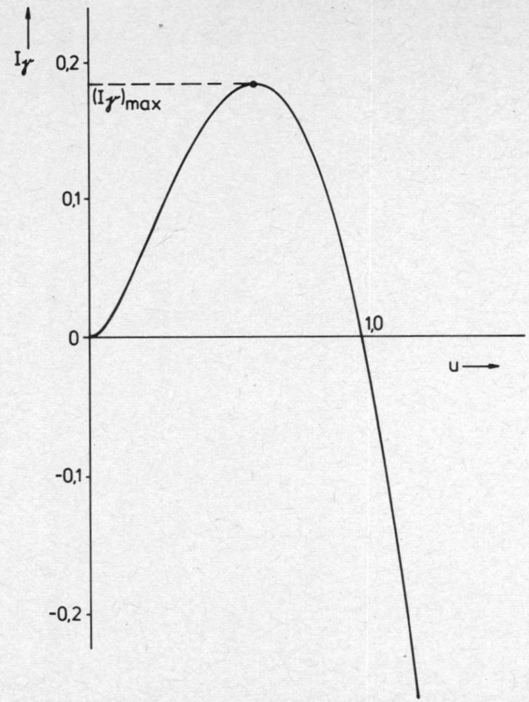

Fig. 3: Quenching condition of a premixed flamelet due to stretch

Fig. 4: A turbulent vertical jet diffusion flame of propane in air (photo by S. Donnerhack)

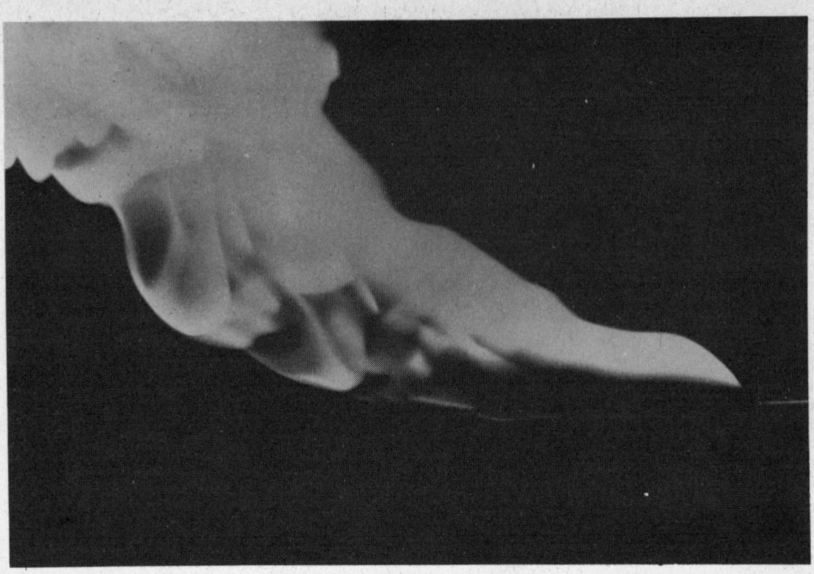

Fig. 5: A horizontal jet diffusion flame in air (photo by S. Donnerhack)

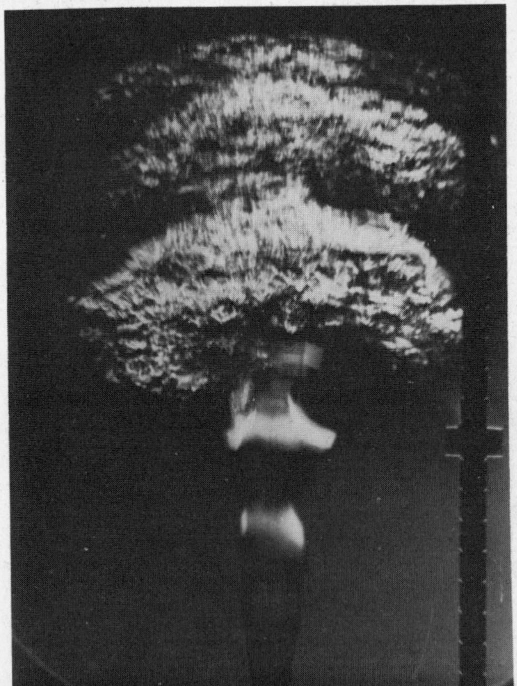

Fig. 6: Schlieren photo of a vertical hydrogen flame by Altgeld

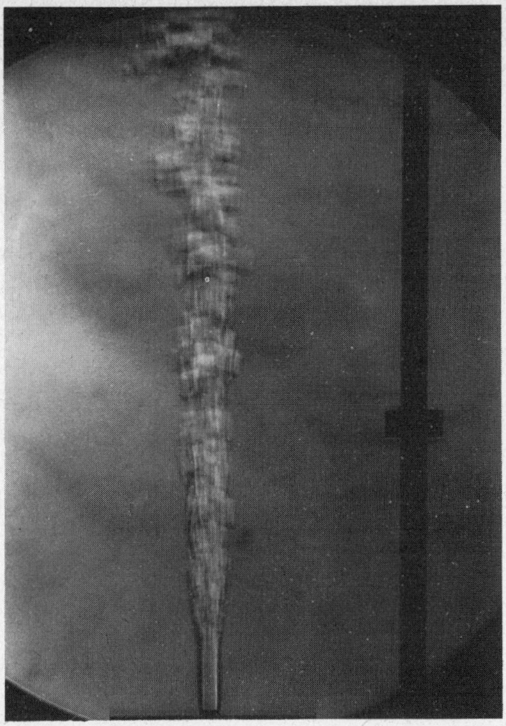

Fig. 7: Schlieren photo of an inert hydrogen jet by Altgeld

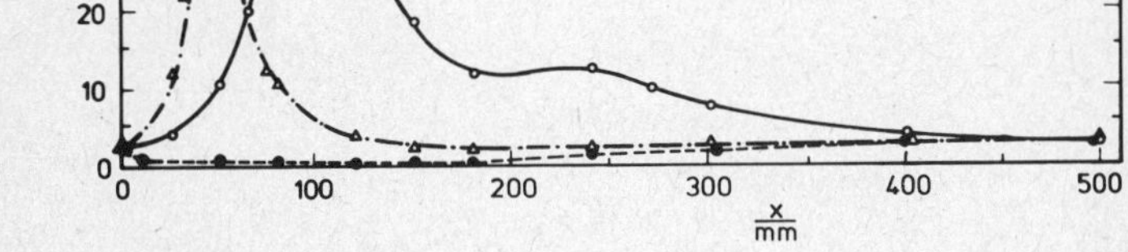

Fig. 8: Comparison of the turbulent kinetic energy of the flame and the inert jet at the axis by Altgeld

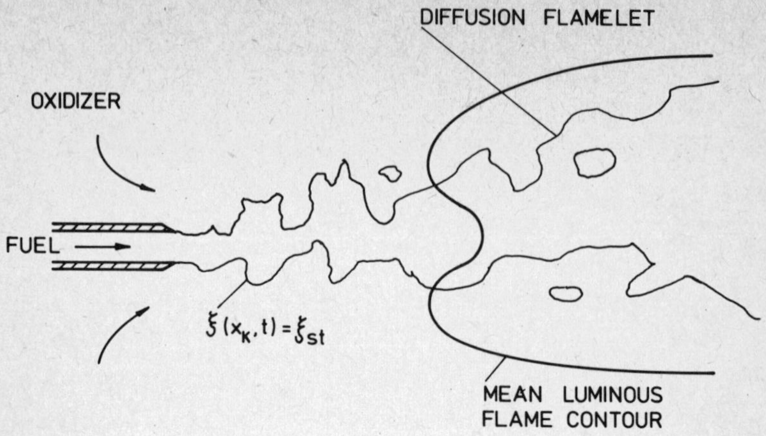

Fig. 9: Schematic of a lifted turbulent diffusion flame

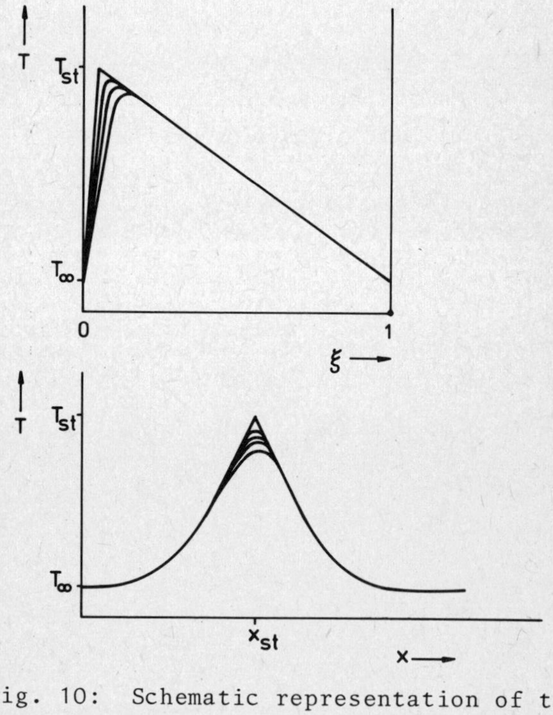

Fig. 10: Schematic representation of the temperature profile across the diffusion flamelet: a) over the conserved scalar, b) over the space coordinate

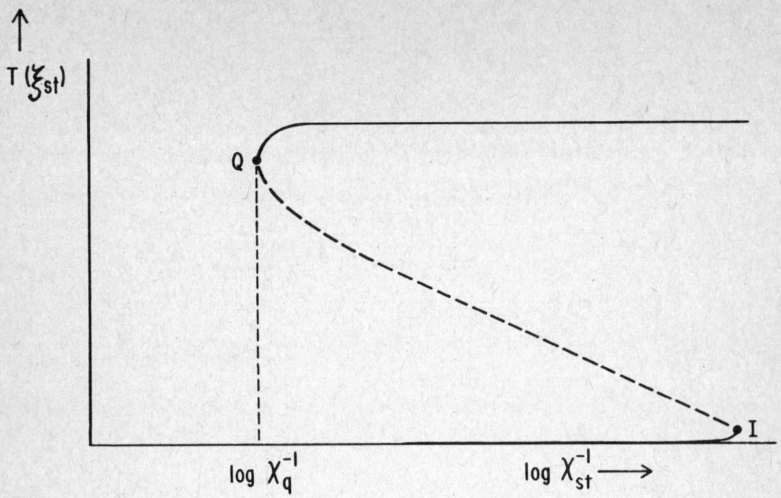

Fig. 11: The S-shaped curve
Q = quenching, I = ignition

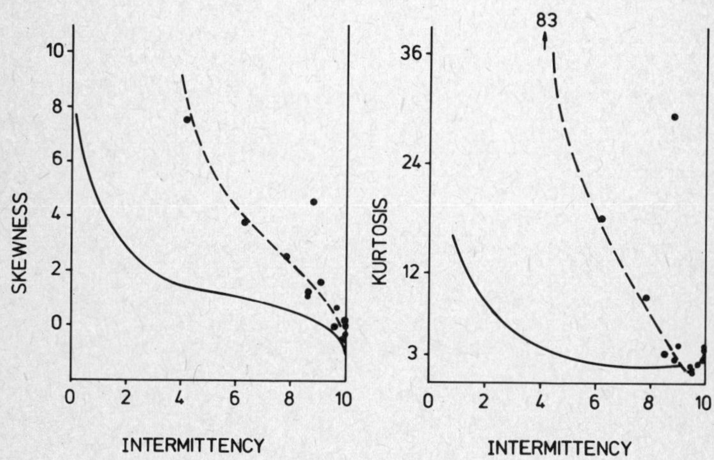

Fig. 12: Comparison of higher moments for flames and inert shear flows:
- - - line through Kennedy and Kent's data points,
-----line through data points for various inert shear flows as reported by Bilger /36/

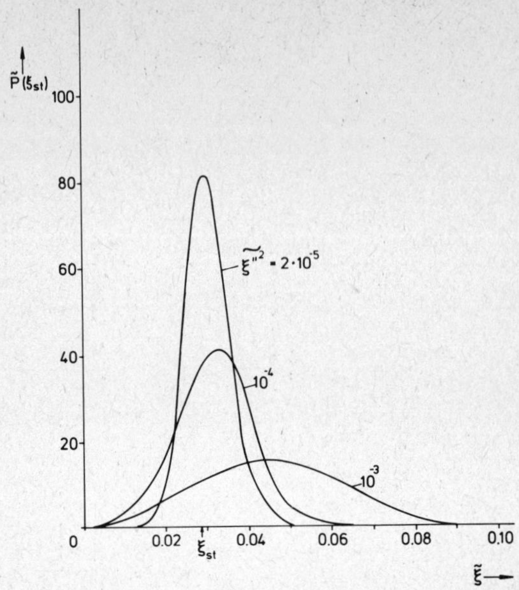

Fig. 13: The value of the p.d.f. of the conserved scalar at stoichiometric conditions

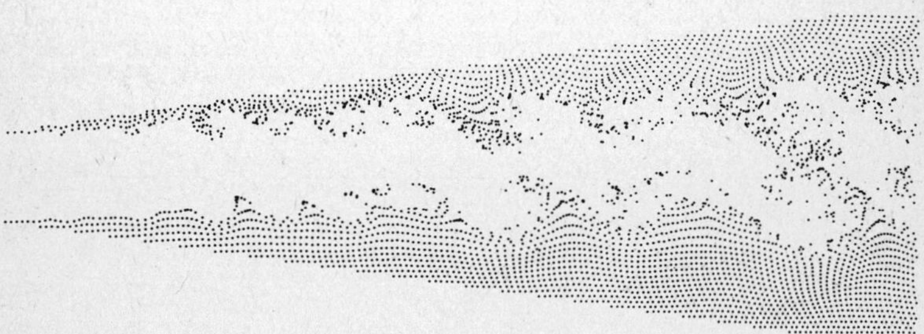

Fig. 14: A numerical simulation of the mixing process in a shear layer, particles from both feeds taken deliberately apart, by Thies

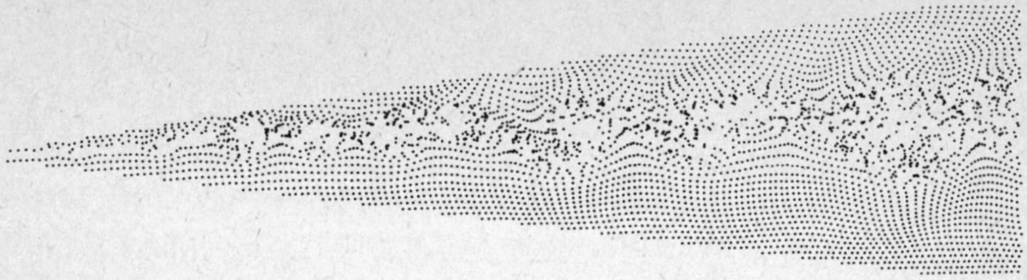

Fig. 15: As fig. 14, both feeds fitted together

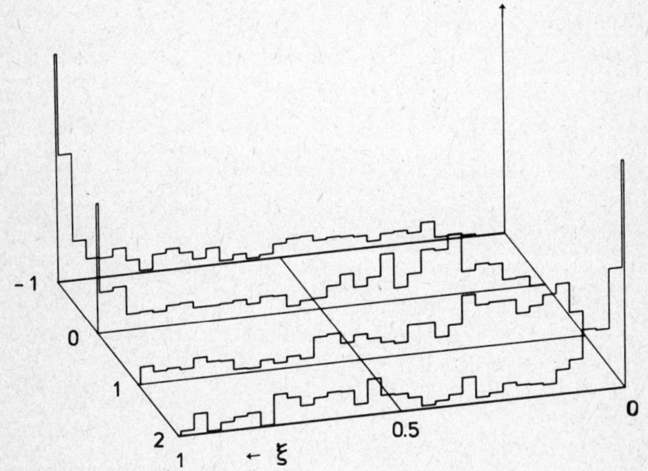

Fig. 16: Probability density function of concentration in a mixing layer, calculated with the random vortex method by Thies

Lecture Notes in Physics

Vol. 114: Stellar Turbulence. Proceedings, 1979. Edited by D. F. Gray and J. L. Linsky. IX, 308 pages. 1980.

Vol. 115: Modern Trends in the Theory of Condensed Matter. Proceedings, 1979. Edited by A. Pekalski and J. A. Przystawa. IX, 597 pages. 1980.

Vol. 116: Mathematical Problems in Theoretical Physics. Proceedings, 1979. Edited by K. Osterwalder. VIII, 412 pages. 1980.

Vol. 117: Deep-Inelastic and Fusion Reactions with Heavy Ions. Proceedings, 1979. Edited by W. von Oertzen. XIII, 394 pages. 1980.

Vol. 118: Quantum Chromodynamics. Proceedings, 1979. Edited by J. L. Alonso and R. Tarrach. IX, 424 pages. 1980.

Vol. 119: Nuclear Spectroscopy. Proceedings, 1979. Edited by G. F. Bertsch and D. Kurath. VII, 250 pages. 1980.

Vol. 120: Nonlinear Evolution Equations and Dynamical Systems. Proceedings, 1979. Edited by M. Boiti, F. Pempinelli and G. Soliani. VI, 368 pages. 1980.

Vol. 121: F. W. Wiegel, Fluid Flow Through Porous Macromolecular Systems. V, 102 pages. 1980.

Vol. 122: New Developments in Semiconductor Physics. Proceedings, 1979. Edited by F. Beleznay et al. V, 276 pages. 1980.

Vol. 123: D. H. Mayer, The Ruelle-Araki Transfer Operator in Classical Statistical Mechanics. VIII, 154 pages. 1980.

Vol. 124: Gravitational Radiation, Collapsed Objects and Exact Solutions. Proceedings, 1979. Edited by C. Edwards. VI, 487 pages. 1980.

Vol. 125: Nonradial and Nonlinear Stellar Pulsation. Proceedings, 1980. Edited by H. A. Hill and W. A. Dziembowski. VIII, 497 pages. 1980.

Vol. 126: Complex Analysis, Microlocal Calculus and Relativistic Quantum Theory. Proceedings, 1979. Edited by D. Iagolnitzer. VIII, 502 pages. 1980.

Vol. 127: E. Sanchez-Palencia, Non-Homogeneous Media and Vibration Theory. IX, 398 pages. 1980.

Vol. 128: Neutron Spin Echo. Proceedings, 1979. Edited by F. Mezei. VI, 253 pages. 1980.

Vol. 129: Geometrical and Topological Methods in Gauge Theories. Proceedings, 1979. Edited by J. Harnad and S. Shnider. VIII, 155 pages. 1980.

Vol. 130: Mathematical Methods and Applications of Scattering Theory. Proceedings, 1979. Edited by J. A. DeSanto, A. W. Sáenz and W. W. Zachary. XIII, 331 pages. 1980.

Vol. 131: H. C. Fogedby, Theoretical Aspects of Mainly Low Dimensional Magnetic Systems. XI, 163 pages. 1980.

Vol. 132: Systems Far from Equilibrium. Proceedings, 1980. Edited by L. Garrido. XV, 403 pages. 1980.

Vol. 133: Narrow Gap Semiconductors Physics and Applications. Proceedings, 1979. Edited by W. Zawadzki. X, 572 pages. 1980.

Vol. 134: $\gamma\gamma$ Collisions. Proceedings, 1980. Edited by G. Cochard and P. Kessler. XIII, 400 pages. 1980.

Vol. 135: Group Theoretical Methods in Physics. Proceedings, 1980. Edited by K. B. Wolf. XXVI, 629 pages. 1980.

Vol. 136: The Role of Coherent Structures in Modelling Turbulence and Mixing. Proceedings 1980. Edited by J. Jimenez. XIII, 393 pages. 1981.